This book presents an integrated treatment of the theory of nonnegative matrices, emphasizing connections with the themes of game theory, combinatorics, inequalities, optimization, and mathematical economics. Some related classes of positive matrices such as positive semidefinite matrices, M-matrices, P-matrices, and distance matrices are also discussed, but the main emphasis is on entrywise nonnegative matrices.

The book begins with the basics of the subject, such as the Perron-Frobenius Theorem. Only a minimal background in linear algebra is assumed, although familiarity with linear programming and statistics will be helpful in following some sections. Each of the later chapters is devoted to an area of applications, including doubly stochastic matrices (price fixing, scheduling, and the fair division problem), combinatorial matroids, and economics. These applications have been carefully chosen both for their elegant mathematical content and for their accessibility. The treatment is rigorous and almost all results are proved completely.

About half of the material in the book presents standard topics in a novel fashion, the remaining portion reports many new results in matrix theory for the first time in a book form.

ENCYCLOPEDIA OF MATHEMATICS AND ITS APPLICATIONS

EDITED BY G.-C. ROTA

Editorial Board

R. S. Doran, M. Ismail, T.-Y. Lam, E. Lutwak, R. Spigler

Volume 64

Nonnegative Matrices and Applications

ENCYCLOPEDIA OF MATHEMATICS AND ITS APPLICATIONS

ENCYCLOPEDIA OF MATHEMATICS AND ITS APPLICATIONS

Nonnegative Matrices and Applications

R. B. BAPAT　　T. E. S. RAGHAVAN

Indian Statistical Institute　　*University of Illinois at Chicago*

CAMBRIDGE
UNIVERSITY PRESS

CAMBRIDGE UNIVERSITY PRESS
Cambridge, New York, Melbourne, Madrid, Cape Town, Singapore, São Paulo, Delhi

Cambridge University Press
The Edinburgh Building, Cambridge CB2 8RU, UK

Published in the United States of America by Cambridge University Press, New York

www.cambridge.org
Information on this title: www.cambridge.org/9780521118668

© Cambridge University Press 1997

First published 1997
This digitally printed version 2009

A catalogue record for this publication is available from the British Library

Library of Congress Cataloguing in Publication data
R. B. Bapat.
   Nonnegative matrices and applications / R. B. Bapat, T. E. S.
Raghavan.
        p.   cm. - - (Encyclopedia of mathematics and its applications;
v. 64)
   ISBN 0-521-57167-7 hardback
   1. Non-negative matrices.   I. Raghavan, T. E. S.   II. Title.
III. Series.
QA188.B355   1996
512.9′434--dc20   96-12844
                                                            CIP

ISBN 978-0-521-57167-8 hardback
ISBN 978-0-521-11866-8 paperback

To the memory of my father,

Bhalachandra S. Bapat

- R. B. Bapat

To the memory of my father,

Eachambadi S. Narasimhachari

- T. E. S. Raghavan

CONTENTS

PREFACE

This book is aimed at first year graduate students as well as research workers with a background in linear algebra. The theory of nonnegative matrices is unfolded in the book using tools from optimization, inequalities and combinatorics. The topics and applications are carefully chosen to convey the excitement and variety that nonnegative matrices have to offer. Some of the applications also illustrate the depth and the mathematical elegance of the theory of nonnegative matrices. The treatment is rigorous and almost all the results are completely proved. While about half of the material in the book presents many topics in a novel fashion, the remaining portion reports many new results in matrix theory for the first time in a book form. Although the only prerequisite is a first course in linear algebra and advanced calculus, familiarity with linear programming and statistics will be helpful in appreciating some sections.

To give some examples, the Perron-Frobenius Theorem and many of its consequences are derived using the theory of matrix games where all rows and columns are essential for optimal play. The chapter on conditionally positive definite matrices and distance matrices has several new results appearing for the first time in a book. A transparent proof of the Alexandroff inequality for mixed discriminants is presented and a characterization of graphs giving rise to a finite Coxeter group is given in the chapter on combinatorial theory. The importance of P-matrices and M-matrices to several areas besides linear economic models is stressed and many of these results are seen via Game theory.

The application topics include, among other things, areas like game theory, Markov chains, probabilistic algorithms, numerical analysis, discrete distributions, categorical data, group theory, matrix scaling and economics. The chapter on doubly stochastic matrices contains applications to the pricing of houses in markets with known expectations, the problem of arriving at a fair division that respects peoples' individual preferences, scheduling jobs in a preemptive environment, and assigning people to jobs to minimize total job completion time. As described in the chapter on matrix scalings, while scaling is a powerful tool to speed up convergence, it is also a useful tool to estimate cell entries of an unknown matrix of the future parametric values from the known entries of the past. Such a procedure could be used to estimate the size of a population in a city or the growth of tumor after an operation and to reconstruct an image based on partial information.

The chapter on topics in economics describes the economic implications of properties of many special classes of matrices. The Perron-Frobenius Theorem can be used to explain both the Leontief and the Sraffa system. In the Leontief system, workers slave with fixed consumption and fixed wage to achieve targeted social output for the future generations. In the Sraffa system, all are entrepreneurs of identical skill who get equal rate of return. In international trade, the famous Hecksher-Ohlin Theorem shows that free trade could very well be a substitute for immigration. Another application for such matrices occurs in the price stability of an economy where any two goods are gross substitutes. In such economies, consumers switch from one brand to another if there is a steep price increase for a particular brand. An increase in price for a brand at a price equilibrium triggers a price increase for other substitute brands due to increased demand. However a dynamic price stability for the brand can be attained, allowing for varying speeds of price adjustments for other brands that are gross substitutes. It may be noted that these topics and results have been cited as major contributions to economic theory by P. Samuelson, Sir John Hicks and B. Ohlin (all Noble prize laureates) by the Noble prize committee.

This book was being written over a period of several years and it has benefitted from the comments, suggestions, and works of a large number of people. We only mention here some names that immediately come to mind. At the outset we wish to acknowledge the influence of the magnificent contributions of John von Neumann and Issai Schur on our approach to the subject at many places. We would also like to mention that the short note by Blackwell on minimax and irreducible matrices and Kaplansky's notion of completely mixed games have given the necessary game theoretic armory for many matrix problems.

We gratefully acknowledge comments and corrections, pertaining to various portions of the manuscript, due to Adi Ben-Israel, Charles Broyden, Gregory M. Constantine, S. Chandrasekaran, John Copas, M. V. Menon, S. R. Mohan, Dale Olesky, S. Panchapakesan, Ashok Ramu, Arunava Sen, Debapriya Sengupta, R. Sridhar, Pauline van den Driessche, and James Weber.

Evangelista Fe, Tamas Solymosi, Murali Srinivasan, Zamir Syed, and Julin Wu took a course based on the first three chapters from Raghavan and made helpful remarks. In particular, Fe made several corrections in spelling, grammer etc., Solymosi suggested improvements on the section on cooperative games and assignment problems and corrected some mathematical typos, while Murali carefully went through some topics on doubly stochastic matrices.

We owe a lot to S. Sankaran who made extensive corrections, particularly in the usage of syntax and grammar.

Dipankar Dasgupta convinced Raghavan of the Sraffa system as a much refined application of the Perron-Frobenius Theorem to Economics, whereas V. K. Chetty spent hours explaining to him the contributions and the depth of the works of Ricardo and Sraffa's formulation of the theory of production of commodities by means of commodities as a clear solution to the problem that Ricardo attempted to solve without success.

We warmly acknowledge the support given by our family members, Ragini, Sudeep, Usha, Deepa, Sampath, Santanu, Tara, and Manu, during the course of writing this book. The theory of Nonnegative matrices is indeed a fascinating and rewarding area. The Perron-Frobenius Theorem, the central result of the theory, was formulated at the beginning of the twentieth century and many significant developments have taken place during the past ninety years. In this book we have tried to touch upon some of these developments and the choice of topics clearly reflects our personal interests. We would consider our efforts amply rewarded even if a single curious and youthful mind is drawn to the subject and happens to share the same fascination after reading this book.

RBB
TESR

1

Perron-Frobenius theory and matrix games

The Perron-Frobenius Theorem is central to the theory of nonnegative matrices. An irreducible nonnegative matrix can be viewed as the payoff matrix of a zero-sum, two-person game with positive value. A matrix game is said to be completely mixed if no row or column is dispensable for optimal play. In this chapter we first exploit the properties of completely mixed matrix games to prove the Perron-Frobenius Theorem. The next few sections deal with certain related topics such as M-matrices, the structure of reducible nonnegative matrices, primitive matrices, and polyhedral sets with a least element. We then describe the basic aspects of finite Markov chains. In the final section we prove the Perron-Frobenius Theorem for operators that leave the Lorentz cone invariant.

1.1. Irreducible nonnegative matrices

We work with real matrices throughout, unless stated otherwise. Let $A = (a_{ij})$ be an $m \times n$ matrix. We say that the matrix A is *nonnegative* and write $A \geq 0$, if $a_{ij} \geq 0$ for all i, j. If $a_{ij} > 0$ for all i, j, then the matrix A is called *positive* and we write $A > 0$. For matrices A, B, we say $A \geq B$ if $A - B \geq 0$. Similar definitions and notation apply for vectors. The Euclidean n-space is denoted by R^n. The identity matrix of the appropriate order is denoted by I. The transpose of the matrix A is denoted by A^T.

An $n \times n$ matrix P is called a *permutation matrix* of order n if P can be obtained from the $n \times n$ identity matrix by permuting its rows and columns. Suppose we permute the rows of a matrix A to get the new matrix B. We can write the matrix B as $B = PA$, where P is the permutation matrix obtained by permuting the rows of the identity matrix, in the same way as B is obtained from A. Similarly, any column permutation of A corresponds to a matrix $C = AP$, where P is a permutation matrix.

A matrix A of order $n \times n$ is said to be *reducible*, either if A is the 1×1 zero matrix or if $n \geq 2$ and there exists a permutation matrix P such that

$$PAP^T = \begin{bmatrix} B & 0 \\ C & D \end{bmatrix},$$

where B and D are square matrices and 0 is a zero matrix. The matrix A is *irreducible* if it is not reducible.

The following lemma is useful in identifying reducible matrices.

Lemma 1.1.1. *Let A be an $n \times n$ matrix with $n \geq 2$. Let $a_{ij} = 0$ for $i \in S$, $j \notin S$ for some nonempty, proper subset S of $\{1, 2, \ldots, n\}$. Then A is reducible.*

Proof. Let $S = \{i_1, i_2, \ldots, i_k\}$, where we assume, without loss of generality, that $i_1 < i_2 < \cdots < i_{k-1} < i_k$. Let $S^c = T$ be the complement of S consisting of the ordered set of elements $j_1 < j_2 < \cdots < j_{n-k}$. Consider the permutation σ of $\{1, 2, \ldots, n\}$ given by

$$\sigma = \begin{pmatrix} 1 & 2 & \cdots & k & k+1 & k+2 & \cdots & n \\ i_1 & i_2 & \cdots & i_k & j_1 & j_2 & \cdots & j_{n-k} \end{pmatrix}.$$

Note that σ can be represented by the permutation matrix $P = (p_{ij})$, where $p_{rs} = 1$ if $\sigma(r) = s$. We prove that

$$PAP^T = \begin{bmatrix} B & 0 \\ C & D \end{bmatrix},$$

where B and D are square matrices and 0 is a $k \times (n-k)$ zero matrix. Consider row α and column β, where $1 \leq \alpha \leq k$ and $k+1 \leq \beta \leq n$. Now

$$(PAP^T)_{\alpha\beta} = \sum_i \sum_j p_{\alpha i} a_{ij} p_{\beta j}.$$

It is enough to show that each term in the summation is zero. Suppose $p_{\alpha i} = p_{\beta j} = 1$. Thus $\sigma(\alpha) = i$ and $\sigma(\beta) = j$. Since $1 \leq \alpha \leq k$, then $i \in \{i_1, i_2, \ldots, i_k\}$; similarly, since $k+1 \leq \beta \leq n$, we have $j \in \{j_1, j_2, \ldots, j_{n-k}\}$. By assumption, for such a pair i, j, we have $a_{ij} = 0$. That completes the proof. ∎

Some important characterizations of irreducible matrices are given in the next result.

Theorem 1.1.2. *Let $A \geq 0$ be an $n \times n$ matrix. Then the following conditions are equivalent:*

(1) *A is irreducible.*

(2) $(I + A)^{n-1} > 0.$

(3) *For any pair* $(i, j), 1 \leq i, j \leq n$, *there is a positive integer* $t = t(i, j) \leq n$ *such that* $(A^t)_{ij} = a_{ij}^{(t)} > 0.$

Proof. $(1) \Rightarrow (2)$: Let $y \geq 0$, $y \neq 0$ be an arbitrary vector in R^n. If a coordinate of y is positive, the same coordinate is positive in $y + Ay = (I + A)y$ as well. We claim that $(I + A)y$ has fewer zero coordinates than y as long as y has a zero coordinate. If the claim is not true, then $y_j = 0 \Rightarrow y_j + (Ay)_j = 0$ for any coordinate j. Let $J = \{j : y_j > 0\}$. For any $j \notin J, r \in J$, we have $(Ay)_j = \sum_k a_{jk} y_k = 0$ and $y_r > 0$. Thus, $a_{jr} = 0$. It follows by Lemma 1.1.1 that A is reducible, which is a contradiction and the claim is proved. Thus $(I + A)y$ has at most $n - 2$ zero coordinates. Continuing in this manner we conclude that $(I + A)^{n-1}y > 0$. We now set y as a column of the identity matrix, so the corresponding column of $(I + A)^{n-1}$ must be positive. Thus (2) holds.

$(2) \Rightarrow (3)$: Since $(I + A)^{n-1} > 0$, $A \geq 0$, then $A \neq 0$ and we have

$$A(I + A)^{n-1} = \sum_{k=1}^{n} \binom{n - 1}{k - 1} A^k > 0.$$

Thus for any i, j, at least one of the matrices $A, A^2, \ldots, A^n$ has its (i, j)-th coordinate positive.

$(3) \Rightarrow (1)$: Suppose A is reducible. Then for some permutation matrix P,

$$PAP^T = \begin{bmatrix} B_1 & 0 \\ C_1 & D_1 \end{bmatrix},$$

where B_1 and D_1 are square matrices. Furthermore, $PAP^T PAP^T = PA^2 P^T$, whence for some square matrices B_2, C_2 we have

$$PA^2 P^T = \begin{bmatrix} B_2 & 0 \\ C_2 & D_2 \end{bmatrix}.$$

More generally, for some matrix C_t and square matrices B_t and D_t,

$$PA^t P^T = \begin{bmatrix} B_t & 0 \\ C_t & D_t \end{bmatrix}.$$

Thus $(PA^t P^T)_{\alpha\beta} = 0$ for $t = 1, 2, \ldots$ and for any α, β corresponding to an entry of the zero submatrix in PAP^T. Now

$$0 = (PA^t P^T)_{\alpha\beta} = \sum_k \sum_l P_{\alpha k} a_{kl}^{(t)} P_{\beta l} \quad \text{for } t = 1, \ldots, n.$$

Choose k, l so that $p_{\alpha k} = p_{\beta l} = 1$. Then $a_{kl}^{(t)} = 0$ for all t, contradicting the hypothesis and thereby completing the proof.   ■

It follows from Theorem 1.1.2 that A^T is irreducible whenever A is irreducible.

We will make use of elementary concepts from graph theory without defining them explicitly. We refer the reader to Lovász (1979) and Bondy and Murty (1976) for these concepts.

If A is a nonnegative $n \times n$ matrix then we may associate a directed graph with A as follows: The graph has n vertices, which we denote by $1, 2, \ldots, n$. There is an edge from vertex i to vertex j if and only if a_{ij} is positive. Denote this graph by $G(A)$. A directed graph is said to be *strongly connected* if there is a path from any vertex to any other vertex. (In contrast with the standard terminology, we will not make any distinction between a walk and a path; thus a path may have a vertex appearing more than once.) Observe that $a_{ij}^{(t)} > 0$ if and only if there is a path of length t from vertex i to vertex j in $G(A)$. It follows from the equivalence of (1) and (3) in Theorem 1.1.2 that A is irreducible if and only if $G(A)$ is strongly connected.

1.2. Perron's Theorem on positive matrices

A set $S \subset R^n$ is said to be *convex* if for any $x, y \in S$ and for any $0 \le \lambda \le 1$, $\lambda x + (1 - \lambda)y \in S$. The empty set and any set with exactly one element are convex. Geometrically, a set S is convex if for any pair of points in S, the line segment joining the pair of points completely lies in S.

Here are some examples of convex sets:

(i) $S_1 = \{(x_1, x_2, \ldots, x_n) : \sum_j x_j^2 \le 1\}$.
(ii) $S_2 = \{(x_1, x_2, \ldots, x_n) : \sum_j a_{ij}x_j \ge b_i, i = 1, 2, \ldots, m\}$, where a_{ij}, b_i are given real numbers.
(iii) $S_3 = \{(x_1, x_2, \ldots, x_n) : \sum_j x_j = 1; x_j \ge 0 \text{ for all } j\}$.

We may think of a nonnegative $n \times n$ matrix A as a linear transformation with respect to a fixed basis. Notice that if $x \ge 0$ in R^n, then $Ax \ge 0$. Thus the set of all nonnegative vectors in R^n is mapped into itself by the matrix A. A set $K \subset R^n$ is called a *cone* if $x, y \in K \Rightarrow x + y \in K$ and $x \in K \Rightarrow \lambda x \in K$ for any $\lambda \ge 0$. A set K in R^n is called a *convex cone* if it is a convex set and if for any $x \in K$ and $\lambda \ge 0$, $\lambda x \in K$. The set of all nonnegative vectors is a convex cone, and a nonnegative matrix leaves this cone invariant.

The Perron-Frobenius Theorem was originally proved by Perron for positive matrices. In this section we prove the main aspects of Perron's Theorem.

The technique is elementary, except for the fact that we will use Brouwer's Fixed Point Theorem, which we now state without proof. For a proof using combinatorial ideas, see Bondy and Murty (1976).

Theorem 1.2.1 (Brouwer's Fixed Point Theorem). *Let S be a nonempty, closed, bounded, convex set in R^n. Let $f : S \to S$ be a continuous map. Then there exists an $x \in S$ such that $f(x) = x$.*

We now recall some elementary facts about multiplicities of eigenvalues, which will be needed in subsequent sections. For any square matrix of order n with real or complex entries, the characteristic polynomial of the matrix can be written as $p(\lambda) = c \prod_{i=1}^{k} (\lambda - \lambda_i)^{m_i}$, where c is a constant and $\lambda_1, \lambda_2, \ldots, \lambda_k$ are distinct. Here m_i is called the *algebraic multiplicity* of the characteristic root $\lambda_i, i = 1, 2, \ldots, k$. We call a root, say λ_1, a *simple root*, if $m_1 = 1$. If $m_1 = 1$, then $\frac{d}{d\lambda} p(\lambda)|_{\lambda = \lambda_1} \neq 0$. If $m_1 > 1$, then $\frac{d}{d\lambda} p(\lambda)|_{\lambda = \lambda_1} = 0$. Thus, a characteristic root λ_1 is simple if and only if the derivative $p'(\lambda_1)$ is nonzero. Another notion of multiplicity is that of the geometric multiplicity of the characteristic root. For any characteristic root λ, let $S_\lambda = \{u : Au = \lambda u\}$. Here S_λ is a vector space in its own right and A, viewed as a linear transformation, leaves the subspace invariant. The dimension of S_λ is called the *geometric multiplicity* of the characteristic root λ. In general the two multiplicities need not be the same. For example, the matrix

$$\begin{bmatrix} 0 & 0 & 0 \\ 1 & 0 & 0 \\ 0 & 2 & 0 \end{bmatrix}$$

has $p(\lambda) = |A - \lambda I| = -\lambda^3$, where $|\cdot|$ denotes determinant. Thus, 0 is a root of A with algebraic multiplicity 3. However, the only characteristic vector for the characteristic root 0 is $(0, 0, 1)^T$, up to a scalar multiple. Hence the geometric multiplicity of λ is 1. In general the algebraic multiplicity is not less than the geometric multiplicity. The following argument can be made precise to show this claim. Fix an eigenvalue λ_0, and think of A as a linear transformation on the vector space S_{λ_0} into itself. This restriction of A to S_{λ_0} can be thought of as another linear transformation A_0 with $|A_0 - \lambda I| = (\lambda - \lambda_0)^m$, where m is the dimension of S_{λ_0}. Clearly, we have $m \leq m_{\lambda_0}$ where m_{λ_0} is the algebraic multiplicity of λ_0 as an eigenvalue of A.

Theorem 1.2.2 (Perron's Theorem). *Let $A > 0$ be an $n \times n$ matrix. Then*

(i) $Ay = \lambda_0 y$ *for some* $\lambda_0 > 0$, $y > 0$.

(ii) *The eigenvalue λ_0 is maximal in modulus among all the eigenvalues of A. That is, for any eigenvalue μ of A, $|\mu| \leq \lambda_0$.*

(iii) *The eigenvalue λ_0 is geometrically simple. That is, any two eigenvectors corresponding to λ_0 are linearly dependent.*

(iv) *Any positive eigenvector of A (corresponding to any eigenvalue) is a scalar multiple of y.*

Proof. Let

$$S = \left\{ (x_1, x_2, \ldots, x_n)^T : \sum_i x_i = 1 \text{ and } x_i \geq 0 \text{ for all } i \right\}.$$

Define the map $f : S \to S$ as follows:

$$f(x) = \left\{ \sum_i (Ax)_i \right\}^{-1} Ax.$$

Here $(Ax)_i$ is the i-th coordinate of Ax. If $x \in S$ then, because $A > 0$, the vector Ax is nonzero and the map f is well defined. It is easily checked that f is continuous and maps S into S. By Brouwer's Fixed Point Theorem, $f(y) = y$ for some $y \in S$. Thus

$$\left\{ \sum_i (Ay)_i \right\}^{-1} Ay = y.$$

If we set $\sum_i (Ay)_i = \lambda_0$, then $\lambda_0 > 0$ and $Ay = \lambda_0 y$. Since $A > 0$, it follows that $y > 0$. Hence (i) is proved.

If we apply (i) to A^T, then we conclude that $A^T z = \lambda_1 z$ for some $\lambda_1 > 0$, $z > 0$. Now

$$\lambda_1 y^T z \doteq y^T A^T z = \lambda_0 y^T z,$$

and since $y^T z > 0$, we have $\lambda_0 = \lambda_1$. Thus $A^T z = \lambda_0 z$. This fact will be used in the rest of the proof. Let $Au = \mu u$ for some real or complex eigenvalue μ. Let u^+ be defined by $u^+ = (|u_1|, |u_2|, \ldots, |u_n|)^T$, where $u = (u_1, u_2, \ldots, u_n)^T$. Without loss of generality, let u^+ be a probability vector. We have

$$\sum_j a_{ij} |u_j| \geq \left| \sum_j a_{ij} u_j \right| = |\mu u_i| = |\mu| |u_i|.$$

Thus $Au^+ \geq |\mu| u^+$. Premultiply this last inequality by z^T to conclude that $|\mu| \leq \lambda_0$. This completes the proof of (ii).

We now prove (iii). Suppose $Av = \lambda_0 v$ for some real, nonzero vector v. We must show that v is a scalar multiple of y. If v and y are linearly independent, then there exists a real number α such that $y - \alpha v$ is a nonnegative, nonzero vector with at least one zero coordinate. Since

$$A(y - \alpha v) = \lambda_0 (y - \alpha v),$$

$y - \alpha v$ is an eigenvector of A. However, since $A > 0$, any nonnegative eigenvector of A must in fact be positive and we get a contradiction. Thus v is a scalar multiple of y. By considering the real and the imaginary parts separately, we can show that any complex eigenvector of A corresponding to λ_0 is a scalar multiple of y.

To prove (*iv*), suppose $Au = \mu u$ for $u > 0$. We have $\mu z^T u = z^T Au = \lambda_0 z^T u$. Since $z^T u > 0$, we have $\mu = \lambda_0$. The result now follows by (*iii*). ∎

1.3. Completely mixed games

The theory of games is concerned with problems of conflict. The simplest form of such games are the so-called *matrix games* played as follows: Players I and II secretly choose a column j and a row i, respectively, of a matrix $A = (a_{ij}), 1 \leq i \leq m, 1 \leq j \leq n$. Their choices are revealed to a referee who then calls Player II to pay Player I the amount a_{ij}. If $a_{ij} < 0$, it is an income to Player II from Player I. (We have slightly deviated from the standard conventions of matrix games in which the rows are usually chosen by Player I.)

Example 1.1. Player I shows 1, 2, or 3 fingers, and simultaneously Player II shows 1 or 2 fingers. The payoff to I from II is the total number of fingers shown by both. The matrix of this game is

$$\text{II's actions} = \begin{matrix} 1f \\ 2f \end{matrix} \begin{pmatrix} 2 & 3 & 4 \\ 3 & 4 & 5 \end{pmatrix}.$$

with I's actions $= 1f \quad 2f \quad 3f$.

Obviously, if they play the game several times, Player I will show 3 fingers and Player II will show 1 finger every time.

Example 1.2. The game is the same as above but the payoff to Player I is 1 if the total number of fingers shown is odd and -1 if the number is even. The payoff matrix is

$$A = \begin{bmatrix} -1 & 1 & -1 \\ 1 & -1 & 1 \end{bmatrix}.$$

In repeated play it is not good for Player II to always show 1 finger, for in that case, Player I will show 2 fingers every time and collect 1 unit from II. Similarly, it is not desirable for I to always show the same number of fingers. It is clear, however, that for Player I showing 1 finger is the same as showing 3 fingers against any choice of the opponent and Player I might as well not bother about showing 3 fingers. Suppose Player I shows 1 or 2 fingers based on the outcome of the toss of a fair coin. If Player II chooses 1 finger, then I loses 1 unit half the time and gains 1 unit half the time. Irrespective of the choice of Player II, the average gain for I is zero. Similarly, Player II's loss on the average is zero if he also uses a fair coin to show 1 or 2 fingers. Thus, tossing a coin to select 1 or 2 fingers is a good strategy for both Players.

Example 1.3. The game is the same as above. We have the following modified payoff to Player I: Player I receives from Player II the total of the number of fingers shown if the total is odd; otherwise he pays Player II the total of the number of fingers shown. The payoff matrix to Player I is

$$A = \begin{bmatrix} -2 & 3 & -4 \\ 3 & -4 & 5 \end{bmatrix}.$$

Suppose, as in the previous example, Player II tosses a fair coin and decides to show 1 or 2 fingers depending on the outcome. By showing 1 finger all the time, Player I can gain on the average at most half a unit. Player II can do better. Suppose he shows 1 finger with chance $7/12$ and 2 fingers with chance $5/12$. The average gain to Player I would be $1/12$ if he shows 1 or 2 fingers and $-3/12$ if he shows 3 fingers. Thus, Player II loses no more than $1/12$ on the average—no matter what Player I does. This is certainly better than tossing a coin. However, it is not clear whether Player II can do better than this. From the point of view of Player I, the strategy that selects 1 finger with chance $7/12$ and 2 fingers with chance $5/12$ guarantees, on the average, $1/12$ to player I, no matter how many fingers Player II shows. Therefore, we say that the mixed strategy $(7/12, 5/12)$ is *optimal* for Player II and the mixed strategy $(7/12, 5/12, 0)$ is optimal for Player I. Further, we say that the *value* of the game is $1/12$.

Example 1.4. The following dialogue takes place between a teacher and a student:

> Student: Professor, will you give us a take home final for the Game Theory course?
> Teacher: I don't believe in them.
> Student: I am nervous in the regular exam. I get only Cs in those exams. However, I can do much better with a take home.
> Teacher: I know.
> Student: I heard that you always recycle old questions!
> Teacher: I would not contradict that. In fact, for the final exam, I have photocopied one among the ten questions that are on my desk.
> Student: Can I pick up all of them to try at home?
> Teacher: No, not really. I can let you pick up only one of them.

Student: I appreciate your help. I am curious. What do you expect from me if I take one of these questions home?

Teacher: It depends. If you pick up question i, and if it also happens to be the final exam question, you could probably score $q_i > 0$; otherwise you can hope for your usual score c.

How shall we compute the expected score for this pessimistic student? Without loss of generality, let $q_1 \geq q_2 \geq \cdots \geq q_{10} > c > 0$. The student keen on maximizing his average score and ensuring a score c on any exam, can use the following payoff matrix for the game between him and the teacher:

$$
\begin{array}{c c}
 & \begin{array}{cccc} 1 & 2 & \cdots & 10 \end{array} \\
\begin{array}{c} 1 \\ 2 \\ \vdots \\ 10 \end{array} &
\left(\begin{array}{cccc}
q_1 - c & 0 & \cdots & 0 \\
0 & q_2 - c & \cdots & 0 \\
\vdots & \vdots & \ddots & \vdots \\
0 & 0 & 0 & q_{10} - c
\end{array} \right)
\end{array}.
$$

If x_i is the chance for the i-th question to be selected by the student, then the expected score for the student is at least $\min_i (q_i - c)x_i$. Temporarily pretending $(q_i - c)x_i = v$ for all i, we get $x_i = v/(q_i - c)$. Summing x_i, we get $v = 1/\sum_i 1/(q_i - c)$. The same strategy is seen to also work for the sadistic teacher.

The existence of such optimal strategies and value is not just a coincidence in these examples. We have the following celebrated theorem of von Neumann. For a proof, see, for example, Parthasarathy and Raghavan (1971).

Theorem 1.3.1 (Minimax Theorem). *Let $A = (a_{ij})$ be an $m \times n$ payoff matrix. Then there exists a unique constant v (called the value) and mixed strategies $x = (x_1, x_2, \ldots, x_m)$ for Player II and $y = (y_1, y_2, \ldots, y_n)^T$ for Player I such that*

$$
\sum_j a_{ij} y_j \geq v, \quad i = 1, 2, \ldots, m, \qquad \sum_i a_{ij} x_i \leq v, \quad j = 1, 2, \ldots, n.
$$

$$(1.3.1)$$

(Here $x_i \geq 0$ for all i and $\sum_i x_i = 1$, $y_j \geq 0$ for all j and $\sum_j y_j = 1$.) The strategy x is called an optimal strategy for Player II. The strategy y is called an optimal strategy for Player I.

A player not knowing the exact choice of his opponent must be ready to take care of himself under every possible course of action by the opponent. The first of these inequalities expresses the fact that when Player I chooses column j in the payoff matrix with chance y_j, then for any choice i by Player II, the expected income $\sum_j a_{ij} y_j$ to Player I is at least v. Similarly, Player II, though not knowing the actual choice of Player I, can safeguard his expected losses by choosing action i (row i) with chance x_i. The expected loss of Player II when Player I chooses column j is $\sum_i a_{ij} x_i$. The second inequality says that it is at most v.

A mixed strategy x for Player II is called *completely mixed* if $x > 0$, that is, if all the rows of the payoff matrix are chosen with positive probability. A completely mixed strategy for Player I is defined similarly. A matrix game is called completely mixed if every optimal mixed strategy x for Player II and y for Player I are completely mixed.

Before we state the next result we recall the following facts on systems of equations. Let A be a real $m \times n$ matrix. The *null space* $\{x : Ax = 0\}$ is a vector space with dimension $n - r$, where r is the rank of A. The dimension of the range space $\{Aw : w \in R^n\}$ is r. We can choose a basis for the null space with $n - r$ elements. Thus, we have $Aw^{(1)} = Aw^{(2)} = \cdots = Aw^{(n-r)} = 0$, where $w^{(1)}, w^{(2)}, \ldots, w^{(n-r)}$ are linearly independent.

If u, v are vectors in R^n, then we denote their inner product by $\langle u, v \rangle$. We denote by $\mathbf{1}$ the column vector of appropriate size with each entry equal to 1.

Theorem 1.3.2. *Let v be the value of the matrix game A. Let some optimal strategy of Player II be completely mixed. Then, for any optimal strategy y of Player I, $Ay = v\mathbf{1}$.*

Proof. If x and y are optimal strategies for Player II and Player I respectively, then it follows from (1.3.1) that $x^T Ay = v$. Now suppose x is completely mixed. We have

$$0 = \langle x, Ay \rangle - v = \langle x, Ay - v\mathbf{1} \rangle. \qquad (1.3.2)$$

Let $u_i = (Ay - v\mathbf{1})_i = \sum_j a_{ij} y_j - v$. Since y is an optimal strategy for Player I, $u_i = \sum_j a_{ij} y_j - v \geq 0$. From (1.3.2) we have $\sum_i x_i u_i = 0$. Since $x_i > 0$ for all i, $u_i = 0$. That completes the proof. ∎

Theorem 1.3.3. *Let the value of the $m \times n$ matrix game $A = (a_{ij})$ be zero and suppose that every optimal strategy for Player II is completely mixed. Then $m - 1 \leq rank A \leq n - 1$. If $rank A = m - 1$, then the optimal strategy for Player II is unique.*

Proof. By Theorem 1.3.2 we have $Ay = 0$ for any optimal strategy y of Player I. Since $y \neq 0$, rank $A \leq n - 1$. In case rank $A \leq m - 2$ there exist at least two linearly independent solutions to $A^T u = 0$. We can assume that one of them is independent of an optimal strategy x of Player II. Let π be such a solution. Without loss of generality, either $\sum_i \pi_i = 0$ or $\sum_i \pi_i = 1$.

Case (i): $\sum \pi_i = 0$.

Since $\pi \neq 0$ and $\sum \pi_i = 0$, then $\pi_i > 0$ for some i. Let $\frac{1}{\theta} = \max_i (\frac{\pi_i}{x_i}) > 0$. The vector $x - \theta\pi$ (≥ 0) has at least one of its coordinates equal to zero. (It is that coordinate i for which the above maximum is attained.) Further, since $\sum \pi_i = 0$, $x - \theta\pi$ is a probability vector. Also, we have $A^T(x - \theta\pi) = A^T x \leq 0$. This shows that the vector $x - \theta\pi$ is optimal for Player II, but is not completely mixed. This is a contradiction to our assumption. Thus rank $A \geq m - 1$.

Case (ii): $\sum \pi_i = 1$.

Let $x > 0$ be any optimal strategy for Player II. Consider the vector $z = (1 + \theta)x - \theta\pi$. For $\theta > 0$ and sufficiently small, the vector z is positive. If θ is chosen such that $\frac{1+\theta}{\theta} = \max_i \frac{\pi_i}{x_i}$, then z is a nonnegative vector with at least one zero coordinate. Since $\sum \pi_i = 1$, z is a probability vector. Furthermore, $A^T z = (1+\theta)A^T x - \theta A^T \pi = (1+\theta)A^T x \leq 0$. Thus, z is an optimal strategy for Player II that is not completely mixed. This contradicts our assumption and therefore rank $A \geq m - 1$.

Lastly, let rank $A = m - 1$. This means that the equation $A^T \pi = 0$ has precisely one solution up to a scalar multiple. Suppose Player II has two distinct optimal strategies. They are necessarily linearly independent, and one of them will be linearly independent of π. We can repeat the above proof verbatim with π and the chosen optimal strategy of Player II, which is independent of π. As proved above, we can again contradict our assumption that the optimal strategies of Player II are completely mixed. Hence, when rank $A = m - 1$, Player II has a unique optimal strategy. ∎

We now wish to obtain an important characterization of completely mixed games due to Kaplansky (1945). We first prove some preliminary results.

Theorem 1.3.4. *Let A be an $m \times n$ payoff matrix. If $m > n$, then Player II can optimally skip a row. If $m < n$, then Player I can optimally skip a column.*

Proof. If v is the value of the matrix game $A = (a_{ij})$, then the matrix $B = (a_{ij} - v)$ has value zero and the optimal strategies of the players remain unchanged. Therefore, we may assume that $v = 0$. Let $m > n$. If every optimal strategy of Player II is completely mixed, then by Theorem 1.3.3, $m - 1 \leq n - 1$. Because this contradicts our assumption there should be an optimal strategy that selects a row with zero probability (that is, it skips a row). A similar argument can be given when $m < n$. ∎

Theorem 1.3.5. *Let A be an n × n payoff matrix. If the game is not completely mixed, then both players have optimal strategies that are not completely mixed.*

Proof. We again assume, without loss of generality, that $v = 0$. Contrary to the assertion, let us suppose that every optimal strategy of Player II is completely mixed. By Theorem 1.3.3 the optimal strategy is unique for Player II. since the game is not completely mixed, Player I has an optimal strategy that is not completely mixed. Let y be such a strategy with, say, $y_1 = 0$. By Theorem 1.3.2, $Ay = 0$. Also, because A is singular,

$$\sum_j a_{ij} A_{kj} = 0, \quad i = 1, 2, \ldots, n \text{ for each fixed } k,$$

where A_{kj} denotes the cofactor of a_{kj}. By Theorem 1.3.3, the rank of A is $n - 1$. Therefore,

$$(A_{k1}, A_{k2}, \ldots, A_{kn}) = \alpha_k (y_1, y_2, \ldots, y_n)$$

for some α_k, and $A_{k1} = \alpha_k y_1 = 0$, for each k. Thus, $A_{11} = A_{21} = \cdots = A_{n1} = 0$ and, therefore,

$$\begin{bmatrix} a_{12} & a_{13} & \cdots & a_{1n} \\ \cdots & \cdots & \cdots & \cdots \\ a_{n2} & a_{n3} & \cdots & a_{nn} \end{bmatrix}$$

has rank $\leq n - 2$. Then the system of equations $\sum_i a_{ij} u_i = 0$, $j = 2, \ldots, n$ has at least two linearly independent solutions. The optimal strategy x for Player II comprises one solution; the other solution, call it π, necessarily satisfies $\sum_i a_{i1} \pi_i \neq 0$, for otherwise $\sum_i a_{ij} u_i = 0$, $j = 1, 2, \ldots, n$ will have x and π as two linearly independent solutions, which contradicts rank $A = n - 1$. As in Theorem 1.3.3 we can assume $\sum_i \pi_i = 0$ or $\sum_i \pi_i = 1$.

Case (i): $\sum_i a_{i1} \pi_i > 0, \sum_i \pi_i = 0$.
As in the proof of Theorem 1.3.3 we can choose $\lambda > 0$ such that the vector $z = (z_1, z_2, \ldots, z_n)^T$ with $z_i = x_i - \lambda \pi_i$, $i = 1, 2, \ldots, n$ satisfies $z_i \geq 0$ for all i, and $z_{i_0} = 0$ for some i_0. However, $\sum_i a_{ij} z_i = \sum_i a_{ij} x_i - \lambda \sum_i a_{ij} \pi_i = -\lambda \sum_i a_{ij} \pi_i$ for all j. For $j = 1$, $\sum_i a_{i1} z_i = -\lambda \sum_i a_{i1} \pi_i < 0$ and, by assumption, $\sum_i a_{ij} z_i = -\lambda \sum_i a_{ij} \pi_i = 0$ for $j = 2, 3, \ldots, n$. Also, $\sum z_i = 1$. Thus, z is optimal for Player II with $z_{i_0} = 0$. This contradicts our assumption on the optimal strategies of Player II.

Case (ii): $\sum_i a_{i1} \pi_i < 0, \sum_i \pi_i = 0$.
The vector $(z_1, z_2, \ldots, z_n)^T$ with $z_i = x_i + \lambda \pi_i$, $i = 1, 2, \ldots, n$ can be used

to contradict our assumption on the optimal strategies of Player II.

Case (iii): $\sum_i a_{i1}\pi_i > 0$, $\sum_i \pi_i = 1$.
The vector $z = (z_1, z_2, \ldots, z_n)^T$ with $z_i = (1 + \lambda)x_i - \lambda\pi_i$ for suitable $\lambda > 0$
can be used to contradict our assumption on the optimal strategies of Player II
as in case (*ii*).

Case (iv): $\sum_i a_{i1}\pi_i < 0$, $\sum \pi_i = 1$.
Use $z_i = (1 - \lambda)x_i + \lambda\pi_i$, $i = 1, 2, \ldots, n$. The proof is similar to the previous
cases. Thus, in all cases there is a strategy for Player II that is not completely
mixed, and the proof is complete. ∎

Theorem 1.3.6. *An $m \times n$ matrix game with value zero is completely mixed if
and only if*

(i) $m = n$ and the rank of the matrix is $n - 1$.
(ii) All cofactors are different from zero and are of the same sign.
(iii) The optimal strategies are unique for the two players.

Proof. (Necessity). By Theorem 1.3.4 and Theorem 1.3.5 we observe that condition (*i*) is necessary. Furthermore, since both players have unique completely
mixed optimal strategies, (*iii*) also follows. Now $\sum_j a_{ij}A_{kj} = 0$, $i = 1, 2, \ldots, n$
and $(A_{k1}, A_{k2}, \ldots, A_{kn}) = \alpha(y_1, y_2, \ldots, y_n)$ for the unique optimal y of Player
I for some α. Since the y_is are positive, the vector $(A_{k1}, A_{k2}, \ldots, A_{kn})$ is either
positive or negative or zero. By using the unique completely mixed strategy of
Player II we have a similar assertion for $(A_{1j}, A_{2j}, \ldots, A_{mj})$ for any j. Thus all
the cofactors A_{ij} are positive or negative or zero. Since the rank of the matrix is
$n - 1$, some cofactor A_{ij} is nonzero and the matrix $C = (A_{ij})$ is either positive
or negative.
(Sufficiency). Conversely, let the matrix A be square with rank $n - 1$ and
with all cofactors A_{ij} of the same sign. Let $\frac{1}{\alpha} = \sum_j A_{1j}$. Observe that $y = \alpha(A_{11}, A_{12}, \ldots, A_{1n})$ is a mixed strategy and that

$$\sum_j a_{ij}y_j = \alpha \sum_j a_{ij}A_{1j} = 0, \quad i = 1, 2, \ldots, n.$$

Thus y is optimal for Player I.
Similarly, for $\frac{1}{\beta} = \sum_i A_{i1}$, the vector $x = \beta(A_{11}, A_{21}, \ldots, A_{n1})^T$ is a mixed
strategy and

$$\sum_i a_{ij}x_i = \beta \sum_i a_{ij}A_{ij} = 0, \quad i = 1, 2, \ldots, n.$$

Hence, x and y are optimal strategies for Players II and I, respectively. From
Theorem 1.3.2 any optimal strategy u of Player I will satisfy $\sum_j a_{ij}u_j = 0$.

Because the rank of A is $n - 1$, $u = y$ and the optimal strategy for Player I is unique. Similarly, the optimal strategy for Player II is unique. Thus, the optimal strategies are completely mixed and, by Theorem 1.3.5, the game is completely mixed. That completes the proof. ∎

A determinantal formula for the value of a completely mixed game is given in the next result.

Theorem 1.3.7. *Let A be a square matrix such that the game A is completely mixed. Then $\sum_{i,j} A_{ij}$ is nonzero and the value v of A is given by*

$$v = \frac{|A|}{\sum_{i,j} A_{ij}}.$$

Proof. If $v = 0$, Theorem 1.3.2 shows that $Ay = 0$ and $|A| = 0$. Furthermore, by Theorem 1.3.6, $\sum_{i,j} A_{ij}$ is nonzero. This proves the theorem for $v = 0$. If, however, $v \neq 0$, then A must be nonsingular, for otherwise there would exist $\pi \neq 0$ with $A\pi = 0$. From Theorem 1.3.2 we get $Ay = v\mathbf{1}$, which is nonzero for the optimal y of Player I. Obviously, π and y are independent. We can use them as in Theorem 1.3.3 to contradict the fact that the game is completely mixed. Thus A cannot be singular. Now, $Ay = v\mathbf{1}$ gives

$$y = A^{-1}Ay = vA^{-1}\mathbf{1} = v\frac{(\text{adj } A)\mathbf{1}}{|A|}.$$

[Here $(\text{adj } A)$ is the transpose of the matrix $C = (A_{ij})$.] Thus,

$$\frac{1}{v} = \frac{\sum_i y_i}{v} = \frac{\sum_{i,j} A_{ij}}{|A|}.$$

This completes the proof of the theorem. ∎

Theorem 1.3.8. *Let A be an $n \times n$ payoff matrix that is not completely mixed. Then the value v of the game satisfies the equation*

$$v = \min_i \max_j v_{ij} = \max_j \min_i v_{ij},$$

where v_{ij} is the value of the $(n - 1) \times (n - 1)$ game obtained by deleting the i-th row and the j-th column in A.

Proof. Let $A_{.j}$ be the payoff matrix obtained from A by deleting column j with value $v_{.j}$. Since the square matrix A is not completely mixed, by Theorem 1.3.5, both players can skip one of their choices (row or column) optimally. Thus, $v = \max(v_{.1}, v_{.2}, \ldots, v_{.n})$. (Player I will skip the column in such a way that the value is not reduced by playing the subgame.) Similarly, if $A_{i.}$ is the matrix obtained from A by deleting row i and with value $v_{i.}$, we have $v = \min(v_{1.}, v_{2.}, \ldots, v_{n.})$. Let v_{ij} be the value of the matrix game $A(i, j)$ obtained from A by deleting row i and column j. Since $A_{.j}$ is a payoff with more rows than columns, by Theorem 1.3.4, Player I can skip a row and play optimally. Therefore, $v_{.j} = \min_i v_{ij}$, which shows that $v = \max_j v_{.j} = \max_j \min_i v_{ij}$. Similarly, we have $v = \min_i v_{i.} = \min_i \max_j v_{ij}$, and the proof is complete. ∎

In a seminal paper Kaplansky (1945) first introduced the notion of completely mixed strategies. The theorems in this section are essentially from this paper.

1.4. The Perron-Frobenius theorem

The results on completely mixed matrix games proved in the previous section will now be used to establish the Perron-Frobenius Theorem on nonnegative matrices. The next result provides the crucial link for such a proof. For each real λ, let $v(\lambda)$ be the value of the matrix game with payoff $A - \lambda I$.

Theorem 1.4.1. *Let A be an $n \times n$ payoff matrix. Then*

(i) $v(\lambda) \to \pm\infty$ as $\lambda \to \mp\infty$.
(ii) $v(\lambda)$ is nonincreasing in λ.
(iii) $v(\lambda)$ is Lipschitz continuous.

Proof. (*i*) Let Player II choose all rows with equal chance $1/n$. Then Player I can expect on the average at most $\max_j \frac{1}{n} \sum_i a_{ij} - (\lambda/n)$ for the payoff $A - \lambda I$. Thus, no matter what Player I does, we have

$$v(\lambda) \leq \max_j \frac{1}{n} \sum_i a_{ij} - \frac{\lambda}{n} \to -\infty \quad \text{as } \lambda \to \infty.$$

(*ii*) Since $A - \lambda I \geq A - \mu I$ when $\lambda \leq \mu$, Player I gains as much by playing $A - \lambda I$ as by playing $A - \mu I$. Thus $v(\lambda) \geq v(\mu)$ and $v(\lambda)$ is nonincreasing in λ.

(*iii*) Let x° and y° be optimal strategies for Player II and Player I in $A - \lambda I$. Let x^*, y^* be optimal strategies for Player II and I in $A - \mu I$. We have for the inner product

$$\langle x^*, (A - \lambda I)y^\circ \rangle \geq v(\lambda) \tag{1.4.1}$$

and

$$\langle (A - \mu I)^T x^*, y^\circ \rangle = \langle x^*, (A - \mu I)y^\circ \rangle \leq v(\mu). \tag{1.4.2}$$

Using (1.4.1) and (1.4.2) we get

$$v(\lambda) - v(\mu) \leq \langle x^*, (A - \lambda I)y^\circ \rangle - \langle x^*, (A - \mu I)y^\circ \rangle$$
$$\leq (\mu - \lambda)\langle x^*, y^\circ \rangle$$
$$\leq |\mu - \lambda|,$$

since $\langle x^*, y^\circ \rangle \leq \sum_i x_i^* \sum_i y_i^\circ = 1$. Similarly,

$$-|\mu - \lambda| \leq (\lambda - \mu)\langle x^\circ, y^* \rangle \leq v(\lambda) - v(\mu)$$

and therefore

$$|v(\lambda) - v(\mu)| \leq |\lambda - \mu|.$$

Thus v is Lipschitz continuous.   ∎

Lemma 1.4.2. *Let $A \geq 0$ be an irreducible matrix of order n. Then there exists $\lambda_0 > 0$ with $v(\lambda_0) = 0$.*

Proof. Suppose $v(0) = $ value $(A) \leq 0$. Thus, for some optimal x of Player II, $A^T x \leq 0$. Since $A \geq 0$ we have $A^T x = 0$. If $x_p > 0$ for some coordinate p of x, then $a_{pj} = 0$ for $j = 1, 2, \ldots, n$. Permuting the first row with the p-th row and the first column with the p-th column we find that the matrix A can be reduced to the form

$$\begin{bmatrix} B & 0 \\ C & D \end{bmatrix},$$

where B, D are square matrices of order 1 and $n - 1$, respectively. This contradicts the irreducibility of A. By Theorem 1.4.1, $v(\lambda) \to -\infty$ as $\lambda \to \infty$ and $v(\lambda)$ is Lipschitz continuous. Since $v(0) > 0$ we have $v(\lambda_0) = 0$ for some $\lambda_0 > 0$.   ∎

Lemma 1.4.3. *Let $A \geq 0$ be irreducible. Let λ_0 be defined as in Lemma 1.4.2. Then the matrix game $A - \lambda_0 I$ is completely mixed.*

Proof. By Theorem 1.3.5 it is enough to prove that every optimal strategy of Player II in the game $A - \lambda_0 I$ is completely mixed. Suppose to the contrary that x is an optimal strategy of Player II with $x_p = 0$ (i.e., row p is skipped optimally by Player II). We have $(A - \lambda_0 I)^T x \leq 0$. This gives $A^T x \leq \lambda_0 x$. Since $A \geq 0$, $(A^T)^k x \leq \lambda_0^k x$ for all $k = 1, 2, \ldots, n$. In particular, $\sum_i a_{ip}^{(k)} x_i \leq \lambda_0^k x_p = 0$. However, at least some $x_q > 0$ and therefore $a_{qp}^{(k)} = 0$ for all k. This contradicts the irreducibility of A by Theorem 1.1.2 and completes the proof.   ∎

We are now ready to prove the Perron-Frobenius Theorem.

Theorem 1.4.4 (Perron-Frobenius Theorem). *Let $A \geq 0$ be an $n \times n$ irreducible matrix. Then*

(i) *$Ay = \lambda_0 y$ for some $\lambda_0 > 0$, $y > 0$.*

(ii) *The eigenvalue λ_0 is geometrically simple.*

(iii) *The eigenvalue λ_0 is maximal in modulus among all the eigenvalues of A. That is, for any eigenvalue μ of A, $|\mu| \leq \lambda_0$.*

(iv) *The only nonnegative, nonzero eigenvectors of A are just the positive scalar multiples of y.*

(v) *The eigenvalue λ_0 is algebraically simple.*

(vi) *Let $\lambda_0, \lambda_1, \ldots, \lambda_{(k-1)}$ be the distinct eigenvalues of A with $|\lambda_i| = \lambda_0$, $i = 1, 2, \ldots, k - 1$. Then they are precisely the solutions of the equation $\lambda^k - \lambda_0^k = 0$.*

Proof. Let λ_0 be defined as in Lemma 1.4.2. Then $v(\lambda_0) = 0$ and, by Lemma 1.4.3, the game $A - \lambda_0 I$ is completely mixed. If y is an optimal strategy for Player I and x is an optimal strategy for Player II, we know that $(A - \lambda_0 I)y = 0$. This proves (*i*). We know from Theorem 1.3.6 that rank $(A - \lambda_0 I) = n - 1$. Thus, any solution u to $(A - \lambda_0 I)u = 0$ is a scalar multiple of y, which gives (*ii*).

Let $Au = \mu u$ for some real or complex eigenvalue μ. Recall that u^+ is defined as

$$u^+ = (|u_1|, |u_2|, \ldots, |u_n|)^T,$$

where $u = (u_1, u_2, \ldots, u_n)^T$. Without loss of generality, let u^+ be a probability vector. Now, for $A = (a_{ij}) \geq 0$, we have

$$\sum_j a_{ij}|u_j| \geq \left| \sum_j a_{ij}u_j \right| = |\mu u_i| = |\mu||u_i|.$$

Thus $Au^+ \geq |\mu|u^+$ and $(A - |\mu|I)u^+ \geq 0$, showing that $v(|\mu|) \geq 0$. In the case where $|\mu| > \lambda_0$, we can use the completely mixed optimal strategy x of Player II in $A - \lambda_0 I$ to get

$$(A - |\mu|I)^T x = (A - \lambda_0 I)^T x + (\lambda_0 - |\mu|)x = 0 + (\lambda_0 - |\mu|)x < 0,$$

which contradicts $v(|\mu|) \geq 0$. Thus $|\mu| \leq \lambda_0$.

Let $Aw = \alpha w$, where $w \geq 0$. We know $A^T x = \lambda_0 x$ for the optimal strategy x of Player II in $A - \lambda_0 I$. We also have $x > 0$. Thus $\alpha \langle w, x \rangle = \langle \alpha w, x \rangle = \langle Aw, x \rangle = \langle w, A^T x \rangle = \langle w, \lambda_0 x \rangle = \lambda_0 \langle w, x \rangle$. Since $\langle w, x \rangle > 0$, $\alpha = \lambda_0$ and, from (*ii*), w is a scalar multiple of y. This proves (*iv*). ∎

Before proving (v) we recall the following:

Lemma 1.4.5. *Let $C(\lambda) = A - \lambda I$. Let $C_{ij}(\lambda)$ be the cofactor of the (i, j)-th entry in $C(\lambda)$. Then*

$$\frac{d}{d\lambda}|C(\lambda)| = -\sum_i C_{ii}(\lambda).$$

Proof. For any continuously differentiable function $F = F(u_1, u_2, \ldots, u_m)$ in $(u_1, \ldots, u_m)$ we have, by the chain rule,

$$dF = \sum_j \frac{\partial F}{\partial u_j} du_j.$$

If the u'_j are functions of a real parameter t, then

$$\frac{dF}{dt} = \sum_j \frac{\partial F}{\partial u_j} \frac{du_j}{dt}.$$

Applying this formula to $F = |C| = F(c_{11}, c_{12}, \ldots, c_{nn})$ we get

$$dF = \sum_i \sum_j \frac{\partial F}{\partial c_{ij}} dc_{ij},$$

where $c_{ij} = c_{ij}(\lambda)$. Thus,

$$\frac{d}{d\lambda}|C(\lambda)| = \sum_i \sum_j \frac{\partial}{\partial c_{ij}}(|C|)\frac{d}{d\lambda}c_{ij}(\lambda).$$

Since $|C| = \sum_j c_{ij}C_{ij}$ we have

$$\frac{\partial}{\partial c_{ij}}|C| = C_{ij} \quad \text{for all } i, j.$$

Therefore,

$$\frac{d}{d\lambda}|C(\lambda)| = \sum_i \sum_j C_{ij}(\lambda)\frac{d}{d\lambda}c_{ij}(\lambda).$$

However, $c_{ij}(\lambda) = a_{ij} - \lambda\delta_{ij}$ (where δ_{ij} denotes the Kronecker δ) and

$$\frac{d}{d\lambda}c_{ij}(\lambda) = -\delta_{ij}.$$

Thus $\frac{d}{d\lambda}|C(\lambda)| = -\sum_i C_{ii}(\lambda)$.

Now continuing with our proof of Theorem 1.4.4, for the characteristic polynomial $p(\lambda) = |A - \lambda I|$,

$$\frac{d}{d\lambda}p(\lambda_0) = \frac{d}{d\lambda}p(\lambda)|_{\lambda=\lambda_0} = -\sum_i C_{ii}(\lambda_0).$$

Because the game $A - \lambda_0 I$ is completely mixed with value zero, by Theorem 1.3.6, all cofactors $C_{ii}(\lambda_0)$ are different from zero and are of the same sign. Thus $\frac{d}{d\lambda} p(\lambda_0) \neq 0$ and, therefore, λ_0 is a simple root of the characteristic polynomial $p(\lambda)$. This proves (v). ∎

Before we prove the last assertion of Theorem 1.4.4 we need the following lemma.

Lemma 1.4.6. *Let $x = (x_1, x_2, \ldots, x_n)$ be a vector with complex coordinates such that $|\sum_i x_i| = \sum_i |x_i|$. Then $x_i = \theta |x_i|$, $i = 1, 2, \ldots, n$ for some complex number θ with $|\theta| = 1$.*

Proof. The proof is easily seen by induction on n, and we verify only the case $n = 2$. Let x_1 and x_2 be two complex numbers. If $|x_1 + x_2| = |x_1| + |x_2|$ and $|x_1 + x_2| = 0$, then in that case $x_1 = x_2 = 0$ and the lemma follows. Let $x_1 + x_2 \neq 0$. We can represent $x_1, x_2, x_1 + x_2$ by points P, Q, R in the complex plane. Our hypothesis says that $|OP| + |OQ| = |OR|$, where $|OP|$, for example, denotes the length of the line segment OP with O as the point 0 of the complex plane. Thus $O, P, Q,$ and R lie on the same line and, therefore,

$$\theta = \frac{x_1}{|x_1|} = \frac{x_2}{|x_2|} = \frac{x_1 + x_2}{|x_1 + x_2|}.$$

This is the same as saying $x_i = \theta |x_i|$, $i = 1, 2$, where $\theta = \frac{x_1 + x_2}{|x_1 + x_2|}$. ∎

The last assertion in Theorem 1.4.4 does not appear to be based on game theoretic ideas. We first prove the last assertion for the special case $\lambda_0 = 1$, from which the general case easily follows. The first step in this direction is the following:

Lemma 1.4.7. *Let $A \geq 0$ be an $n \times n$ irreducible matrix with the maximal eigenvalue (in modulus) $\lambda_0 = 1$. Then the set of eigenvalues of absolute value unity form a group under multiplication.*

Proof. Let λ and μ be two eigenvalues (not necessarily distinct) of A with $|\lambda| = |\mu| = 1$. Let $Au = \lambda u$ and $Av = \mu v$. As before, let u^+ denote the vector $u^+ = (|u_1|, |u_2|, \ldots, |u_n|)^T$, where $u = (u_1, u_2, \ldots, u_n)$. Without loss of generality, we can assume that u^+ and v^+ are probability vectors. We have

$$\sum_j a_{ij} u_j = \lambda u_i, \quad i = 1, 2, \ldots, n.$$

Taking absolute values on both sides of the above equation, we get

$$|u_i| = |\lambda||u_i| = |\lambda u_i| = \left| \sum_j a_{ij} u_j \right| \le \sum_j a_{ij} |u_j|, \quad i = 1, 2, \ldots, n.$$

$$(1.4.3)$$

We thus have $(A-I)u^+ \ge 0$. Since $\lambda_0 = 1$, the game $A-I$ is completely mixed with value zero by Lemma 1.4.3, and so $(A-I)u^+ = 0$. That is, $Au^+ = u^+$ and, similarly, $Av^+ = v^+$. From (1.4.3) we see that

$$\sum_j a_{ij} |u_j| = \left| \sum_j a_{ij} u_j \right|, \quad i = 1, 2, \ldots, n.$$

Applying Lemma 1.4.6 n times we see that, for some θ_i, π_i of absolute value unity,

$$a_{ij} u_j = \theta_i a_{ij} |u_j|, \qquad a_{ij} v_j = \pi_i a_{ij} |v_j| \quad \text{for all } i, j. \qquad (1.4.4)$$

However, the probability vector u^+ is the unique eigenvector for the maximal eigenvalue 1 and, therefore, $u^+ = v^+$. Now consider the vector $(\pi_1 u_1, \ldots, \pi_n u_n)$. We show that it is an eigenvector. We have

$$\sum_j a_{ij} \pi_j u_j = \sum_j a_{ij} u_j \pi_j$$

$$= \sum_j \pi_j \theta_i a_{ij} |u_j| \quad \text{[from (1.4.4)]}$$

$$= \sum_j \pi_j \theta_i a_{ij} |v_j| \quad \text{(since } |u_j| = |v_j|\text{)}$$

$$= \theta_i \sum_j \pi_j a_{ij} \left(\sum_k a_{jk} |v_k| \right) \quad \text{(since } Av^+ = v^+\text{)}$$

$$= \theta_i \sum_j a_{ij} \left(\sum_k \pi_j a_{jk} |v_k| \right)$$

$$= \theta_i \sum_j a_{ij} \left(\sum_k a_{jk} v_k \right) \quad \text{[from (1.4.4))]}$$

$$= \theta_i \sum_j a_{ij} \mu v_j \quad \text{(since } Av = \mu v\text{)}$$

$$= \mu \theta_i \sum_j a_{ij} v_j$$

$$= \mu \theta_i \sum_j \pi_i a_{ij} |v_j| \quad \text{[from 1.4.4]}$$

$$= \mu\theta_i \sum_j \pi_i a_{ij} |u_j| \quad (\text{since } u^+ = v^+)$$

$$= \mu\pi_i \sum_j \theta_i a_{ij} |u_j|$$

$$= \mu\pi_i \sum_j a_{ij} u_j \quad [\text{from (1.4.4)}]$$

$$= \mu\lambda\pi_i u_i.$$

Thus $(\pi_1 u_1, \ldots, \pi_n u_n)^T$ is an eigenvector for the eigenvalue $\mu\lambda$. This shows that the set of eigenvalues of absolute value unity is closed under multiplication. Since there are only a finite number of distinct eigenvalues, we have, for any arbitrary eigenvalue λ of absolute value unity, that $\lambda, \lambda^2, \ldots, \lambda^l$ are eigenvalues and $\lambda^s = \lambda^l$ for some $s > l$. In particular, $\lambda^{s-l} = 1$. If $s - l = 1$, then $\lambda = 1$ is the only eigenvalue of absolute value unity. If $s - l > 1$, then λ^{s-l-1} is also an eigenvalue of A and, further, since $\lambda^{s-l-1} = \lambda^{-1}$, the multiplicative inverse of λ, the set of eigenvalues of absolute value unity form a group under multiplication. ∎

The proof of Lemma 1.4.7 is based on Karlin (1959), who attributes it to Bohnenblust. Rota (1961) extended the technique to positive operators on L_1 and L_∞ spaces.

We recall the following property of finite groups:

Lemma 1.4.8. *Let id denote the identity element of a finite group G with N elements. Let $x \neq id$ be any arbitrary element of G with $x^k \neq id$ for $1 \leq k < h$ and $x^h = id$. Then h divides N.*

Proof. This is essentially Lagrange's Theorem. [See, Herstein (1964), p. 35.] ∎

Now, continuing with our proof of Theorem 1.4.4, by a direct application of Lemmas 1.4.7 and 1.4.8, $\lambda^k - 1 = 0$ has $\lambda_0, \lambda_1, \lambda_2, \ldots, \lambda_{(k-1)}$ as the solution set. The proof for the general case in assertion (vi) of Theorem 1.4.4 follows by considering the matrix $\frac{1}{\lambda_0} A$. This completes the proof of Theorem 1.4.4.

We refer to λ_0 as the *Perron eigenvalue* (or the *Perron root*) and to y as a *(right) Perron eigenvector* of A. A *left Perron eigenvector* of A is simply a right Perron eigenvector of A^T.

We will often refer to basic concepts from the theory of linear programming [see, for example, Gale (1960), Chvátal (1983)] and, in particular, the Duality

Theorem will be used several times. The next result, which is a version of Farkas' Lemma, is derived using the Duality Theorem.

Theorem 1.4.9. *Let A be an $m \times n$ matrix and let b be a column vector of order m. Then exactly one of the following statements is true.*

(i) *The matrix equation $Ax = b$ has a solution $x \geq 0$.*

(ii) *The inequalities $A^T y \geq 0$ and $(b, y) < 0$ have a solution.*

Proof. Consider the dual linear programming problems:

primal: $\max \langle 0, x \rangle$ subject to $Ax = b$, $x \geq 0$

dual: $\min \langle b, y \rangle$ subject to $A^T y \geq 0$.

Clearly, $y = 0$ is feasible for the dual. If the dual problem has an optimal solution, then by the Duality Theorem, the primal has an optimal, and hence a feasible solution, and the objective function has value zero at the optimal solution. Thus $A^T y \geq 0 \Rightarrow \langle b, y \rangle \geq 0$ and hence (ii) does not hold but (i) holds. Conversely, if the dual has no optimal solution, the primal cannot have any optimal solution. However, for the primal, any feasible solution is optimal and thus the primal cannot even be feasible. Thus, when the dual has no optimal solution, then (ii) is true. That completes the proof of the theorem. ∎

Theorem 1.4.10. *Let $Ay = v\mathbf{1}$ for all optimal strategies y of Player I for a payoff matrix A with value v. Then there exists an optimal strategy x for Player II that is completely mixed.*

Proof. We assume, without loss of generality, that $v = 0$. Let $e_1 = (1, 0, \ldots, 0)^T$. Consider the equation

$$\begin{bmatrix} A^T \\ e_1^T \\ -e_1^T \end{bmatrix} x \leq \begin{bmatrix} 0 \\ 1 \\ -1 \end{bmatrix}. \tag{1.4.5}$$

If (1.4.5) has a nonnegative solution x, then x can be normalized to give an optimal strategy for Player II with $x_1 > 0$. Otherwise, by Theorem 1.4.9,

$$\begin{bmatrix} A & e_1 & -e_1 \end{bmatrix} \begin{bmatrix} y \\ \alpha \\ \beta \end{bmatrix} \geq 0, \quad y \geq 0, \alpha \geq 0, \beta \geq 0, \alpha - \beta < 0$$

has a solution. That is, $(Ay) + (\alpha - \beta)e_1 \geq 0$. Thus, $(Ay)_1 > 0$ and y can be normalized to give an optimal strategy for Player I. However, $(Ay)_1 > 0$, which is a contradiction to our assumption. Hence, there exists an optimal strategy x

for Player II with $x_1 > 0$. A similar argument shows that Player II has an optimal x with $x_i > 0$ for any i. The average of such optimal strategies is an optimal completely mixed strategy for Player II. That completes the proof. ∎

Theorem 1.4.11. *Let A be a positive symmetric payoff matrix with no completely mixed optimal strategy for Player II. Then the convex sets of optimal strategies for the two players are disjoint and can be strictly separated by a hyperplane $\langle x^0, x \rangle = c$ for some c, where x^0 is a Perron eigenvector.*

Proof. Let $Ax^0 = \lambda_0 x^0, \lambda_0 > 0$, and $x^0 > 0$. Let y be any optimal strategy for Player I. By Theorem 1.4.10 we can assume, without loss of generality, that $(Ay)_1 = \sum_j a_{1j} y_j > v$ and $\sum_j a_{ij} y_j \geq v$ for all i. Since $A = A^T$ we have

$$\langle x^0, y \rangle = \lambda_0^{-1} \langle Ax^0, y \rangle = \lambda_0^{-1} \langle x^0, Ay \rangle > v\lambda_0^{-1} \sum_i x_i^0.$$

For any optimal strategy x of Player II we have

$$\langle x^0, x \rangle = \lambda_0^{-1} \langle Ax^0, x \rangle = \lambda_0^{-1} \langle x^0, Ax \rangle \leq v\lambda^{-1} \sum_i x_i^0.$$

Thus, x^0 strictly separates the optimal strategy sets. ∎

Theorem 1.4.12. *Let A, B be two positive square payoff matrices that commute. Let Player I in each game have a completely mixed optimal strategy and let Player II have an optimal strategy common to both payoffs A and B. Then the ratio of the values of the games coincides with the corresponding ratio of the Perron eigenvalues.*

Proof. Let $Ay = \lambda y, y > 0$. We have $BAy = A(By) = \lambda By$. Because λ is a simple root of A, By is a scalar multiple of y; let $By = \mu y$. Let x^0 be a common optimal strategy for Player II for the games A and B. Since Player I has a completely mixed strategy we have, by Theorem 1.3.2, $A^T x^0 = v_1 \mathbf{1}$ and $B^T x^0 = v_2 \mathbf{1}$, where v_1, v_2 are the values of A and B, respectively. Thus $\langle x^0, Ay \rangle = v_1 \langle \mathbf{1}, y \rangle = \lambda \langle x^0, y \rangle$. Similarly, $\langle x^0, By \rangle = v_2 \langle \mathbf{1}, y \rangle = \mu \langle x^0, y \rangle$. Since $y > 0$, it follows that

$$\frac{v_1}{v_2} = \frac{\lambda}{\mu},$$

which completes the proof. ∎

The Perron-Frobenius Theorem has been proved and reproved in many different ways. Perron (1907) originally proved the theorem for positive matrices while studying the theory of continued fractions. The important extension to the irreducible case was carried out by Frobenius (1912). Alexandroff and Hopf (1935) used topological methods to prove the existence of a positive eigenvalue and a nonnegative eigenvector for a nonsingular, nonnegative matrix (see the proof of Perron's Theorem given in Section 1.2). Putnam (1958) and Karlin (1959) gave alternative proofs using the following theorem due to Pringshiem: *Let $f(z) = \sum a_n z^n$ be a power series with $a_n \geq 0$ for all n. Let r be the radius of convergence. Then the analytic function f is singular at $z = r$. In addition, if this singularity is a pole, it is of maximal order on $|z| = r$.*

Ostrowski (1963) and Householder (1964) use matrix norms and series expansions to prove parts of the Perron-Frobenius Theorem. Numerical analysts have generally followed the proof due to Wielandt (1950). Blackwell (1961) first used the Minimax Theorem of von Neumann to give an elegant proof, which forms the basis of the proof given here. For a comprehensive account of the Perron-Frobenius theory, see Gantmacher (1959).

1.5. Nonsingular M-matrices

A matrix A is called a Z-matrix if all the off-diagonal entries (those not on the main diagonal) are nonpositive. A Z-matrix A is called an M-matrix if A can be represented as $A = sI - B$, where $B \geq 0$ and $s \geq \rho(B)$, the spectral radius of B. The following game theoretic observation is the key to our study of M-matrices.

Theorem 1.5.1. *Let A be a square payoff matrix with nonpositive off-diagonal entries. Then the following assertions hold:*

 (i) *If the value of the game is positive, the game is completely mixed.*
 (ii) *If the value of the game is nonnegative and the off-diagonal entries negative, the game is completely mixed.*

Proof. To prove the first part of the theorem, suppose that the game is not completely mixed. Then, by Theorem 1.3.5, Player I has an optimal strategy x°, with some coordinate, say $x_n^0 = 0$. The $n \times (n-1)$ submatrix, with the last column deleted from A, has as its last row, a vector with nonpositive entries. Player II can constantly choose this row, so clearly the value of the new matrix is nonpositive. However, with $x_n^0 = 0$, the matrices have the same value. This contradicts our assumption that the value of A is positive. The second part of the theorem is similarly proved. ∎

We denote the value of the payoff matrix A by $v(A)$. In the following theorem we have chosen only those equivalent conditions that have game theoretic import, even though several other equivalent conditions are available in the literature; see Berman and Plemmons (1994).

Theorem 1.5.2. *Let $A = (a_{ij})$ be an $n \times n$ Z-matrix. Then the following conditions are equivalent to the statement that A is a nonsingular M-matrix:*

(i) *There exists an $x \geq 0$ such that $Ax > 0$.*

(ii) *There exists an $x > 0$ such that $Ax > 0$.*

(iii) *Any Z-matrix C satisfying $C \geq A$ is nonsingular.*

(iv) *Each real eigenvalue of A is positive.*

(v) *All principal minors of A are positive.*

(vi) *The inverse matrix A^{-1} exists and $A^{-1} \geq 0$.*

(vii) *The matrix A reverses the sign only for the zero vector. That is, if $x_i(Ax)_i \leq 0$ for all i, then $x = 0$.*

(viii) *Every real eigenvalue of each principal submatrix of A is positive.*

(ix) *Every regular splitting of A is convergent. That is, if $A = M - N$, $M^{-1} \geq 0$, $N \geq 0$, then the spectral radius of $M^{-1}N$ is less than unity.*

(x) *There exists a positive diagonal matrix D such that $AD + DA^T$ is positive definite.*

Proof. $(i) \Rightarrow (ii)$: If we normalize x to a probability vector, condition (i) says $v > 0$ and, by Theorem 1.5.1, the game is completely mixed. Thus the optimal x^0 for Player I is positive and $Ax^0 > 0$.

$(ii) \Rightarrow (iii)$: Since $C \geq A$, $v(C) \geq v(A)$. By condition (ii), $v(A) > 0$. Thus $v(C) > 0$. In addition, the off-diagonal entries of C are nonpositive and so C is completely mixed by Theorem 1.5.1. By Theorem 1.3.7, C is nonsingular.

$(iii) \Rightarrow (iv)$: Let, if possible, $Ax = \alpha x$ for some $\alpha \leq 0$ so that $A - \alpha I$ is singular. However, $A - \alpha I$ is a Z-matrix and $A - \alpha I \geq A$. By condition (iii), $A - \alpha I$ is nonsingular and we have a contradiction.

$(iv) \Rightarrow (v)$: Let the real eigenvalues of A be positive. We first assume that the off-diagonal entries of A are negative. By assumption, the product of the real eigenvalues of A is positive. The product of the complex eigenvalues of the real matrix A is positive anyway. Hence $|A| > 0$. By condition (iv), the polynomial $p(\lambda) = |A - \lambda I|$ has no negative root and so has the same sign for all $\lambda \leq 0$. In particular, $|A + \alpha I| > 0$ for all $\alpha \geq 0$. We first show that $v(A) > 0$. Let, if possible, $v(A) \leq 0$. Then, by Theorem 1.4.1, $A + \alpha I$ has value 0 for some $\alpha \geq 0$. Since the off-diagonal entries of A are negative, Player I cannot skip any column in any optimal strategy and, by Theorem 1.3.5, the game is completely mixed. By Theorem 1.3.6, the matrix $A + \alpha I$ is singular so that $-\alpha \leq 0$ is an eigenvalue of A. This contradicts (iv) and we have $v(A) > 0$.

Now, given any Z-matrix A with positive real eigenvalues (if any), by continuity we can find a Z-matrix $B \leq A$ such that its real eigenvalues are positive. Clearly, $v(B) \leq v(A)$. By the above argument, $v(B) > 0$ and $v(A) > 0$. This means that $a_{ii} > 0$ for all i; otherwise some row would have only nonpositive entries, in which case, $v(A) \leq 0$. Furthermore, every principal submatrix shares the properties of A. For example, the principal submatrix $A(M) = (a_{ij})$, $i, j \in M \subset \{1, 2, \ldots, n\}$ is a Z-matrix. We show that the real eigenvalues of $A(M)$ (if any) are positive. Suppose, on the contrary, that $A(M)$ has a nonpositive eigenvalue of $-\beta$. Then $[A(M) + \beta I]z = 0$ for some $z \neq 0$. Let $u = (u_1, u_2, \ldots, u_n)^T$ with $u_i = z_i$ if $i \in M$ and $u = 0$ otherwise. This shows that the matrix $C = (c_{ij})$, with

$$
\begin{aligned}
c_{ij} &= a_{ij} \quad i, j \in M \\
c_{ii} &= a_{ii} \quad \text{for all } i \\
&= 0 \quad \text{for } i \neq j \text{ and } i \text{ or } j \notin M,
\end{aligned}
$$

has $(C + \beta I)u = 0$ and C has a nonpositive eigenvalue. But $C \geq A$ and hence $v(C) > 0$. Also, since C has nonpositive off-diagonal entries, it satisfies (i) and hence (iv) and, therefore, the real eigenvalues of C and of $A(M)$ are positive. However,

$$
|C| = \prod_{i \notin M} a_{ii} |A(M)| > 0.
$$

This shows that $|A(M)| > 0$, and the proof is complete.

(v) $\Rightarrow$ (vi): Let the principal minors of A be positive. As in the previous argument, by continuity we have a matrix $B \leq A$, where all the off-diagonal entries of B are negative and all the principal minors of B are positive. By continuity we can also assume that the cofactors B_{ij} of B have the same sign as the cofactors A_{ij} of A whenever $A_{ij} \neq 0$. Since the principal minors of B are positive, $|B + \alpha I| > 0$ for all $\alpha \geq 0$. Further, we claim that $v(B) > 0$, for otherwise, by Theorem 1.4.1, $v(B + \alpha I) = 0$ for some $\alpha \geq 0$. Since $B + \alpha I$ is a Z-matrix, by Theorem 1.5.1, the game $B + \alpha I$ is completely mixed. Now, by Theorem 1.3.6, $B + \alpha I$ is singular, which is a contradiction. Thus $v(B) > 0$ and hence $v(A) > 0$. By Theorem 1.4.1, A is completely mixed. Now consider the matrix

$$
B(\alpha) = \begin{bmatrix}
b_{11} & b_{12} & \cdots & b_{1n-1} & b_{1n} - \alpha \\
b_{21} & b_{22} & \cdots & b_{2n-1} & b_{2n} - \alpha \\
\cdots & \cdots & \cdots & \cdots & \cdots \\
b_{n1} & b_{n2} & \cdots & b_{nn-1} & b_{nn} - \alpha
\end{bmatrix}.
$$

Using an argument as in the proof of Theorem 1.4.1, we get $v[B(\alpha^*)] = 0$ for some $\alpha^* > 0$. Since the off-diagonal entries of $B(\alpha^*)$ are negative, by

Theorem 1.4.1, the matrix $B(\alpha^*)$ is completely mixed. Now, using Theorem 1.3.6, all the cofactors of $B(\alpha^*)$ are different from zero and are of the same sign. However, the cofactors $B_{in}(\alpha^*)$ are equal to B_{in}, $i = 1, 2, \ldots, n$. And, since $B_{nn} > 0$, $B_{in} > 0$ for $i = 1, 2, \ldots, n$. Similarly, we have $B_{ij} > 0$ for all i, j. By assumption this implies $A_{ij} \geq 0$. Further, $|A| > 0$. Thus, $A^{-1} \geq 0$, which completes the proof.

$(vi) \Rightarrow (i)$: Let $v(A) \leq 0$. Then there exists an optimal strategy y for Player II such that $A^T y \leq 0$. Since, by assumption, $(A^T)^{-1} \geq 0$ we have $(A^T)^{-1} A^T y \leq 0$. This gives $y \leq 0$, which is a contradiction, and shows that $v(A) > 0$ and $Ax > 0$ for the optimal x of player I. This completes the proof.

Having established the equivalence of (i) through (vi), we next prove that they are equivalent to the remaining conditions.

$(i) \Rightarrow (vii)$: Suppose $x_i (Ax)_i \leq 0$ for all i, with $x \neq 0$; consider the matrix $J = \{i : x_i \neq 0\}$. Our condition is equivalent to $My + Uy = 0$, where M is the principal submatrix of A with row and column indices from J, y is the vector x retaining coordinates corresponding to indices in J, and U is a nonnegative diagonal matrix. Now, since $M + U \geq M$ and since $(i) \Rightarrow (v)$, then $v(M + U) \geq v(M) > 0$. Thus $M + U$ satisfies (i) and, by (iii), is nonsingular. This contradicts $(M + U)y = 0$ for some $y \neq 0$. Thus $(i) \Rightarrow (vii)$.

$(vii) \Rightarrow (i)$: Suppose $x_i (Ax)_i \leq 0$ for all i implies that $x = 0$. Since A^T is also a Z-matrix, then by the symmetry of condition (v), it is sufficient to prove that $v(A^T) > 0$. In the case where $v(A^T) \leq 0$, any optimal strategy y in A^T of Player II gives $(A^T)^T y = Ay \leq 0$. In particular, $y_i (Ay)_i \leq 0$ for all i and $y \neq 0$. This contradicts our assumption. Hence $v(A^T) > 0$ and therefore $(A^T)^{-1} \geq 0$, $A^{-1} \geq 0$, and $v(A) > 0$ by the equivalence of (i) through (vi).

The proof of $(iv) \Rightarrow (viii)$ is contained in the proof of $(iv) \Rightarrow (v)$, whereas $(viii)$ trivially implies (iv).

$(ii) \Rightarrow (ix)$: Let $A = M - N$, $M^{-1} \geq 0$, and $M^{-1}N = C \geq 0$. By assumption, $Ay > 0$ for some $y > 0$. Since $M^{-1} \geq 0$, $Cy < y$. Let $C^T u = \alpha u$, where α is the spectral radius of C. By Theorem 1.4.4, $u \geq 0$, $\alpha \geq 0$. Further, since $u \neq 0$, $\langle u, y \rangle > 0$. Thus $\alpha \langle u, y \rangle = \langle u, Cy \rangle < \langle u, y \rangle$. This shows that $\alpha < 1$, and hence the implication is proved.

$(ix) \Rightarrow (iv)$: By Theorem 1.4.1, $v(A + \alpha I) > 0$ for α sufficiently large. Since $(i) \Rightarrow (vi)$, $(A + \alpha I)^{-1} \geq 0$. Thus $A = A + \alpha I - \alpha I$ is a regular splitting and, therefore, we have at least one regular splitting. Let $A = M - N$ be any regular splitting. By assumption, $C = M^{-1}N$ has spectral radius less than unity. By continuity, we have a positive matrix D with $0 \leq C < D$ such that the spectral radius of D is less than unity. Let $D^T u = \alpha u$, where α is the spectral radius and $u > 0$. We claim that $v(A^T) > 0$; otherwise $Ay \leq 0$ for some probability vector y. However, this means that $My \leq Ny$ and $y \leq Cy < Dy$. Further,

$\alpha \langle u, y \rangle = \langle u, Dy \rangle > \langle u, Cy \rangle \geq \langle u, y \rangle$. Since $\langle u, y \rangle > 0$, we have $\alpha > 1$, which is a contradiction. Thus $v(A^T) > 0$. Now, $(i) \Leftrightarrow (iv)$ applied to A^T implies that the real eigenvalues of A^T, and hence the real eigenvalues of A, are positive. Thus the implication is proved.

$(i) \Rightarrow (x)$: As in the above argument, when $Ay > 0$ for some $y > 0$, $A^T x > 0$ for some $x > 0$. Define the diagonal matrix D with the j-th diagonal element as $\frac{y_j}{x_j}$, where y_j, x_j are the j-th coordinates of y and x, respectively. We get $(AD + DA^T)x > 0$ and condition (ii) is satisfied for $AD + DA^T$. However, $AD + DA^T$ as a symmetric matrix has only real eigenvalues; in addition, it has nonpositive off-diagonals. Since $(ii) \Rightarrow (iv)$ the symmetric matrix $AD + DA^T$ has only positive eigenvalues and hence it is positive definite.

$(x) \Rightarrow (i)$: Suppose $v(A) \leq 0$. We have $A^T y \leq 0$ for some probability vector y. Since D is a positive diagonal matrix, $DA^T y \leq 0$ and

$$\langle y, (AD + DA^T)y \rangle = 2 \langle y, DA^T y \rangle \leq 0.$$

This contradicts the positive definiteness of $AD + DA^T$. Thus $v(A) > 0$ and we have the required implication. ∎

Nonsingular M-matrices were introduced by Ostrowski (1937–1938) and many of their properties were discovered independently by Fiedler and Ptak (1962), Fan (1958), and Ostrowski (1937–1938). The game theoretic proof of Theorem 1.5.2 given here is adapted from Raghavan (1978).

We now indicate an application of Theorem 1.5.2 to the method of successive approximations and iterative methods in obtaining the solutions to matrix equations. Suppose we wish to solve the matrix equation $Ax = b$, that is,

$$\begin{bmatrix} a_{11} & a_{12} & \cdots & a_{1n} \\ a_{21} & a_{22} & \cdots & a_{2n} \\ \vdots & \vdots & \ddots & \vdots \\ a_{n1} & a_{n2} & \cdots & a_{nn} \end{bmatrix} \begin{bmatrix} x_1 \\ x_2 \\ \vdots \\ x_n \end{bmatrix} = \begin{bmatrix} b_1 \\ b_2 \\ \vdots \\ b_n \end{bmatrix}.$$

If A is known to be a nonsingular M-matrix, then, by conditions (v) and (ix) of Theorem 1.5.2, the diagonal entries of A are positive, furthermore,

$$\begin{bmatrix} a_{11} & 0 & \cdots & 0 \\ 0 & a_{22} & \cdots & 0 \\ 0 & \vdots & \ddots & 0 \\ 0 & 0 & \cdots & a_{nn} \end{bmatrix} - \begin{bmatrix} 0 & -a_{12} & \cdots & -a_{1n} \\ -a_{21} & 0 & \cdots & -a_{2n} \\ \vdots & \vdots & \ddots & \vdots \\ -a_{n1} & -a_{n2} & \cdots & 0 \end{bmatrix}$$

is a convergent regular splitting. Denoting the above splitting by $A = M - N$, we have $M^{-1}N \geq 0$, $M \geq 0$, and so $C = M^{-1}N$ is easily found. Let $u = M^{-1}b$.

Consider the iterative scheme

$$x^{(r)} = Cx^{(r-1)} + u, \quad r = 1, 2, \ldots.$$

By a repeated application of the above iterative scheme, we get

$$x^{(r)} = C^{(r)}x^{(0)} + (I + C + C^2 + \cdots + C^{r-1})u.$$

Let $r(\cdot)$ denote the spectral radius. Since $r(C) < 1$, $(I - C)^{-1}$ exists. Thus

$$
\begin{aligned}
x^{(r)} &= C^r x^{(0)} + (I - C)^{-1}(I - C)(I + C + \cdots + C^{r-1})u \\
&= C^r x^{(0)} + (I - C)^{-1}(I - C^r)u.
\end{aligned}
$$

The matrix C is nonnegative. Choose $H > 0$ and $H \geq C$ such that $\beta = r(H) < 1$. We have $Hw = \beta w$ for some $w > 0$. Now $\beta^r \to 0$. This shows that $\sum_i \sum_j h_{ij}^{(r)} w_j \to 0$. Since $w > 0$, $h_{ij}^{(r)} \to 0$ for all i, j. In particular, $c_{ij}^{(r)} \to 0$ for all i, j and $x^{(r)} \to (I - C)^{-1}u = A^{-1}b$. Thus the iterative scheme determines the solution for the equation $Ax = b$ when A is a nonsingular M-matrix.

Theorem 1.5.3. *Let $A = M_1 - N_1 = M_2 - N_2$ be two regular splittings of a nonsingular M-matrix A. Let $a_{ij} < 0$ for all $i \neq j$. If $N_2 \geq N_1 \geq 0$, then $r(M_2^{-1}N_2) \geq r(M_1^{-1}N_1)$.*

Proof. If $A = M - N$ is a regular splitting, then $M = A + N$ and $M^{-1}N = (I + G)^{-1}G$, where $G = A^{-1}N$. However, G and $M^{-1}N$ have common eigenvectors. In fact, if $Gx = \alpha x$, then $(I + G)^{-1}Gx = \frac{\alpha}{1+\alpha}x$. Conversely, if $M^{-1}Nz = \mu z$, then $Gz = (\frac{\mu}{1-\mu})z$ and $\mu \neq 1$. Thus $r(M^{-1}N) = \frac{r(A^{-1}N)}{1+r(A^{-1}N)}$ and $r(M^{-1}N)$ is a monotonic function of $r(A^{-1}N)$. Also, $A^{-1} \geq 0$ since A is a nonsingular M-matrix. Thus, if $N_2 \geq N_1$, then $r(A^{-1}N_2) \geq r(A^{-1}N_1)$, which yields $r(M_2^{-1}N_2) \geq r(M_1^{-1}N_1)$. ∎

Example 1.5. Suppose we wish to solve the system of equations

$$6x - 4y - z = -5$$

$$-3x + 7y - 2z = 5.$$

$$-5x - 4y + 10z = 17$$

Define

$$
M = \begin{bmatrix} 6 & 0 & 0 \\ 0 & 7 & 0 \\ 0 & 0 & 10 \end{bmatrix}, \quad
N = \begin{bmatrix} 0 & 4 & 1 \\ 3 & 0 & 2 \\ 5 & 4 & 0 \end{bmatrix}, \quad \text{and} \quad
b = \begin{bmatrix} -5 \\ 5 \\ 17 \end{bmatrix}.
$$

Starting with the initial solution

$$x^0 = \begin{bmatrix} 1 \\ 1 \\ 1 \end{bmatrix},$$

the iterative algorithm yields

$$x^{(1)} \simeq \begin{bmatrix} 0.001 \\ 1.429 \\ 2.600 \end{bmatrix}, \quad x^{(2)} \simeq \begin{bmatrix} 0.554 \\ 1.462 \\ 2.272 \end{bmatrix}, \quad x^{(3)} \simeq \begin{bmatrix} 0.521 \\ 1.602 \\ 2.886 \end{bmatrix},$$

$$x^{(4)} \simeq \begin{bmatrix} 0.718 \\ 2.226 \\ 2.601 \end{bmatrix}, \quad x^{(5)} \simeq \begin{bmatrix} 1.086 \\ 1.766 \\ 2.949 \end{bmatrix},$$

$$x^{(6)} \simeq \begin{bmatrix} 0.837 \\ 2.023 \\ 2.949 \end{bmatrix}, \quad \text{and} \quad x^{(7)} \simeq \begin{bmatrix} 1.008 \\ 1.916 \\ 2.928 \end{bmatrix}.$$

The solutions $x^{(7)}$ and $x^{(8)}$ are almost the same and $(1, 2, 3)^T$ is a good approximate solution.

The above iterative method is an improvement over the Gaussian elimination algorithm if the matrix is sparse (i.e., has many zero entries).

1.6. Polyhedral sets with least elements

In this section we discuss a generalization of the notion of an M-matrix in a geometric setting. The next few results on polyhedral sets with least elements are due to Cottle and Veinott (1972).

Let $X \subset R^n$. We say that X admits a *least element* if for some $\tilde{x} \in X, x \geq \tilde{x}$ for all $x \in X$. Clearly, when a least element exists, it is unique. As an example consider the box $X = \{(x_1, x_2, x_3) : 0 \leq x_1, x_2, x_3 \leq 1\}$. Then $(0, 0, 0)$ is the least element in X.

A set, in general, may not possess a least element. The existence of least elements in polyhedra is closely related to the Duality Theorem of linear programming.

Let A be an $m \times n$ matrix and let b be a vector of order m. We consider polyhedra X_b and X_b^+ defined by

$$X_b = \{x : Ax \geq b\},$$
$$X_b^+ = \{x : Ax \geq b, x \geq 0\}.$$

Given A and b, X_b or X_b^+ may be empty.

In case A has rank n, we say that an $n \times n$ submatrix B of A determines an element x of X_b if and only if (1) B is nonsingular and (2) $Bx = b_B$, where b_B is the restriction of the vector b to the rows corresponding to the rows of B. The matrix B is called a basis for determining x. Given the matrix A, the vector b, and the nonnegative vector c, we consider two dual linear programming problems:

Problem I. Minimize $\langle c, x \rangle$ subject to $Ax \geq b$.
Problem II. Maximize $\langle b, y \rangle$ subject to $A^T y = c$, $y \geq 0$.

With the above preliminaries, we can state the following theorem, characterizing polyhedra with least elements.

Theorem 1.6.1. *An element $\tilde{x}$ of the polyhedron X_b is the least element if and only if the matrix A has a nonnegative left inverse A^- and $\tilde{x} = A^- b$.*

Proof. When $\tilde{x}$ is the least element of X_b, then, for any $c \geq 0$, $\langle c, x \rangle \geq \langle c, \tilde{x} \rangle$ for all $x \in X_b$. Thus Problem I has an optimal solution for any $c \geq 0$. Choosing $c = e_j$, the j-th column of the $n \times n$ identity matrix, we have, by the Duality Theorem, an optimal solution $y^{(j)}$ to Problem II. Thus

$$A^T y^{(j)} = e_j, \quad y^{(j)} \geq 0, j = 1, 2, \ldots, n,$$
$$\langle e_j, \tilde{x} \rangle = \langle y^{(j)}, b \rangle, j = 1, 2, \ldots, n. \tag{1.6.1}$$

If $(A^-)^T = (y^{(1)}, \ldots, y^{(n)})$, we get $A^- \geq 0$, $A^T (A^-)^T = I$, and $\tilde{x}_j = (A^- b)_j$ from 1.6.1). Conversely, let $\tilde{x} \in X_b$ and let $A^- \geq 0$, satisfying $A^- A = I$ and $A^- b = \tilde{x}$. Then for any $x \in X_b$, $Ax \geq b \Rightarrow A^- Ax \geq A^- b$. Thus $A^- Ax = x \geq A^- b = \tilde{x}$. This completes the proof of the theorem. ∎

As a consequence we have the following theorem:

Theorem 1.6.2. *Let $\tilde{x}$ be the least element of X_b and let $A\tilde{x} - b$ have at most n zero coordinates. Then there exists a basis B determining $\tilde{x}$ such that $B^{-1} \geq 0$.*

Proof. From Theorem 1.6.1 we know that $A^- A = I$ and $A^- b = \tilde{x}$ for some $A^- \geq 0$. Since $A^- A = I$, rank $A = $ rank $A^- = n$. Without loss of generality let $(A\tilde{x} - b)_i = 0$ if $i \leq p$ and $(A\tilde{x} - b) > 0$ if $i > p$, where $p \leq n$. Let $A^- = [P, Q]$, $A = \begin{bmatrix} B \\ D \end{bmatrix}$, where P has p columns and B has p rows. We write $A\tilde{x} - b = (0, u^T)^T$, where 0 is a zero vector of order p and $u > 0$. Since $A^- b = \tilde{x}$, $A^- (A\tilde{x} - b) = 0$. Thus $Qu = 0$. Since $Q \geq 0$ and $u > 0$, we have $Q = 0$. The matrix $A^- = [P, 0]$ and $PB = I$. Therefore, rank $P = $ rank $B = n$.

Since P has p columns with $p \leq n$, and rank $P = n$, we have $p = n$ and $P = B^{-1} \geq 0$. Then $B\tilde{x} = b_B$, and the proof is complete. ∎

The following theorem uses dual feasibility to characterize polyhedra with least elements.

Theorem 1.6.3. *Let A and b be defined as in Problems I and II above. Let P be an $n \times m$ matrix such that $y = P^T c$ is feasible for Problem II for any $c \geq 0$ and optimal for some $\tilde{c} > 0$. Then X_b admits a least element.*

Proof. Let $\tilde{x} = Pb$. We show that $\tilde{x} \in X_b$ and further that $\tilde{x}$ is the least element in X_b. Since for all $c \geq 0$, $y = P^T c \geq 0$ is feasible for Problem II, then $A^T(P^T c) = c$ for all $c \geq 0$. Thus $P \geq 0$ and $PA = I$. If we show that $\tilde{x} \in X_b$, then, from Theorem 1.6.1, our assertion follows by identifying $P = A^-$. Let $\tilde{c} > 0$ and let $\tilde{y} = P^T \tilde{c}$ be optimal for Problem II. Let x^0 be optimal for Problem I (x^0 exists as a consequence of the Duality Theorem).

In the case in which $(P^T \tilde{c})_j = 0$, by the positivity of $\tilde{c}$, we have $(P^T c)_j = 0$ for any $c \geq 0$. By complementary slackness we see that $(Ax^0 - b)_i = 0$ or $P^T \tilde{c}_i = 0$ for each i. Thus x^0 and $P^T c$ are optimal for the dual problems for any $c \geq 0$. That is, $\langle P^T c, Ax^0 - b \rangle = 0$ for all $c \geq 0$. Thus $\langle c, P(Ax^0 - b) \rangle = 0$ for all $c \geq 0$. Substituting $PA = I$ we get $x^0 - Pb = 0$. Thus $x^0 = \tilde{x} \in X_b$, and the proof is complete. ∎

Theorem 1.6.4. *Let A and b be defined as in Problems I and II above. Let $x^* \in X_b$ be determined by B, an $n \times n$ nonsingular submatrix of A with a nonnegative inverse. Then x^* is the least element of X_b.*

Proof. The polyhedron X_b admits a least element if and only if $Q(X_b)$ admits a least element for every permutation matrix Q, that is, if $\{x : QAx \geq Qb\}$ has a least element for every permutation matrix Q. We can therefore assume without loss of generality that $A = \begin{bmatrix} B \\ D \end{bmatrix}$, where B is the given basis determining x^*. The vector $y = (c^T B^{-1}, 0)^T = (y_B^T, 0)^T$ satisfies $A^T y = c$ for any $c \geq 0$. Furthermore, $y \geq 0$ since $B^{-1} \geq 0$ and $c \geq 0$. Thus y is feasible for Problem II. Also,

$$\langle y, b \rangle = \langle y_B, b_B \rangle = \langle (B^{-1})^T c, Bx^* \rangle = \langle c, x^* \rangle.$$

Thus x^* is optimal for the primal Problem I for any $c \geq 0$. That is, $x \geq x^*$ for all $x \in X_b$, and the proof is complete. ∎

A matrix A of order $m \times n$ is called a *Leontief matrix* if the following two conditions hold: (i) each column has at most one positive element and (ii) $Ax > 0$ for some $x \geq 0$. The notion of a Leontief matrix stems from the theory of input-output matrices in economics. Incidentally, it also generalizes the notion of a nonsingular M-matrix to rectangular matrices.

Theorem 1.6.5. *Let A be an $m \times n$ matrix. The following conditions are equivalent:*

(i) *For any vector b, if the set $X_b = \{x : Ax \geq b\}$ is nonempty, then it has a least element.*

(ii) *There is an $n \times n$ nonsingular submatrix B of A with $(B^{-1})^T c \geq 0$ for some $c > 0$. Further, each such basis has a nonnegative inverse.*

Furthermore, if the $n \times n$ identity matrix is a submatrix of A, then conditions (i) and (ii) are equivalent to A^T being Leontief.

Proof. (i) $\Rightarrow$ (ii): Since $X_0 = \{x : Ax \geq 0\}$ has a least element, from Theorem 1.6.1 it follows that $A^- A = I$, $A^- \geq 0$ for some A^-. Moreover, since $A^- 0 = 0$, from Theorem 1.6.1 it follows that 0 is the least element of X_0. Let $c > 0$. Clearly, $y = (A^-)^T c$ is feasible for the dual Problem II. Therefore, there exists a basic feasible solution to the dual problem. We know from $A^- A = I$ that rank $A = $ rank $A^- = n$. Thus the basic feasible solution can be taken to be

$$\bar{y} = \begin{pmatrix} (B^{-1})^T c \\ 0 \end{pmatrix}, \quad \text{where } A = \begin{pmatrix} B \\ D \end{pmatrix}.$$

Let b be a vector of order m with the first n coordinates 0 and the last $m - n$ coordinates negative. The vectors 0 and $\bar{y}$ are feasible to the dual programming Problems I and II, respectively, with b defined as above. In addition, the complementary slackness is fulfilled and therefore $0, \bar{y}$ are optimal to the dual problem. Since X_b has a least element $\bar{x}$, we know that $\langle c, x \rangle$ attains its minimum at $\bar{x}$, besides 0. Since $c > 0$, we should have $\bar{x} = 0$. Now $A\{0\} - b$ has exactly n zero components and, by Theorem 1.6.2, B has a nonnegative inverse.

(ii) $\Rightarrow$ (i): Suppose X_b is nonempty. Without loss of generality, let $A = \begin{bmatrix} B \\ D \end{bmatrix}$. By assumption, $y = \begin{bmatrix} (B^{-1})^T c \\ 0 \end{bmatrix}$ is feasible for the dual Problem II. By the Duality Theorem we have a basic optimal $\bar{y}$ with the associated basis Q for Problem II for some $c > 0$. Since $(Q^{-1})^T c \geq 0$, we have, by assumption, $Q^{-1} \geq 0$. Let $\bar{x} = Q^{-1} b_Q$. We can pretend that $A = \begin{bmatrix} Q \\ R \end{bmatrix}$. The matrix $A^- = (A^{-1}, 0)$ satisfies $A^- A = I$, $A^- \geq 0$. Also, $(A^-)^T d$ is feasible for Problem II for all $d \geq 0$ and optimal for some $d > 0$. Since $A^- b = \bar{x}$, the result follows from Theorem 1.6.3.

Now suppose A^T contains the $n \times n$ identity matrix as a submatrix. Without loss of generality, let $A^T = (B^T, I)$. As usual, let $\mathbf{1}$ be the n-vector with all entries unity. The vector $x = (0, \mathbf{1}^T)^T$ of the appropriate size satisfies $A^T x > 0$.

It suffices to prove that any row of A that is not a row of the identity matrix has at most one positive entry. Without loss of generality, let $a_{11} > 0$. We have

$$B = \begin{bmatrix} a_{11} & a_{12} & a_{13} & \cdots & a_{1n} \\ 0 & 1 & 0 & \cdots & 0 \\ 0 & 0 & 1 & \cdots & 0 \\ \vdots & \vdots & \vdots & \ddots & \vdots \\ 0 & 0 & 0 & \cdots & 1 \end{bmatrix},$$

$$B^{-1} = \begin{bmatrix} a_{11}^{-1} & -a_{12}a_{11}^{-1} & \cdots & \cdots & -a_{1n}a_{11}^{-1} \\ 0 & 1 & 0 & \cdots & 0 \\ 0 & 0 & 1 & \cdots & 0 \\ \vdots & \vdots & \vdots & \ddots & 0 \\ 0 & 0 & 0 & \cdots & 1 \end{bmatrix}.$$

Let $c = (\epsilon, \mathbf{1}^T)^T$. For a sufficiently small ϵ, $(B^{-1})^T c \geq 0$. Thus, by assumption, $B^{-1} \geq 0$. This shows that a_{11} is the only positive entry in the first row of B. By permuting the rows and columns if necessary, we can always bring any row of A with a positive entry to the first row with first entry positive, which leads to the implication.

Finally, let A^T be Leontief. Thus $A^T x > 0$ for some $x \geq 0$. We may assume after a normalization that x is a mixed strategy for the payoff A. Thus the value of A is positive. Then no column of A can be nonpositive, and therefore A has at least one positive entry in each column. We know by assumption that A has at most m positive entries. Thus $n \leq m$. We leave it as an exercise (see Exercise 10) to show that both players can restrict themselves to an $n \times n$ submatrix B and that B has the same value as A. Since B has positive value, it has exactly n positive entries. Rearranging the rows of B we can bring the positive entries to the main diagonal. Thus, without loss of generality, $A = \begin{bmatrix} B \\ D \end{bmatrix}$, where B is a Z-matrix with a positive value. We know from Theorem 1.5.2 that $B^{-1} \geq 0$. For any $c \geq 0$, the vector $y = \begin{bmatrix} (B^{-1})^T c \\ 0 \end{bmatrix}$ is feasible for the dual Problem II and B is the required basis. Now suppose Q is a basis for A and $(Q^{-1})^T c \geq 0$ for some $c > 0$. Then for $u = (Q^{-1})^T c$ we have $Q^T u = c > 0$ and $u \geq 0$. Thus Q has positive value and Q consists of n rows of A. For some permutation P, PQ is again a Z-matrix with the main diagonal positive. Thus $(PQ)^{-1} = Q^{-1}P^{-1} \geq 0$ and $Q^{-1} = Q^{-1}P^{-1}P \geq 0$. That completes the proof.   ∎

1.7. Reducible nonnegative matrices

If $C = (c_{ij})$ is a complex matrix, then we define $C^+ = (|c_{ij}|)$. The purpose of this section is to indicate how some aspects of the Perron-Frobenius Theorem continue to hold without the assumption of irreducibility. We first prove some preliminary results.

Theorem 1.7.1. *Let $A \geq 0$ be an irreducible $n \times n$ matrix and let C be an $n \times n$ matrix such that $C^+ \leq A$. If λ is an eigenvalue of C, then $|\lambda| \leq \lambda_0$, where λ_0 is the Perron eigenvalue of A.*

Proof. Let $Cz = \lambda z$. Then $C^+ z^+ \geq |\lambda| z^+$; this in turn gives $Az^+ \geq |\lambda| z^+$. By Theorem 1.4.4 there exists $u > 0$ with $A^T u = \lambda_0 u$. Also,

$$\lambda_0 \langle u, z^+ \rangle = \langle A^T u, z^+ \rangle = \langle u, Az^+ \rangle \geq |\lambda| \langle u, z^+ \rangle.$$

Since $z^+ \neq 0$ and $u > 0$, we get $\langle u, z^+ \rangle > 0$ and $\lambda_0 \geq |\lambda|$. That completes the proof. ∎

Theorem 1.7.2. *Let $A \geq 0$ be irreducible and let λ_0 be the Perron eigenvalue of A. Then for any $\lambda > \lambda_0$, $|\lambda I - A| > 0$ and $(\lambda I - A)^{-1} > 0$.*

Proof. For $\lambda > \lambda_0$, the matrix $\lambda I - A$ is nonsingular and has off-diagonal entries nonpositive. Thus, by Theorem 1.5.2, $(\lambda I - A)^{-1} \geq 0$. Let $v \geq 0$, $v \neq 0$. There exists $u \geq 0$, $u \neq 0$ such that $(\lambda I - A)u = v$. Thus $\lambda u \geq Au$. Suppose $u_1 > 0$. Since A is irreducible, $(A^m u)_2 > 0$ for some m. It follows that $\lambda^m u \geq A^m u$ and thus $u_2 > 0$. Similarly, we can show that all the coordinates of u are positive. Thus $(\lambda I - A)^{-1} v > 0$. Since v is an arbitrary nonnegative, nonzero vector, it follows that $(\lambda I - A)^{-1} > 0$. ∎

Observe that in the next few results we do not impose the assumption that the matrix is irreducible.

Theorem 1.7.3. *Let $A \geq 0$ be a square matrix. Then the spectral radius λ_0 of A is an eigenvalue of A. There is an eigenvector $y \geq 0$ for λ_0. Furthermore, $(\lambda I - A)^{-1} \geq 0$ for $\lambda > \lambda_0$.*

Proof. We use induction on the order of the matrix. Let A be an $n \times n$ matrix. For $n = 1$, the theorem is obvious. If A is irreducible, the assertions are contained in Theorems 1.4.4 and 1.7.2. If A is reducible, then without loss of generality, we can take

$$A = \begin{bmatrix} B & 0 \\ C & D \end{bmatrix},$$

where B and D are square matrices. The eigenvalues of A are precisely the eigenvalues of B and D taken together. Thus λ_0 is either $r(B)$ or $r(D)$, where r denotes spectral radius.

Case (i). $\lambda_0 = r(D)$. By induction, $Dy_2 = \lambda_0 y_2$ for some $y_2 \geq 0$, $y_2 \neq 0$ and $(\lambda I - D)^{-1} \geq 0$ for $\lambda > \lambda_0$. We have

$$\begin{bmatrix} B & 0 \\ C & D \end{bmatrix} \begin{bmatrix} 0 \\ y_2 \end{bmatrix} = \lambda_0 \begin{bmatrix} 0 \\ y_2 \end{bmatrix},$$

that is, λ_0 is an eigenvalue of A with a nonnegative eigenvector. By induction, $(\lambda I - B)^{-1} \geq 0$ and $(\lambda I - D)^{-1} \geq 0$. Thus, given $u_1 \geq 0$, $u_2 \geq 0$, there exist $v_1 \geq 0$, $v_2 \geq 0$ such that

$$\begin{bmatrix} \lambda I - B & 0 \\ -C & \lambda I - D \end{bmatrix} \begin{bmatrix} v_1 \\ v_2 \end{bmatrix} = \begin{bmatrix} u_1 \\ u_2 \end{bmatrix}.$$

It should be noted that $Cv_1 \geq 0$ because $C \geq 0$. Thus $(\lambda I - A)^{-1}$ maps nonnegative vectors to nonnegative vectors and $(\lambda I - A)^{-1} \geq 0$ for $\lambda > \lambda_0$.

Case (ii). $\lambda_0 = r(B) > r(D)$. In this case, by induction, we have $By_1 = \lambda_0 y_1$ for some $y_1 \geq 0$, $y_1 \neq 0$. Also by induction $(\lambda_0 I - D)^{-1} \geq 0$, and there exists $y_2 \geq 0$ with $(\lambda_0 I - D)y_2 = Cy_1$. Consequently,

$$\begin{bmatrix} B & 0 \\ C & D \end{bmatrix} \begin{bmatrix} y_1 \\ y_2 \end{bmatrix} = \lambda_0 \begin{bmatrix} y_1 \\ y_2 \end{bmatrix}.$$

This shows that λ_0 is an eigenvalue of A with a nonnegative eigenvector. Also by induction, given any $u_1 \geq 0$, $u_2 \geq 0$ for $\lambda > \lambda_0$, we have

$$(\lambda I - B)v_1 = u_1 \quad \text{for some } v_1 \geq 0$$

and

$$(\lambda I - D)v_2 = Cv_1 + u_2 \quad \text{for some } v_2 \geq 0.$$

Therefore, $(\lambda I - A)^{-1} \geq 0$. That completes the proof. ∎

Theorem 1.7.4. *Let $A \geq 0$ be a square matrix with spectral radius λ_0. Then for every proper principal submatrix of A the maximal eigenvalue is at most λ_0. If A is irreducible, the inequality is strict; otherwise, equality occurs for some proper principal submatrix.*

Proof. If A is reducible, we can reduce it to

$$PAP^T = \begin{bmatrix} B & 0 \\ C & D \end{bmatrix},$$

where B and D are principal submatrices of A. As noted in the proof of Theorem 1.7.3, $\lambda_0 = r(B)$ or $r(D)$. This in fact is the last assertion in the present theorem. Let $\lambda > \lambda_0$ be the spectral radius of A. By Theorem 1.7.3, $(\lambda I - A)^{-1} \geq 0$.

Thus, by Theorem 1.5.2, $\lambda I - A$ is a nonsingular M-matrix and all the principal minors of $\lambda I - A$ are positive. Thus λ is not an eigenvalue of any principal submatrix of A for any $\lambda > \lambda_0$ and, consequently, the spectral radius of any principal submatrix of A is at most λ_0. Now suppose A is irreducible and let M be a proper principal submatrix of A. Without loss of generality,

$$A = \begin{bmatrix} M & A_{12} \\ A_{21} & A_{22} \end{bmatrix}, \qquad B = \begin{bmatrix} M & 0 \\ 0 & 0 \end{bmatrix}.$$

We know that $A \geq B \geq 0$. If $C = (A + B)/2$, then $A \geq C \geq B \geq 0$. Since $C \geq A/2$, C is therefore irreducible. Let $C^T u = \beta u$, $\beta > 0$, $u > 0$. Let $Ay = \lambda_0 y$, $y > 0$. Since $A_{12} \geq 0$, $A_{12} \neq 0$, we have

$$\beta \langle u, y \rangle = \langle C^T u, y \rangle = \langle u, Cy \rangle < \langle u, Ay \rangle = \lambda_0 \langle u, y \rangle.$$

It follows that the spectral radius of M is less than λ_0. ∎

Theorem 1.7.5. *Let $A \geq 0$ be a square matrix and let λ_0 be the spectral radius of A. Then the following statements hold:*

(i) $B(\lambda) = adj(\lambda I - A) \geq 0$ for $\lambda > \lambda_0$.

(ii) $\frac{d}{d\lambda} B(\lambda) \geq 0$ for $\lambda > \lambda_0$.

(iii) $B(\lambda_0) \geq 0$.

(iv) $\frac{d}{d\lambda}(\lambda I - A)^{-1}|_{\lambda=\lambda_0} \geq 0$.

Proof. Since $|\lambda I - A| \to \infty$ as $\lambda \to \infty$, and since $|\lambda I - A| \neq 0$ for $\lambda > \lambda_0$, we have $|\lambda I - A| > 0$. Statement (i) follows by Theorem 1.7.3. By continuity, we get statement (iii). Let $B_{ik}(\lambda)$ be the (i, k)-th entry of $B(\lambda)$ and let $B^{(j)}(\lambda)$ be the matrix obtained from $B(\lambda)$ by deleting the j-th row and the j-th column. By (i), $B^{(j)}(\lambda) \geq 0$ for all j when $\lambda > \lambda_0$. Since $\frac{d}{d\lambda} B_{ik}(\lambda) = \Sigma_j B_{ik}^{(j)}(\lambda)$, statement (ii) is obtained. Statement (iv) follows by continuity. ∎

We now introduce some definitions. Let $A \geq 0$ be a reducible matrix. Then, for some permutation matrix P_1,

$$P_1 A P_1^T = \begin{bmatrix} B & 0 \\ C & D \end{bmatrix}.$$

If D is reducible, we have, for some permutation matrix P_2,

$$P_2 P_1 A P_1^T P_2^T = \begin{bmatrix} E & 0 & 0 \\ F & G & 0 \\ H & K & L \end{bmatrix}.$$

After a finite number of such reductions, we get

$$
QAQ^T = \begin{bmatrix}
A_{11} & 0 & \cdots & 0 \\
A_{21} & A_{22} & \cdots & 0 \\
\vdots & \vdots & \ddots & \vdots \\
A_{s1} & A_{s2} & \cdots & A_{ss}
\end{bmatrix},
$$

where Q is a permutation matrix. Here each A_{ii} is either a 1×1 zero matrix or is irreducible. If $A_{ij} = 0$ for all $j \neq i$, we call A_{ii} an *isolated block*. By suitable permutations, if all the isolated blocks are placed in the leading position along the main diagonal, we get the form

$$
\begin{bmatrix}
A_1 & 0 & \cdots & 0 & \cdots & 0 \\
0 & A_2 & & 0 & \cdots & 0 \\
& & \ddots & & & \\
0 & 0 & \cdots & A_t & \cdots & 0 \\
A_{t+1,1} & A_{t+1,2} & \cdots & & A_{t+1} & 0 \\
\cdots & \cdots & \cdots & \cdots & \cdots & \cdots \\
A_{s1} & A_{s2} & \cdots & & A_{s,s-1} & A_s
\end{bmatrix}. \qquad (1.7.1)
$$

This form is known as the *Frobenius Normal Form* of the matrix.

It is instructive to interpret the Frobenius Normal Form in terms of the graph $G(A)$ associated with A, which was introduced in Section 1.1. Suppose A is $n \times n$ and let the vertices of $G(A)$ be denoted by $1, 2, \ldots, n$. We say that i is equivalent to j if $i = j$ or if in $G(A)$ there is a path from i to j as well as a path from j to i. This is clearly an equivalence relation. Let $C_i, i = 1, 2, \ldots, s$ denote the corresponding equivalence classes. Then, after a renumbering of the vertices if necessary, the elements in C_i correspond to the row indices of the block A_i in the Frobenius Normal Form. The class corresponding to an isolated block is called a *final* class. The class C_i is called a *basic* class if the Perron root of A_i equals that of A.

Theorem 1.7.6. *Let $A \geq 0$ be reducible and suppose A is in the Frobenius Normal Form 1.7.1). The spectral radius λ_0 has a positive eigenvector y if and only if λ_0 is an eigenvalue of $A_1, \ldots, A_t$ but is not an eigenvalue of $A_{t+1}, \ldots, A_s$ (i.e., the basic and final classes are the same).*

Proof. If $\lambda_0 = 0$, then A is the zero matrix and the result is trivial. So suppose $\lambda_0 > 0$. Let $Ay = \lambda_0 y$ with $y > 0$. We have

$$
\begin{bmatrix}
A_1 & 0 & \cdots & 0 & \cdots & 0 \\
0 & A_2 & & 0 & \cdots & 0 \\
& & \ddots & & & \\
0 & 0 & \cdots & A_t & \cdots & 0 \\
A_{t+1,1} & A_{t+1,2} & \cdots & & A_{t+1} & 0 \\
\cdots & \cdots & \cdots & \cdots & & \cdots \\
A_{s1} & A_{s2} & \cdots & & A_{s,s-1} & A_s
\end{bmatrix}
\begin{bmatrix}
y_1 \\ y_2 \\ \vdots \\ y_t \\ y_{t+1} \\ \vdots \\ y_s
\end{bmatrix}
= \lambda_0
\begin{bmatrix}
y_1 \\ y_2 \\ \vdots \\ y_t \\ y_{t+1} \\ \vdots \\ y_s
\end{bmatrix}
$$

for the conformal partition $y_1, y_2, \ldots, y_s$ of y. Obviously, λ_0 is an eigenvalue of $A_1, \ldots, A_t$. Furthermore,

$$
\sum_{j=1}^{t} A_{t+1,j} y_j + A_{t+1} y_{t+1} = \lambda_0 y_{t+1}.
$$

We can show that the spectral radius of A_{t+1} is less than λ_0. This is obvious if A_{t+1} is the 1×1 zero matrix. Therefore, we assume that A_{t+1} is irreducible. Since $y_j > 0$, $j = 1, 2, \ldots, t$, and since $A_{t+1,j} \neq 0$ for some $j \leq t$, we have $A_{t+1} y_{t+1} \leq \lambda_0 y_{t+1}$ with strict inequality in some coordinate. There exists $z > 0$ with $A_{t+1}^T z = \mu z$, where μ is the Perron root of A_{t+1}. Thus $\mu \langle z, y_{t+1} \rangle = \langle z, A_{t+1} y_{t+1} \rangle < \lambda_0 \langle z, y_{t+1} \rangle$. This shows that $\mu < \lambda_0$. Similarly, the Perron root of A_m is less than λ_0, for $t + 1 < m \leq s$.

Conversely, let λ_0 be an eigenvalue of $A_1, A_2, \ldots, A_t$ but not an eigenvalue of $A_{t+1}, \ldots, A_s$. Let $y_1, y_2, \ldots, y_t$ be the components of y that are the eigenvectors for λ_0 for the matrices $A_1, \ldots, A_t$, respectively. Since A_i is irreducible, then $y_i > 0$, $i = 1, 2, \ldots, t$. Observe that λ_0 is greater than the Perron root of A_{t+1}. Also, $A_{t+1,j} \neq 0$ for some $j \leq t$. Therefore,

$$
\sum_{j=1}^{t} A_{t+1,j} y_j > 0.
$$

If A_{t+1} is the 1×1 zero matrix, then set

$$
y_{t+1} = \frac{1}{\lambda_0} \sum_{j=1}^{t} A_{t+1,j} y_j.
$$

If A_{t+1} is irreducible, then by Theorem 1.7.2, $(\lambda_0 I - A_{t+1})^{-1} > 0$. Thus, for some $y_{t+1} > 0$,

$$
\sum_{j=1}^{t} A_{t+1,j} y_j + A_{t+1} y_{t+1} = \lambda_0 y_{t+1}.
$$

Inductively, we can construct the $y_j > 0$ for $j = t + 2, \ldots, s$ and then $y = (y_1^T, \ldots, y_s^T)^T$ constitutes a positive eigenvector of A.   ∎

1.8. Primitive matrices

Let $\lambda_0, \lambda_1, \ldots, \lambda_{k-1}$ be the set of eigenvalues of an irreducible nonnegative matrix A with $|\lambda_i| = \lambda_0$, $i = 1, 2, \ldots, k - 1$, where λ_0 is the spectral radius of A. The matrix A is called *primitive* if $k = 1$. The irreducible matrix A is called *cyclic of index k* if $k > 1$.

Theorem 1.8.1. *Let $A > 0$ be an $n \times n$ matrix. Then A is primitive.*

Proof. Without loss of generality let the spectral radius $\lambda_0 = 1$. We have $Ay = y$ and $A^T u = u$ for some $y, u > 0$. Let μ be any eigenvalue of A with $|\mu| = 1$. We can prove that $\mu = 1$. If $Az = \mu z$ for an eigenvector z, then by following the line of proof of Lemma 1.4.7, we can conclude that for some θ, $\theta_i a_{ij} z_j^+ = a_{ij} z_j$ and that $\theta_i z_j^+ = z_j$ for all i, j; i.e., $\theta z^+ = \theta y = z$. Thus

$$\langle \mu z, u \rangle = \langle Az, u \rangle = \langle z, A^T u \rangle = \langle z, u \rangle = \theta \langle y, u \rangle,$$

and hence, $\mu \theta \langle y, u \rangle = \theta \langle y, u \rangle$. Therefore, either $\mu = 1$ or $\theta = 0$. Since $z \neq 0$, then $\theta \neq 0$, and therefore $\mu = 1$. That completes the proof.   ∎

The next result is stronger than Theorem 1.8.1.

Theorem 1.8.2. *Let $A \geq 0$ be an irreducible matrix with spectral radius 1. Then the following conditions are equivalent:*

  (i) A is primitive.
 (ii) $A^k \to P$ as $k \to \infty$ for some matrix P.
(iii) For some positive integer p, $A^p > 0$.
 (iv) A^k is irreducible for each $k \geq 1$.

Proof. $(i) \Rightarrow (ii)$: By Theorem 1.7.2, $(\lambda I - A)^{-1} > 0$ for $\lambda > 1$. Let $B(\lambda) = \text{adj}(\lambda I - A)$. Then

$$(\lambda - 1)(\lambda I - A)^{-1} = (\lambda - 1)\frac{B(\lambda)}{|\lambda I - A|}.$$

Since A is irreducible, by Theorem 1.4.4, 1 is a simple root of A. Thus $\lim_{\lambda \downarrow 1} \frac{\lambda - 1}{|\lambda I - A|}$ exists, because the factor $(\lambda - 1)$ in the denominator has multiplicity 1. Therefore,

$$\lim_{\lambda \downarrow 1} (\lambda - 1)(\lambda I - A)^{-1}$$

exists. Let this limit be P. Then, clearly, $P \geq 0$. Furthermore,

$$AP = [(A - \lambda I) + \lambda I] \lim_{\lambda \downarrow 1} (\lambda - 1)(\lambda I - A)^{-1}$$

$$= \lim_{\lambda \downarrow 1} \lambda(\lambda - 1)(\lambda I - A)^{-1}$$

$$= P.$$

Similarly, we can prove $PA = P$. Also, $P(\lambda I - A) = (\lambda - 1)P$. Thus for $\lambda > 1$, $P = (\lambda - 1)P(\lambda I - A)^{-1}$. Letting $\lambda \downarrow 1$, we see that $P = P^2$. Thus $PA = AP = P = P^2$. Next, we claim that $B = A - P$ has spectral radius less than unity. To see this suppose that, for $x \neq 0$, $Bx = \alpha x$, $|\alpha| \geq 1$. Since $PB = P(A - P) = P - P^2 = 0$ we get $PBx = \alpha Px = 0$. Thus $Px = 0$. That is, $Bx = Ax = \alpha x$. It follows that $\alpha = 1$ and hence $Ax = x$. Therefore $(\lambda I - A)x = (\lambda - 1)x$ and $x = (\lambda - 1)(\lambda I - A)^{-1}x$. Letting $\lambda \downarrow 1$, we see that $x = Px$. Since $Bx = x$ and $Px = x$ we have $Bx + Px = Ax = 2x$. This is a contradiction since the spectral radius of A is 1. Thus the spectral radius of B is less than 1. We have

$$A^k = (P + B)^k = P + B^k.$$

(Notice that $P^k = P$, $PB = BP = 0$.) Since the spectrum of B lies in the interior of the unit circle, it follows [see, for example, Varga (1962)] that $B^k \to 0$. Thus $A^k \to P$.

(ii) $\Rightarrow$ (iii): Let u be the first column of P. We know from $AP = P$ that $Au = u$. Thus u is an eigenvector of A corresponding to the spectral radius 1 of A. It follows from Theorem 1.4.4 that $u > 0$. Thus $P > 0$. Since $A^k \to P$, for a sufficiently large p, $A^p > 0$.

(iii) $\Rightarrow$ (iv): If A^k is reducible for some k, then $(A^k)^p$ is reducible. However, $(A^k)^p = (A^p)^k > 0$, which is a contradiction.

(iv) $\Rightarrow$ (i): Let if possible $Ax = \alpha x$, $\alpha \neq 1$, $|\alpha| = 1$. Let $Ay = y$. Then $A^h x = x$, where $\alpha^h = 1$. Also, $A^h y = y$. Since A^h is irreducible, $x = \theta y$, for some $\theta \neq 0$. Thus $\alpha \theta y = \alpha x = Ax = A\theta y = \theta y$. This contradicts $\alpha \neq 1$. This completes the proof of the theorem. ∎

If A is a primitive matrix, then the smallest power q for which $A^q > 0$ is called the *index of primitivity*. It can be shown that the index of primitivity is at most $(n - 1)^2 + 1$, where n is the order of the matrix. [See Holladay and Varga (1958) and Perkins (1961).]

Theorem 1.8.3. *Let $A \geq 0$ be irreducible with spectral radius λ_0. Let $\lambda_0, \lambda_1, \ldots, \lambda_{k-1}$ be all the eigenvalues of A with absolute value λ_0. Then the set of eigenvalues of A, as a subset of the complex plane, remains invariant* *under a rotation of the complex plane by an angle $2\pi/k$. Furthermore, there is a permutation matrix P such that*

$$PAP^T = \begin{bmatrix} 0 & A_{12} & 0 & \cdots & 0 \\ 0 & 0 & A_{23} & \cdots & 0 \\ \vdots & \vdots & \vdots & \ddots & \vdots \\ 0 & 0 & 0 & \cdots & A_{k-1,k} \\ A_{k1} & 0 & 0 & \cdots & 0 \end{bmatrix},$$

where the main diagonal blocks are square matrices.

Proof. Consider the smallest angle $\varphi_1 > 0$ with the property that $\lambda_0 e^{i\varphi}$ is an eigenvalue of A. By Theorem 1.4.4, $\varphi_1 = 2\pi/k$. Let $\lambda = \lambda_0 e^{i\varphi_1}$ and suppose $Au = \lambda u$. From 1.4.3) and 1.4.4), $Au^+ = \lambda_0 u^+$ and

$$\theta_i a_{ij} u_j^+ = a_{ij} u_j \text{ for all } i, j \text{ for some } \theta_i \text{s with } |\theta_i| = 1.$$

Summing with respect to j, we get

$$\sum_j \theta_i a_{ij} u_j^+ = \lambda u_i = \lambda_0 e^{i\varphi_1} u_i.$$

Therefore, $\lambda_0 \theta_i u_i^+ = \lambda_0 e^{i\varphi_1} u_i$ and $u_i = e^{-i\varphi_1} \theta_i u_i^+$. Thus $a_{ij} u_j = \theta_i a_{ij} u_j^+ = e^{i\varphi_1} \theta_i a_{ij} \theta_j^{-1} u_j$. Since $u_j \neq 0$ we get $a_{ij} = e^{i\varphi_1} \theta_i a_{ij} \theta_j^{-1}$. The matrix A can then be expressed as

$$A = e^{i\varphi_1} D_1 A D_1^{-1}, \tag{1.8.1}$$

where $D_1 = \text{diag}(\theta_1, \theta_2, \ldots, \theta_n)$. Since A and $D_1 A D_1^{-1}$ have the same set of eigenvalues, (1.8.1) shows that the spectrum of A remains invariant under the rotation of the complex plane by an angle $2\pi/k$. Because there are precisely k eigenvalues that are roots of $\lambda^k - \lambda_0^k = 0$, no smaller angle can retain this rotational invariance. This completes the proof of the first part. To see the second part, let $\theta_1, \theta_2, \ldots, \theta_s$ be distinct among the θ_is. There exists a permutation matrix P such that

$$PD_1 P^T = \begin{bmatrix} \theta_1 I_1 & 0 & \cdots & 0 \\ 0 & \theta_2 I_2 & \cdots & 0 \\ \vdots & \vdots & \ddots & \vdots \\ 0 & 0 & \cdots & \theta_s I_s \end{bmatrix},$$

where the I_js are identity matrices. Let

$$PAP^T = \begin{bmatrix} A_{11} & A_{12} & \cdots & A_{1s} \\ A_{21} & A_{22} & \cdots & A_{2s} \\ \vdots & \vdots & \ddots & \vdots \\ A_{s1} & A_{s2} & \cdots & A_{ss} \end{bmatrix}.$$

Clearly, $s > 1$ and we can assume $\theta_1 = 1$. We have

$$A_{jk} = e^{i\varphi_1}\theta_j A_{jk}\theta_k^{-1}.$$

In particular, $A_{11} = e^{i\varphi_1}\theta_1 A_{11}\theta_1^{-1} = e^{i\varphi_1}A_{11}$. Since $e^{i\varphi_1} \neq 1$, $A_{11} = 0$. Also, $A_{12} = e^{i\varphi_1}\theta_1 A_{12}\theta_2^{-1}$, $A_{13} = e^{i\varphi_1}\theta_1 A_{13}\theta_3^{-1}, \ldots, A_{1s} = e^{i\varphi_1}\theta_1 A_{1s}\theta_s^{-1}$. Since A is irreducible, at least one of $A_{12}, A_{13}, \ldots, A_{1s}$ is not zero. Further, $e^{i\varphi_1}/\theta_2 = 1$ or $e^{i\varphi_1}/\theta_3 = 1$ or $\ldots e^{i\varphi_1}/\theta_s = 1$. Assuming that the arguments of $\theta_1, \theta_2, \ldots, \theta_s$ are ordered increasingly, we can say that $e^{i\varphi_1}/\theta_2 = 1$ and the rest are not 1. Thus $A_{11}, A_{13}, \ldots, A_{1s}$ are all zero matrices. Similarly, we can show that $A_{21}, A_{22}, A_{24}, \ldots, A_{2s}$ are zero, and so on. Finally, $A_{sj} = 0$, $j = 2, \ldots, s$. Therefore,

$$PAP^T = \begin{bmatrix} 0 & A_{12} & 0 & \cdots & 0 \\ 0 & 0 & A_{23} & \cdots & 0 \\ \vdots & \vdots & \vdots & \ddots & \vdots \\ 0 & 0 & 0 & \cdots & A_{s-1s} \\ A_{s1} & 0 & 0 & \cdots & 0 \end{bmatrix}.$$

From this, we also see that $s = k$, and the proof is complete. ∎

Theorem 1.8.4. *Let $A \geq 0$ be an $n \times n$ irreducible cyclic matrix of index k. Then its characteristic polynomial $p(\lambda) = |\lambda I - A|$ is of the form*

$$p(\lambda) = \lambda^m \left(\lambda^k - \lambda_0^k\right) \prod_{i=1}^{r} \left(\lambda^k - \lambda_i^k\right),$$

where $|\lambda_i| < \lambda_0$, $i = 1, 2, \ldots, r$.

Proof. By Theorem 1.8.3 we know that for any eigenvalue λ_i of A, the solutions of $\lambda^k - \lambda_i^k = 0$ are also eigenvalues of A. ∎

In an attempt to generalize the various aspects of the Perron-Frobenius Theorem, a wide class of matrices have been considered. Since the eigenvalues of a square matrix vary continuously with the entries of the matrix, one expects that a few negative entries of small order and many positive entries of large order might still lead to a positive eigenvalue and positive eigenvectors. For example, the matrix

$$A = \begin{bmatrix} 10 & 2 & 1 \\ -1 & 1 & 4 \\ 3 & 1 & 0 \end{bmatrix}$$

with a negative entry has a positive eigenvalue and a positive eigenvector. This phenomenon can be explained by the fact that $A^2 > 0$. We now introduce a more general definition.

A square matrix A is called *polynomially positive* if for some real polynomial $p(t) = \alpha_0 + \alpha_1 + \cdots + \alpha_k t^k$, $p(A) = \alpha_0 I + \alpha_1 A + \cdots + \alpha_k A^k > 0$. If $A^k > 0$ for some k, then A is called *power-positive* and such matrices are clearly polynomially positive.

The following is an extension of the Perron-Frobenius Theorem to polynomially positive matrices.

Theorem 1.8.5. *Let A be a polynomially positive matrix of order n. Then there exists a real eigenvalue μ with positive eigenvectors for A and A^T. Furthermore, μ is a simple root of the characteristic equation.*

Proof. Consider the payoff $A - \alpha I$. By Theorem 1.4.1, we know that $v(\alpha)$, the value of $A - \alpha I$, vanishes for some α. Suppose $v(\mu) = 0$. Let y be optimal for Player II with $(A - \mu I)^T y = v \le 0$. If $v \ne 0$, then since $p(A^T) > 0$, we have $p(A^T)(A - \mu I)^T y = (A - \mu I)^T p(A^T) y < 0$. Normalizing the positive vector $p(A^T) y$ we get another mixed strategy u for Player II such that $(A - \mu I)^T u < 0$. This contradicts $v(\mu) = 0$. Thus $v = 0$. Since a similar argument applies for Player I we have an optimal strategy x for Player I such that $(A - \mu I)x = 0$. Moreover, since $p(A)x = p(\mu)x > 0$, then $x > 0$. Similarly, $y > 0$. Therefore, the game is completely mixed. By Theorem 1.3.6, all the cofactors of $A - \mu I$ are of the same sign and are different from zero. This shows that μ is a simple root of A. That completes the proof. ∎

The concept of a power-positive matrix seems to be due to Brauer (1961). Polynomially positive matrices have been termed *Perron matrices* in Seneta (1973, p. 43). The proof of Theorem 1.8.5, which is based on game theoretic arguments, is from Raghavan (1979).

1.9. Finite Markov chains

Consider a random (stochastic) process $X_n, n = 0, 1, 2, \ldots$ that takes on a finite number of possible values. We assume the set of possible values to be the integers $0, 1, 2, \ldots, k$ for convenience. If $X_n = i$, the process is said to be in state i at time n. We assume that the process moves from state i to state j with a fixed transition probability p_{ij}. More formally,

$$P_n(X_{n+1} = j \mid X_n = i, X_{n-1} = i_{n-1}, \ldots, X_1 = i_1, X_0 = i_0) = p_{ij}$$

for all states $i_0, i_1, \ldots, i_{n-1}, i, j$ and for all $n \ge 0$. Note that the transition probability p_{ij} does not involve $i_0, i_1, \ldots, i_{n-1}$. Clearly, the matrix (p_{ij}) is stochastic, i.e., has all entries nonnegative and the sum of the elements in each row is unity.

Such a random process is called a *stationary Markov chain* with finitely many states. The adjective stationary relates to the fact that the transition probabilities depend only on the current state.

Example 1.6. [A random walk model]. A particle moving on a straight line jumps every second. It moves either one unit to the right with probability p or one unit to the left with probability $1 - p$. The particle starts at the origin. When it reaches the points -3 or 2, it is absorbed permanently in those positions.

The transition probabilities can be described by the transition matrix

$$
\begin{array}{c c c c c c c}
 & -3 & -2 & -1 & 0 & 1 & 2 \\
\begin{array}{c}
-3 \\ -2 \\ -1 \\ 0 \\ 1 \\ 2
\end{array}
&
\left(
\begin{array}{c c c c c c}
1 & 0 & 0 & 0 & 0 & 0 \\
q & 0 & p & 0 & 0 & 0 \\
0 & q & 0 & p & 0 & 0 \\
0 & 0 & q & 0 & p & 0 \\
0 & 0 & 0 & q & 0 & p \\
0 & 0 & 0 & 0 & 0 & 1
\end{array}
\right)
\end{array}.
$$

Example 1.7. [An inventory model]. In a store that stocks a certain commodity, the weekly demands are independent and identically distributed. Let a_j be the probability that j units are demanded, where $j = 0, 1, 2, \ldots, k$. Suppose that the store employs the following policy: At the beginning of the week, if its supply is s or higher, it does not order; if it is below s, then the store orders enough to bring the supply to $S \le k$. We also assume that the order is filled instantaneously. If a demand is not met, we assume that it is lost.

We can formulate the problem as follows: Let X_n be the inventory at the end of the n-th week. Let Y_n be the demand for the n-th week. Then

$$
\begin{aligned}
X_{n+1} &= \max(X_n - Y_{n+1}, 0) \quad \text{if } X_n \ge s \\
&= \max(S - Y_{n+1}, 0) \quad \text{if } X_n < s.
\end{aligned}
$$

Hence X_n is a Markov chain with the transition matrix $P = (p_{ij})$ given by

$$
\begin{aligned}
p_{ij} &= \sum_{r=i}^{k} a_r \quad &\text{if } j = 0, i \ge s \\
&= a_{i-j} \quad &\text{if } 0 < j \le i, i \ge s \\
&= \sum_{r=S}^{k} a_r \quad &\text{if } j = 0, i < s \\
&= a_{S-j} \quad &\text{if } 0 < j \le s, i < s \\
&= 0 \quad &\text{otherwise.}
\end{aligned}
$$

A Markov chain is said to be irreducible if its transition matrix is irreducible. Clearly, this is equivalent to saying that with positive probability the process moves from any state to any other state in finitely many steps. Example 1.7 illustrates a chain that is not irreducible. Once the process reaches states -3 or 2 it gets absorbed. One can modify that example to get an irreducible chain as follows.

Example 1.8. A particle moves as in Example 1.7 except for the fact that when it reaches -3, it jumps to -2 the next second and when it reaches 2, it jumps to 1 the next second. Now the transition matrix is

$$
\begin{array}{c c}
& \begin{array}{c c c c c c} -3 & -2 & -1 & 0 & 1 & 2 \end{array} \\
\begin{array}{c} -3 \\ -2 \\ -1 \\ 0 \\ 1 \\ 2 \end{array} &
\left(\begin{array}{c c c c c c}
0 & 1 & 0 & 0 & 0 & 0 \\
q & 0 & p & 0 & 0 & 0 \\
0 & q & 0 & p & 0 & 0 \\
0 & 0 & q & 0 & p & 0 \\
0 & 0 & 0 & q & 0 & p \\
0 & 0 & 0 & 0 & 1 & 0
\end{array}\right).
\end{array}
$$

If $p_{ij}^{(n)}$ denotes the (i, j)-th entry of P^n, we have

$$
p_{ij}^{(n)} = \sum_t p_{it}^{(n-r)} p_{tj}^{(r)}, \quad r = 1, 2, \ldots, n-1.
$$

These equations are known as the *Kolmogorov-Chapman equations.*

Example 1.9. There are two boxes. Box 1 has two white balls and box 2 has two black balls. A ball is picked at random from each box and transferred to the other box. If X denotes the number of black balls in box 1, what is the chance that after three such transfers the first box has two white balls? We can formulate this problem using Markov chains as follows.

Let X_n denote the number of black balls in the first box after n transfers. Clearly, X_n takes one of the values 0, 1, or 2 and by assumption the system starts at $X_0 = 0$. The transition matrix is

$$
P = \begin{bmatrix}
p_{00} & p_{01} & p_{02} \\
p_{10} & p_{11} & p_{12} \\
p_{20} & p_{21} & p_{22}
\end{bmatrix} = \begin{bmatrix}
0 & 1 & 0 \\
\frac{1}{4} & \frac{1}{2} & \frac{1}{4} \\
0 & 1 & 0
\end{bmatrix}.
$$

Observe that the number of transitions has no influence on the transition probabilities, which depend only on the current state. Hence the process is a stationary Markov chain. We now observe that the entries of any integer power of P have a nice interpretation. We first note that

$$
\begin{aligned}
p_{ik}\,p_{kl}\,p_{lj} &= P\{X_1 = k \mid X_0 = i\}P\{X_2 = l \mid X_1 = k, X_0 = i\} \\
&\quad \times P\{X_3 = j \mid X_2 = l, X_1 = k, X_j = i\} \\
&= P\{X_3 = j, X_2 = l, X_1 = k \mid X_0 = i\}.
\end{aligned}
$$

Thus

$$
\sum_k \sum_l p_{ik}\,p_{kl}\,p_{lj} = P\{X_3 = j \mid X_0 = i\}
$$

$$
= (P^3)_{ij}.
$$

We have

$$
P^3 = \frac{1}{16}\begin{bmatrix} 2 & 12 & 2 \\ 3 & 10 & 3 \\ 2 & 12 & 2 \end{bmatrix}.
$$

We started at state 0. Thus, from the first row of P^3, we can conclude that there is a 1/8 chance we will have two white balls in box 1 after three such transfers. Just as a curiosity, if we calculate P^6 and P^{12}, we find

$$
P^6 = \frac{1}{256}\begin{bmatrix} 44 & 168 & 44 \\ 42 & 172 & 42 \\ 44 & 168 & 44 \end{bmatrix}, \qquad
P^{12} = \frac{1}{8192}\begin{bmatrix} 1366 & 5460 & 1366 \\ 1365 & 5462 & 1365 \\ 1366 & 5460 & 1366 \end{bmatrix}.
$$

The three rows are getting closer to the fixed vector $\left(\frac{1}{6}, \frac{2}{3}, \frac{1}{6}\right)$. In fact, we can prove that the matrix P^n converges to the matrix

$$
\begin{bmatrix} \frac{1}{6} & \frac{2}{3} & \frac{1}{6} \\ \frac{1}{6} & \frac{2}{3} & \frac{1}{6} \\ \frac{1}{6} & \frac{2}{3} & \frac{1}{6} \end{bmatrix}.
$$

Thus no matter where we start, in the long run, we have a 1/6 chance for the first box to contain two white balls or two black balls. It is more likely that the first box has one white and one black ball.

Let P be the transition matrix of a stationary Markov chain with k states. State j is said to be *accessible* from state i if $(P^n)_{ij} > 0$ for some $n \geq 1$. State i is said to *communicate* with state j $(i \leftrightarrow j)$ if $(P^n)_{ij} > 0$ for some n and $(P^m)_{ji} > 0$ for some m. It is straightforward to verify that the binary relation $\leftrightarrow$ is an equivalence relation. In fact, this is precisely the same equivalence relation introduced earlier in connection with the Frobenius Normal Form (see Section 1.7). An irreducible Markov chain has all of its states in one equivalence class. State i is said to be *recurrent* if starting from state i we will eventually return to state i with certainty. State i is said to be *transient* if starting from state i there is a positive probability that the system will never return to state i. State i is said to be *absorbing* if $p_{ii} = 1$. State i is said to have *period* $d = d(i)$ if d is the largest integer with the property that $p_{ii}^n = 0$ except when $n = kd$, where $k = 1, 2, \ldots$. Here we assume that d is the greatest common divisor of all $n \geq 1$ such that $p_{ii}^{(n)} > 0$. If $p_{ii}^{(n)} = 0$ for all n, then we say that the period $d = d(i) = 0$. State i is *aperiodic* if $d = d(i) = 1$. We call a Markov chain aperiodic if all states are aperiodic. We need the following result from elementary number theory for the proof of the next theorem.

Lemma 1.9.1. *Let n_1 and n_2 be positive integers and let d be the greatest common divisor between them. Then there exist integers x and y, such that $d = n_1 x + n_2 y$.*

Proof. We assume, without loss of generality, that $n_1 > n_2$. If n_2 is a factor of n_1, the result is trivial. Otherwise, $n_1 = q_2 n_2 + n_3$, where $0 < n_3 < n_2$. Dividing n_2 by n_3 we get $n_2 = q_3 n_3 + n_4$, where $0 \leq n_4 < n_3$. For some k, we will have $n_{k-1} = q_k n_k$ in such an inductive step. Note that n_k divides n_1 and n_2, whereas d divides n_1 and n_2 and hence n_3. Proceeding this way, we show that d divides n_k. Therefore, $d = n_k$. Furthermore, we have $n_k = n_{k-2} - q_{k-1} n_{k-1}$. Substituting $n_{k-1} = n_{k-3} - q_{k-2} n_{k-2}$, we get $n_k = n_{k-2} - q_{k-1}(n_{k-3} - q_{k-2} n_{k-2}) = \alpha n_{k-2} + \beta n_{k-3}$ for integers α and β. By induction, we see that the process terminates with $d = n_1 x + n_2 y$. ∎

The following is a straightforward generalization of Lemma 1.9.1.

Lemma 1.9.2. *Let $n_1, n_2, \ldots, n_k$ be positive integers with 1 as the greatest common divisor. Then there exist integers $\pi_1, \pi_2, \ldots, \pi_k$ such that*

$$\pi_1 n_1 + \pi_2 n_2 + \cdots + \pi_k n_k = 1.$$

Lemma 1.9.3. *Let d be the greatest common divisor of the positive integers $n_1, n_2, \ldots, n_k$. Then there exists a positive integer h such that for any integer $m \geq h$, $md = \sum_{j=1}^{k} c_j n_j$ for some nonnegative integers $c_1, c_2, \ldots, c_k$.*

Proof. By Lemma 1.9.2 we know that

$$1 = \sum \pi_j \left(\frac{n_j}{d} \right), \quad \text{or equivalently,} \quad d = \sum \pi_j n_j.$$

(Here the π_js are not necessarily nonnegative.) Let $d^* = \sum |\pi_j| n_j$ and define $h = (d^*)^2$. Now for any $m \geq h$, $m = qd^* + r$ and, clearly, $q \geq d^*$; $d^* > r \geq 0$.

Therefore,

$$m = q \sum |\pi_j| n_j + r \sum \pi_j \left(\frac{n_j}{d} \right).$$

Since $q > r$ and $d \geq 1$, $md = \sum n_j c_j$, where $c_j \geq 0$. Hence the result is proved. ∎

Theorem 1.9.4. *If states i and j of a Markov chain communicate $(i \leftrightarrow j)$, then they have the same period.*

Proof. Let $p_{ij}^{(m)} > 0$ and $p_{ji}^{(n)} > 0$. Let $d(i)$ and $d(j)$ be the periods of states i and j, respectively. Let $p_{ii}^{n_s} > 0$, for $s = 1, 2, \ldots, t$. Without loss of generality $d(i)$ is the greatest common divisor of $n_1, n_2, \ldots, n_t$. For a sufficiently large c we can write, by Lemma 1.9.3, $cd(i) = \sum_{r=1}^{t} \lambda_r n_r$, where $\lambda_r \geq 0$. Thus

$$p_{jj}^{[m+n+cd(i)]} \geq p_{ji}^{(n)} p_{ii}^{[cd(i)]} p_{ij}^{(m)} = p_{ji}^{(n)} p_{ii}^{(\Sigma \lambda_r n_r)} p_{ij}^{(m)} > 0.$$

Similarly, $p_{jj}^{[m+n+(c+1)d(i)]} > 0$. Thus $d(j)$ divides

$$(m + n + (c + 1)d(i)) - (m + n + cd(i)) = d(i).$$

By symmetry, $d(i)$ divides $d(j)$ and hence $d(i) = d(j)$. ∎

Theorem 1.9.5. *The transition matrix P is irreducible and aperiodic if and only if $P^N > 0$ for some N.*

Proof. First suppose P is irreducible and aperiodic. Then for each i, j there exists $m = m(i, j)$ such that $p_{ij}^{(m)} > 0$. Let $p_{ii}^{(n_s)} > 0$, for $s = 1, 2, \ldots, r$. Since each state is aperiodic, without loss of generality, the greatest common divisor of $n_1, n_2, \ldots, n_t$ is 1. By Lemma 1.9.3, for any sufficiently large n, we can write $n = \sum \lambda_r n_r$, where $\lambda_r \geq 0$. Now

$$p_{ij}^{(m+n)} \geq p_{ii}^{(n)} p_{ij}^{(m)} = \sum p_{ii}^{(\Sigma \lambda_i n_i)} p_{ij}^{(m)}.$$

Thus $p_{ij}^{(m+n)} > 0$ for all sufficiently large n, say for $n \geq n(i)$. Choosing N as the maximum of $n(i)$ over i we see that $p_{ij}^N > 0$ for all i, j. By Theorem 1.8.2 P is primitive.

Conversely, when $P^N > 0$, P is irreducible. Moreover, $P^{N+m} > 0$ for any $m \geq 1$. This shows that P is aperiodic and completes the proof. ∎

Now we are ready to prove the following important theorem.

Theorem 1.9.6. *Let P be an irreducible aperiodic transition probability matrix of a stationary Markov chain. Then $\lim_k P^k = Q$ exists. Furthermore, $PQ =$*

$QP = Q = Q^2$. *The matrix Q has identical rows. Any row of Q is given by the unique probability vector u satisfying uP = u.*

Proof. By Theorem 1.9.5 we know that P is primitive and 1 is the spectral radius of P. The first two assertions now follow by Theorem 1.8.2. Since $QP = Q$, then $uP = u$, where u is any row of Q. Becuase u is unique, all the rows of Q are identical, and the proof is complete. ∎

Example 1.10. A dermatologist knows that a certain skin disease is only tem-porarily arrested, but never cured, by any of the three available skin creams: A, B, and C. The skin reacts only to the cream used in the previous week, and the creams used in the past have no influence at all on the current condition. If cream A is used for a week, then 70% of the time it continues to be effective for another week. When it fails to be effective, the only effective cream is creme C. Similarly, when cream B is used, 60% of the time it is effective for a week. When it fails, again cream C is prescribed. Cream C is effective for a week 80% of the time. When it fails, both A and B are equally effective and one of them is prescribed at random. What is the chance that a chronic skin patient with the particular skin disease uses cream B on New Year's day after five years?

Clearly, the transition matrix is

$$P = \begin{array}{c} \\ A \\ B \\ C \end{array} \begin{array}{ccc} A & B & C \\ \begin{pmatrix} .7 & 0 & .3 \\ 0 & .6 & .4 \\ .1 & .1 & .8 \end{pmatrix} \end{array}.$$

We want Q, if it exists, which approximates P^k for large k. Note that P is irreducible and that, the diagonal entries are positive. By Theorem 1.8.2, P is primitive and therefore, by Theorem 1.9.5, Q exists. We know that the rows of Q are identical. Let x be a row of Q. Since $xP = x$, we get $x = (x_1, x_2, x_3) = (\frac{4}{19}, \frac{3}{19}, \frac{12}{19})$. Thus the approximate chance of using creams A, B, or C after several years is $\frac{4}{19}$, $\frac{3}{19}$, or $\frac{12}{19}$, respectively.

For a stochastic matrix P, even if P^k may not converge to a limiting matrix, the following result holds.

Theorem 1.9.7. For any stochastic matrix P, $\frac{1}{k}(I + P + P^2 + \cdots + P^{k-1}) \to Q$ for some matrix Q. Also, $QP = PQ = Q = Q^2$.

Proof. Let $Q_k = \frac{1}{k}(I + P + P^2 + \cdots + P^{k-1})$. Since the entries of Q_k lie in the interval $[0,1]$, we have $Q_{k_l} \to Q$ for some subsequence $\{k_l\}$. Since each Q_{k_l} is a stochastic matrix, Q is also stochastic. We show that every subsequence of $\{Q_k\}$

converges to the same limit Q. Let $Q_{k_s} \to Q^*$ for another subsequence $\{k_s\}$. We have $P Q_{k_l} = Q_{k_l} + \frac{1}{k_l}(P^{k^l} - I)$. Taking limits on both sides we get $PQ = Q$. Similarly, $Q_{k_l} P = Q_{k_l} + \frac{1}{k_l}(P^{k^l} - I)$. Thus $QP = Q$. By induction, we get $Q_k Q = Q Q_k = Q$. In the same manner, we deduce $Q_k Q^* = Q^* Q_k = Q^*$ and we have $Q_{k_l} Q^* = Q^* = Q Q^*$. Similarly $Q^* Q_{k_l} = Q^* = Q^* Q$ and $Q Q^* = Q^* Q = Q^*$. By symmetry, $Q = Q^*$, which completes the proof. ∎

Markov chains were initiated by A. A. Markov at the turn of the century. He applied this theory to study the pairings of vowels and consonants in the Russian literature. Kolmogorov (1935) and his school in Moscow enriched the subject with many deep contributions. For a thorough discussion of Markov chains, see Feller (1968), Kemeny and Snell (1960), and Romanovsky (1970). Lemma 1.9.3 is due to Schur; see Holladay and Varga (1958). The modern treatment of Markov chains uses the group generalized inverse; see Meyer (1975) for a survey.

1.10. Self maps of the Lorentz cone

Recall that a set $K \subset R^n$ is called a *cone* if $x, y \in K \Rightarrow x + y \in K$ and $x \in K \Rightarrow \lambda x \in K$ for any $\lambda \geq 0$. We assume that the origin is an extreme point of the cone, i.e., if $x + y = 0$ and $x, y \in K$, then $x = y = 0$. If this property holds, then K is said to be *pointed*. The *Lorentz cone* in R^n is the cone defined as

$$K = \left\{ (x_1, x_2, \ldots, x_n) : x_1^2 \geq \sum_{i=2}^{n} x_i^2, x_1 \geq 0 \right\}.$$

Let K^* be the set of all linear functionals on R^n that are nonnegative on K. Then K^* is called the *dual cone* of K. Note that

$$K^* = \{(f_1, f_2, \ldots, f_n) : f_1 x_1 + f_2 x_2 + \cdots + f_n x_n \geq 0, \text{ for any } x \in K\}.$$

The Lorentz cone has the following interesting property.

Lemma 1.10.1. *If K is the Lorentz cone in R^n, then $K = K^*$.*

Proof. Let $f = (f_1, f_2, \ldots, f_n)$ be an element of K^*. Choose $x = (1, 0, \ldots, 0)$. Since $x \in K$, $f_1 \geq 0$. Let $y_1 = \sqrt{\sum_2^n f_j^2}$, $y_2 = -f_2, \ldots, y_n = -f_n$. Then $y = (y_1, y_2, \ldots, y_n) \in K$. Thus by assumption, $f_1 y_1 + f_2 y_2 + \cdots + f_n y_n \geq 0$. Therefore, $f_1 \sqrt{\sum_2^n f_j^2} \geq \sum_2^n f_j^2$. This shows that $f \in K$. Thus $K^* \subset K$. It is obvious that $K \subset K^*$, and the proof is complete. ∎

Consider the bilinear functional $[,]$ defined by $[x, y] = x_1 y_1 - x_2 y_2 - \cdots - x_n y_n$ for any $x = (x_1, x_2, \ldots, x_n)$, $y = (y_1, y_2, \ldots, y_n)$ in R^n. For any x in the Lorentz cone, $[x, x] \geq 0$. Further, $[x, x] = 0$ if x is a boundary point of the cone.

Lemma 1.10.2. *Let x and y be in the Lorentz cone K. Then*

$$[x, y] \geq \sqrt{[x, x][y, y]}.$$

Furthermore, the inequality is strict except for the case when x and y are linearly dependent.

Proof. We have

$$x_1^2 = [x, x] + \sum_{j=2}^{n} x_j^2$$

$$y_1^2 = [y, y] + \sum_{j=2}^{n} y_j^2.$$

Define $u = (u_1, u_2, \ldots, u_n)$ with $u_1 = \sqrt{[x, x]}, u_2 = x_2, u_3 = x_3, \ldots, u_n = x_n$ and $v = (v_1, v_2, \ldots, v_n)$ with $v_1 = \sqrt{[y, y]}, v_2 = y_2, v_3 = y_3, \ldots, v_n = y_n$. By applying the Cauchy-Schwarz inequality to vectors u, v we get

$$\sum_{j=1}^{n} u_j v_j \leq \sqrt{\sum_{j=1}^{n} u_j^2} \sqrt{\sum_{j=1}^{n} v_j^2}.$$

Thus

$$x_1^2 y_1^2 \geq \left([x, x] + \sum_{j=2}^{n} x_j^2\right) \left([y, y] + \sum_{j=2}^{n} y_j^2\right)$$

$$\geq \left(\sqrt{[x, x]}\sqrt{[y, y]} + \sum_{j=2}^{n} x_j y_j\right)^2.$$

Therefore,

$$x_1 y_1 \geq \sqrt{[x, x]}\sqrt{[y, y]} + \sum_{j=2}^{n} x_j y_j.$$

That is, $[x, y] = x_1 y_1 - \sum_{j=2}^{n} x_j y_j \geq \sqrt{[x, x]}\sqrt{[y, y]}$, and the inequality is proved. If equality holds, then u, v must be linearly dependent and then x, y are linearly dependent as well.   ∎

Let $A : R^n \to R^n$ be a unitary transformation with respect to the bilinear form $[\,,\,]$, i.e.,

$$[Ax, Ay] = [x, y]$$

for all $x, y \in R^n$. Such a map A is called a *Lorentz transformation*.

Theorem 1.10.3. *Let $A : R^n \to R^n$ be a linear transformation that maps the Lorentz cone K into itself. Suppose $[Ax, Ax] > [x, x]$ for $x \in K \setminus \{0\}$. Then A has an eigenvalue $\lambda > 1$ with an eigenvector $u \in K$.*

Proof. Consider the set $T = \{(x_1, x_2, \ldots, x_n) : x_1 = 1, \sum_{i=2}^{n} x_i^2 \leq 1\}$. Then T is a closed, bounded, convex set in R^n. Further, $0 \notin T$. Let $B : x \to \{\frac{1}{(Ax)_1} Ax\}$. [Here $(Ax)_1$ is the first coordinate of (Ax).] By Brouwer's Fixed Point Theorem (see Theorem 1.2.1), $Bu = u$ for some $u \in T$. Thus $Au = \lambda u$, where $\lambda = (Au)_1 > 0$. Now $\lambda^2 [u, u] = [Au, Au] > [u, u]$. Since $u \in T, u_1 = 1$ and thus u is an eigenvector in the cone K. Moreover, $\lambda > 1$ by the assumption on A. ∎

Theorem 1.10.4. *Let $A : R^n \to R^n$ be a Lorentz transformation. Let A map some nonzero element of the Lorentz cone K into itself. Then A has an eigenvector $u \in K$ such that $Au = \lambda u, \lambda \geq 1$. In case $\lambda > 1$, A has another eigenvector $v \in K$ such that $Av = \frac{1}{\lambda} v$. Further, in this case both u and v are on the boundary of K.*

Proof. For any $x \in K, [Ax, Ax] = [x, x] \geq 0$ and therefore either Ax or $-Ax \in K$. Suppose $y \neq 0, z \neq 0 \in K$ and $Ay \in K, -Az \in K$. If $Az \in K$, then we have nothing to prove. By Lemma 1.10.2,

$$[y, z] \geq \sqrt{[y, y][z, z]} \geq 0.$$

The inequality is strict; otherwise, y and z would be linearly dependent and therefore Az would be in K. Since Ay and $-Az \in K$ we have $0 \leq [Ay, -Az] = -[y, z] < 0$, which is a contradiction. Thus A maps K into itself. Let

$$V_\epsilon : (x_1, x_2, \ldots, x_n) \to ((1 + \epsilon)x_1, x_2, \ldots, x_n).$$

We have

$$[AV_\epsilon x, AV_\epsilon x] = [V_\epsilon x, V_\epsilon x] > [x, x]$$

for all $x \in K \setminus \{0\}$. By Theorem 1.10.3,

$$AV_\epsilon u_\epsilon = \lambda_\epsilon u_\epsilon \quad \text{for some } \lambda_\epsilon > 1, u_\epsilon \in T.$$

By going to a subsequence of ϵs we have $V_\epsilon \to I$, (the identity transformation), $AV_\epsilon \to A, u_\epsilon \to u \in T$, and $AV_\epsilon u_\epsilon \to Au$. Thus $\lambda_\epsilon \to \lambda \geq 1$. This proves the first part of the theorem. Suppose $\lambda > 1$. Since A^{-1} is also a Lorentz transformation, we have $A^{-1} v = \mu v$, and $\mu \geq 1$ has a solution $v \in K \setminus \{0\}$. Thus $\frac{1}{\mu}$ is an eigenvalue of A with $\frac{1}{\mu} \leq 1 < \lambda$. Since $\lambda^2 [u, u] = [Au, Au] = [u, u]$, we have $[u, u] = 0$. Similarly, $[v, v] = 0$. We also have

$$[u, v] = [Au, Av] = \frac{\lambda}{\mu}[u, v].$$

If $[u, v] = 0$, then equality will occur in Lemma 1.10.2 and u and v will be linearly dependent, which is not possible. Consequently, $[u, v] \neq 0$ and therefore $\lambda = \mu$. Thus $\frac{1}{\lambda}$ is an eigenvalue of A with an eigenvector $v \in K$. That completes the proof.   ∎

It was Krein who first noticed that any nonnegative matrix can be thought of as a linear transformation that maps the cone of nonnegative vectors to nonnegative vectors. Positive vectors can be thought of as interior points of a cone that induces a partial order on a real vector space of finite dimension. It was the study of cones in general Banach spaces by Krein that motivated Krein and Rutman (1948) to extend the Perron-Frobenius Theorem to linear operators that leave a cone invariant in Banach spaces.

Among cones in R^n, the Lorentz cone as a self-conjugate cone has many mathematically interesting properties. Theorem 1.10.4 is due to Frobenius and our proof is adopted from Krein (1950) and Krein and Rutman (1948) for the case of R^n. The Lorentz cone induces an indefinite metric with one positive and $n - 1$ negative terms in expanding the bilinear form associated to its quadratic form. More generally, one can consider the general bilinear form $\langle x, y \rangle$ with p positive and $n - p$ negative terms given by $\langle x, y \rangle = \sum_1^p x_i y_i - \sum_{(p+1)}^n x_i y_i$ in R^n. One could extend the above arguments to prove the following theorem due to Frobenius (1908): *If the indefinite quadratic form $\langle x, x \rangle$ has exactly p positive squares, then a transformation U, unitary with respect to this indefinite metric, has a p-dimensional subspace L in which the form $\langle x, x \rangle$ is nonnegative and in which all eigenvalues of U are not less than unity in modulus. Furthermore, if A is a Hermitian transformation with respect to the indefinite metric $\langle \cdot, \cdot \rangle$, then A has a p-dimensional invariant subspace L in which the quadratic form $\langle \cdot, \cdot \rangle$ is nonnegative and A has eigenvalues with nonnegative imaginary parts.* The proof given for $p = 1$ can be extended to cover this more general case via Brouwer's Fixed Point Theorem; see Krein (1950).

Exercises

1. Find a permutation matrix P such that if

$$A = \begin{bmatrix} 3 & 1 & 7 & 4 \\ 1 & 6 & 3 & 2 \\ 1 & 4 & 7 & 0 \\ 0 & 1 & 3 & 9 \end{bmatrix},$$

then

$$P A P^T = \begin{bmatrix} 9 & 3 & 1 & 0 \\ 0 & 7 & 4 & 1 \\ 2 & 3 & 6 & 1 \\ 4 & 7 & 1 & 3 \end{bmatrix}.$$

2. Let $A = (a_{ij})$ be a square matrix of order n. Let J be a nonempty proper subset of $\{1, 2, \ldots, n\}$ with $a_{ij} = 0$ if $i \notin J, j \in J$. Prove that A is reducible.

3. If A is reducible, prove that A^2 is reducible.

4. If A is irreducible, is A^2 irreducible?

5. Let A be an irreducible nonnegative matrix of order n. If the main diagonal entries of A are positive, prove that $A^{n-1} > 0$.
[Hint: For some $\epsilon > 0$, $\epsilon(I + A) \leq A$.]

6. A matrix $A = (a_{ij})$ has a *saddle point* at the entry a_{pq} if a_{pq} is the smallest entry in column q and the largest entry in row p. Let A be the 2×2 matrix

$$\begin{bmatrix} a & b \\ c & d \end{bmatrix}.$$

If A has no saddle point, prove that the game is completely mixed and the value of the game is

$$v = \frac{ad - bc}{a - b + d - c}.$$

7. If a payoff matrix $A = (a_{ij})$ is skew-symmetric (i.e., if $A = -A^T$), prove that the value of the game is zero.
[Hint: Check that any optimal strategy of one player is also optimal for the other player.]

8. Show that the skew-symmetric payoff matrix

$$A = \begin{bmatrix} 0 & 5 & -4 & 7 \\ -5 & 0 & 6 & -2 \\ 4 & -6 & 0 & 4 \\ -7 & 2 & -4 & 0 \end{bmatrix}$$

is not completely mixed.
(Hint: From the previous exercise, the value is zero. Evaluate the cofactor A_{11} and use Theorem 1.3.6.)

9. Consider the payoff matrix

$$A = \begin{bmatrix} 5 & -1 & -2 \\ -3 & 8 & -2 \\ -5 & -3 & 9 \end{bmatrix}.$$

Prove that the game is completely mixed.

10. Let A be an $m \times n$ payoff matrix with value v. Prove that there exist optimal mixed strategies for the players with at most $\min(m, n)$ positive coordinates. (Hint: We may assume $m \leq n$. Let y be optimal for Player I. First show that Ay is a boundary point of the convex hull (i.e., the set of all convex combinations) of the columns of A. Then use Carathéodory's Theorem, which asserts that any point in a polyhedral set in R^k can be expressed as a linear combination of at most $k + 1$ vectors from the defining set and furthermore, for a boundary point, at most k vectors suffice.)

11. Let G be a directed graph with vertex set V. Consider the following game: Player I chooses an edge and Player II chooses a vertex of G. If the chosen vertex and the edge are not incident, then the payoff is zero. Otherwise, the payoff to Player II from Player I is 1 or -1, depending on whether the edge originates or terminates at the vertex chosen. (a) Show that the value of the game is nonnegative and is zero if and only if the graph has a directed cycle. (b) Suppose the graph is acyclic and, for each vertex v, let $\ell(v)$ be the maximum possible length of a path originating at v. Show that the value of the game is $(\sum_{v \in V} \ell(v))^{-1}$. (c) Find optimal strategies for the two players. If the graph is acyclic, show that Player II has a unique optimal strategy. [See Bapat and Tijs (1995).]

12. Let A be an $m \times n$ matrix and consider the following game: Player I and Player II choose a nonempty subset of $\{1, \ldots, m\}$ and $\{1, \ldots, n\}$, respectively. The payoff to Player II from Player I is the value of the matrix game determined by the submatrix of A corresponding to the row and column indices chosen by the respective players. Show that this game has a saddle point. (See Exercise 6 for the definition of a saddle point.)

13. Let A and B be $n \times n$ matrices with $A \geq B \geq 0$. Prove that the Perron root of A is greater than or equal to that of B.
(Hint: We may assume A and B to be irreducible, as the general case then follows by a continuity argument. We have $y^T A x \geq y^T B x$ for all nonnegative vectors x and y. Choose x to be a right Perron eigenvector of A and y to be a left Perron eigenvector of B.)

14. If a square matrix $A \geq 0$ has each row sum less than unity, prove that $A^k \to 0$ as $k \to \infty$.

15. Let A be a positive matrix. If the function $v(\mu) = $ value $(A - \mu I)$ has the first two derivatives continuous and strictly convex in a neighborhood of the Perron root, how can you use this to improve the initial guess about the Perron root μ? Assuming the validity of these hypotheses for the matrix

$$\begin{bmatrix} 1 & 6 & 8 \\ 3 & 5 & 2 \\ 4 & 9 & 1 \end{bmatrix}$$

and $\mu = 10$ as the initial guess for the Perron root μ, find an improved estimate for μ.

16. If λ is an eigenvalue of A and $\epsilon > 0$, prove that, for some $B < A$, the matrix B has an eigenvalue μ with $|\mu - \lambda| < \epsilon$.

17. Let A be a nonnegative, lower triangular $n \times n$ matrix. Then, clearly, $a_{11}, \ldots, a_{nn}$ are the eigenvalues of A and we assume that these are distinct. Let $i \in \{1, \ldots, n\}$ be fixed. Show that the following conditions are equivalent:

(i) A has a nonnegative eigenvector corresponding to a_{ii}.

(ii) For any $j \neq i$, if $a_{ji} > 0$ then $a_{ii} > a_{jj}$.

18. Let A be a nonnegative, lower triangular matrix. Characterize the eigenvalues of A that admit a nonnegative eigenvector. Then solve the same problem for an arbitrary nonnegative $n \times n$ matrix by first putting it in the Frobenius Normal Form. [See Schneider (1986).]

19. Let A be a nonnegative, irreducible $n \times n$ matrix. Suppose, for $k = 1, 2, \ldots, n - 1$, the sum of all principal minors of A of order k is zero. Show that there exists a permutation matrix P such that

$$PAP^T = \begin{bmatrix} 0 & x_1 & \cdots & \cdots & 0 \\ 0 & 0 & x_2 & \cdots & 0 \\ \vdots & \vdots & \vdots & \ddots & \vdots \\ 0 & 0 & 0 & \cdots & x_{n-1} \\ x_n & 0 & 0 & \cdots & 0 \end{bmatrix},$$

for some positive numbers $x_1, \ldots, x_n$.

20. Let $X \subset R^n$. A set-valued map $\phi(x) : X \to 2^X$ (all closed, nonempty subsets of X) is called *upper semicontinuous* if for any sequence $\{x_n\} \subset X, x_n \to x$, $y_n \in \phi(x_n)$ and $y_n \to y$ then $y \in \phi(x)$. If $\sigma(A)$ denotes the set of all eigenvalues of an $n \times n$ matrix A, is the map $\phi : A \to \sigma(A)$ uppersemicontinuous?

21. Let C be an $n \times n$ matrix with all its eigenvalues in the set $\{\lambda : |\lambda| < 1\}$. Prove that $\lim_k C^k = 0$.

22. A function $f : R^n \to R$, the set of real numbers, is called *additive* if $f(x + y) = f(x) + f(y)$ for all $x, y \in R^n$. Let K be a cone in R^n with nonempty interior. If an additive function f is nonnegative on the cone (i.e., $f(x) \geq 0$, when $x \in K$), prove that f is continuous.

23. Let K be a pointed cone in R^n. We write $x \succeq y$ if $x - y \in K$. Show that $\succeq$ is a partial order on R^n.

24. A pointed cone $K \in R^n$ *is a lattice* if and only if the partial order $\succeq$ induces a lattice, i.e., given $x \succeq 0, y \succeq 0$, there exists $z \in K$ such that $z \succeq x, z \succeq y$, and further $w \succeq x, w \succeq y \Rightarrow w \succeq z$. The element z is denoted by sup (x, y). Similarly, inf (x, y) can be defined. Check that the positive orthant in R^n is a lattice cone with respect to the partial order $\succeq$ induced by the cone.

25. Prove that the Lorentz cone in R^n has a nonempty interior.

26. Show that when a cone K has nonempty interior in R^n then $R^n = K - K$, where $K - K = \{x - y : x, y \in K\}$

27. Let $f : R^n \to R^n$ be a one-to-one map with $[f(x), f(y)] = [x, y]$ for all $x, y \in R^n$, where $[x, y] = x_1 y_1 - x_2 y_2 - \cdots - x_n y_n$. Prove that f is linear and hence a Lorentz transformation. Also show that f has an inverse that is also linear.

28. Let A and B be $n \times n$ matrices and, for any $n \times n$ matrix U, let $\phi(U) = AUA^T + BUB^T$. Show that ϕ preserves the cone of positive semidefinite matrices. (See Chapter 3 for elementary properties of positive semidefinite matrices.) If ϕ is expressed as a map from R^{n^2} to itself, then the corresponding matrix of this transformation is $A \otimes A + B \otimes B$, where $\otimes$ denotes the Kronecker product. Prove the basic assertions of the Perron-Frobenius Theorem as well as an analogue of Exercise 13 for the map ϕ, concluding that

$$r(A \otimes A + B \otimes B) \geq r(A \otimes A),$$

where r denotes the spectral radius. [This problem has applications in time series analysis; see Liu (1992) and Tigelaar (1991).]

2

Doubly stochastic matrices

A square matrix is called *doubly stochastic* if all entries of the matrix are nonnegative and the sum of the elements in each row and each column is unity. Among the class of nonnegative matrices, stochastic matrices and doubly stochastic matrices have many remarkable properties. Whereas the properties of stochastic matrices are mainly spectral theoretic and are motivated by Markov chains, doubly stochastic matrices, besides sharing such properties, also have an interesting combinatorial structure. In this chapter we first consider the combinatorial properties of the polytope of doubly stochastic matrices. The Birkhoff–von Neumann Theorem, the Frobenius-König Theorem, and related results are proved. An extension of the Frobenius-König Theorem involving matrix rank is given. We then describe a probabilistic algorithm to find a positive diagonal in a nonnegative matrix. Such algorithms are of relatively recent origin. The next several sections focus on diagonal products and permanents of nonnegative as well as doubly stochastic matrices. The proof of the van der Waerden conjecture due to Egorychev is given. We also give an elementary alternative proof of the Alexandroff Inequality, which is along the lines of the proof of the van der Waerden conjecture due to Falikman. The last few sections are concerned with various problems in game theory, scheduling, and economics.

2.1. The Birkhoff–von Neumann Theorem

In this section we develop the basic properties of the polytope of doubly stochastic matrices. We denote the set of $n \times n$ doubly stochastic matrices by Ω_n.

Lemma 2.1.1. *The set Ω_n, viewed as a subset of R^{n^2}, is a closed, bounded, convex set.*

Proof. Let $A, B \in \Omega_n$. For any $0 \le \lambda \le 1$, the matrix $\lambda A + (1 - \lambda)B$ has all entries nonnegative and for any i,

$$\lambda \sum_j a_{ij} + (1 - \lambda) \sum_j b_{ij} = 1.$$

Thus the row sums of $\lambda A + (1 - \lambda) B$ are unity. Similarly, the column sums are unity, and thus Ω_n is a convex set. It is easily proved that Ω_n is bounded and closed.   ∎

The set Ω_2 is given by

$$\Omega_2 = \left\{ A : A = \begin{bmatrix} p & 1 - p \\ 1 - p & p \end{bmatrix} : 0 \le p \le 1 \right\}.$$

Moreover,

$$\Omega_2 = \left\{ p \begin{bmatrix} 1 & 0 \\ 0 & 1 \end{bmatrix} + (1 - p) \begin{bmatrix} 0 & 1 \\ 1 & 0 \end{bmatrix} : 0 \le p \le 1 \right\}.$$

Thus Ω_2 is the convex set generated by

$$\begin{bmatrix} 1 & 0 \\ 0 & 1 \end{bmatrix} \quad \text{and} \quad \begin{bmatrix} 0 & 1 \\ 1 & 0 \end{bmatrix}.$$

A point x_0 of a convex set S is called an *extreme point* of S if $S \setminus \{x_0\}$ is also convex.

Lemma 2.1.2. *A point x_0 of a convex set S is not an extreme point of S if and only if there exists a pair of distinct points $u, w \in S$ such that $x_0 = \frac{1}{2}u + \frac{1}{2}w$; that is, x_0 is the midpoint of a nontrivial line segment in S.*

Proof. Suppose that $x_0 \in S$ is not an extreme point of S. Then $S \setminus \{x_0\}$ is not convex. By convention the empty set and sets with just one element are convex. Therefore, there exist two distinct elements $u, v \in S \setminus \{x_0\}$ such that $\lambda u + (1 - \lambda)v \notin S \setminus \{x_0\}$ for some $\lambda \in (0, 1)$. Thus $\lambda u + (1 - \lambda)v = x_0$. Without loss of generality, we can take $\frac{1}{2} < \lambda < 1$. If $\lambda = \frac{\alpha+1}{2}$, then $0 < \alpha < 1$ and $x_0 = \frac{1}{2}u + \frac{1}{2}w$, where $w = \alpha u + (1 - \alpha)v$. Conversely, if $x_0 = \frac{1}{2}u + \frac{1}{2}w$ and $u \ne w$, then $u, w \in S \setminus \{x_0\}$ but $x_0 \notin S \setminus \{x_0\}$. Thus $S \setminus \{x_0\}$ is not convex.
                                                                            ∎

The next result gives a clue to identifying the extreme points of Ω_n. We will eventually show (in Theorem 2.1.6) that permutation matrices are the *only* extreme points of Ω_n.

Lemma 2.1.3. *Any $n \times n$ permutation matrix is an extreme point of Ω_n.*

Proof. Let $\Pi = (\pi_{ij})$ be a permutation matrix in Ω_n. If Π is not an extreme point of Ω_n, then by Lemma 2.1.2, $\Pi = \frac{A+B}{2}$, $A \neq B$ and $A, B \in \Omega_n$. Now, for all i, j, $\pi_{ij} = 0$ or 1 and $A = (a_{ij})$, $B = (b_{ij})$ satisfy $0 \leq a_{ij}, b_{ij} \leq 1$. Thus if $\pi_{ij} = 0$, then $a_{ij} = b_{ij} = 0$ and if $\pi_{ij} = 1$, then $a_{ij} = b_{ij} = 1$. This contradicts the assumption that $A \neq B$, and the proof is complete. ∎

Given a square matrix $A = (a_{ij})$ of order n, the *diagonal* associated with a permutation σ of $\{1, 2, \ldots, n\}$ is the set $\{a_{1\sigma(1)}, a_{2\sigma(2)}, \ldots, a_{n\sigma(n)}\}$. A diagonal is said to be *positive* if each entry in the diagonal is positive. The product $\prod_{i=1}^{n} a_{i\sigma(i)}$ is called the *diagonal product* of A associated with the permutation σ.

For example, in the 4×4 matrix

$$A = \begin{bmatrix} 3 & 5 & 2^* & 7 \\ 6^* & 0 & 1 & 4 \\ 3 & 2^* & 5 & 1 \\ 2 & 9 & 1 & 3^* \end{bmatrix}$$

the diagonal corresponding to $\sigma : \{1, 2, 3, 4\} \to \{3, 1, 2, 4\}$ consists of the starred entries. The corresponding diagonal product is 72.

The sum of all diagonal products of a matrix is called the *permanent* of the matrix. Thus if $A = (a_{ij})$ is of order $n \times n$, the permanent of A is defined by

$$\text{per } A = \sum_{\sigma} \prod_{i=1}^{n} a_{i\sigma(i)},$$

where the sum is taken over all $n!$ permutations σ of $\{1, 2, \ldots, n\}$.

A *transposition* is a permutation σ that interchanges two indices, leaving the others fixed. Thus $\sigma : \{1, 2, 3, 4\} \to \{3, 2, 1, 4\}$ is a transposition. We can represent σ by the permutation matrix

$$\Pi_1 = \begin{bmatrix} 0 & 0 & 1 & 0 \\ 0 & 1 & 0 & 0 \\ 1 & 0 & 0 & 0 \\ 0 & 0 & 0 & 1 \end{bmatrix}.$$

Recall that it is possible to associate one of the numbers -1 or $+1$ to each permutation σ through a function $\epsilon(\cdot)$ defined as follows:

(1) $\epsilon(id) = 1$, where $id : \{1, 2, \ldots, n\} \to \{1, 2, \ldots, n\}$ is the identity permutation.

(2) $\epsilon(\sigma) = -1$ if σ is a transposition.

(3) $\epsilon(\sigma \circ \tau) = \epsilon(\sigma) \cdot \epsilon(\tau)$, where $\sigma \circ \tau$ denotes the composition of σ and τ.

Then, for any square matrix A,

$$|A| = \sum_{\sigma} \epsilon(\sigma) \prod_{i=1}^{n} a_{i\sigma(i)},$$

where $|A|$ denotes the determinant of A. Thus the permanent and the determinant differ only in the usage of the appropriate sign $\epsilon(\sigma)$ for each term in the sum. While determinants are easily computed by Gaussian elimination, there is no such counterpart for computing permanents. Recently Girko (1995) proved the following theorem: Let $A = (a_{ij})$ be an $n \times n$ matrix with positive entries. Let $\xi_{ij}, i, j = 1, \ldots, n$ be independently and identically distributed random variables with $E\xi_{ij} = 0$, and $E\xi_{ij}^2 = 1$. Then

$$\operatorname{per} A = E(\det(a_{ij}\xi_{ij}) \det(\xi_{ij})).$$

Thus permanents can be computed approximately by simulation.

The permanent, like the determinant, admits a Laplace expansion along any row or column. If A is an $n \times n$ matrix, then for any i, j, let $A(i, j)$ denote the submatrix obtained by deleting the i-th row and the j-th column of A. Then, expansion of the permanent along the i-th row yields

$$\operatorname{per} A = \sum_{j=1}^{n} a_{ij} \operatorname{per} A(i, j).$$

We now prove an important result that is central in the study of zero-nonzero patterns of nonnegative matrices. The result is equivalent to the well-known theorem of Hall (1935) on systems of distinct representatives; see Mirsky (1971) and Marcus and Minc (1964) for details.

Theorem 2.1.4 (Frobenius-König). *The permanent of a nonnegative $n \times n$ matrix A is zero if and only if A has an $r \times s$ zero submatrix with $r + s = n + 1$.*

Proof. We first prove the "only if" part. If $A = 0$, the result obviously holds. Let $A = (a_{ij})$ and $a_{ij} \neq 0$ for some i, j. By assumption, the permanent of A is zero and hence every diagonal product is zero. Thus per $A(i, j) = 0$. By induction assumption $A(i, j)$, and hence A, has a zero submatrix of order $u \times v$ such that $u + v = n$. We can find permutation matrices Π_1, Π_2 such that

$$\Pi_1 A \Pi_2 = \begin{bmatrix} B & 0 \\ C & D \end{bmatrix}.$$

Here B and D are square matrices of order u and $n - u$, respectively. Since per $A = 0$, either per $B = 0$ or per $D = 0$. Suppose per $B = 0$. By induction B has a $p \times q$ zero submatrix with $p + q = u + 1$. Without loss of generality, we can assume it to be the matrix formed by the first p rows and the last q columns of B. Thus A has a $p \times (q + n - u)$ zero submatrix and $p + q + n - u = u + 1 + n - u = n + 1$. A similar argument applies when per D is zero.

To prove the "if" part, let us assume, without loss of generality, that the matrix formed by the first r rows and the first s columns of A is the zero matrix, where $r + s = n + 1$. Let σ be any permutation of $\{1, 2, \ldots, n\}$. We must show that $a_{i\sigma(i)} = 0$ for some i. Suppose that $a_{i\sigma(i)}$ is nonzero for all i. Then we must have

$$\{\sigma(1), \sigma(2), \ldots, \sigma(r)\} \cap \{1, 2, \ldots, s\} = \emptyset,$$

which contradicts the fact that $r + s = n + 1$, and the proof is complete. ∎

Lemma 2.1.5. *If A is a doubly stochastic matrix, then A has a positive diagonal.*

Proof. It is sufficient to show that per $A > 0$. Suppose per $A = 0$. Then, by the Frobenius-König Theorem, A has an $r \times s$ zero submatrix with $r + s = n + 1$. Thus we may assume, after permuting the rows and columns of A if necessary, that

$$A = \begin{bmatrix} B & 0 \\ C & D \end{bmatrix},$$

where the zero matrix is $r \times s$, with $r + s = n + 1$. Thus B is $r \times (n - s)$ and D is $(n - r) \times s$. Since A is doubly stochastic the row sums of B are unity and hence the entries of B add up to r. The column sums of A are also equal to 1 and therefore the entries of B together with the entries of C add up to $n - s$. Thus we must have $r \leq n - s$, which is a contradiction because $r + s = n + 1$. That completes the proof. ∎

Theorem 2.1.6 (Birkhoff–von Neumann). *Any doubly stochastic matrix A can be written as a convex combination of finitely many permutation matrices; that is,*

$$A = \lambda_1 \Pi_1 + \lambda_2 \Pi_2 + \cdots + \lambda_m \Pi_m,$$

where $\Pi_1, \Pi_2, \ldots, \Pi_m$ are permutation matrices and $0 \leq \lambda_1, \lambda_2, \ldots, \lambda_m \leq 1$, $\sum_{i=1}^{m} \lambda_i = 1$.

Proof. If A is a permutation matrix, then the proof is obvious. Otherwise, by Lemma 2.1.5, A has a positive diagonal. Let Π_1 be the permutation corresponding to this diagonal, namely, if $\{a_{1r_1}, a_{2r_2}, \ldots, a_{nr_n}\}$ is the positive diagonal then

$$(\Pi_1)_{ij} = 1 \quad \text{if} \quad (i, j) = (1, r_1), (2, r_2), \ldots, (n, r_n)$$

$$= 0 \quad \text{otherwise.}$$

Let $\lambda_1 = a_{tr_t} = \min\{a_{1r_1}, a_{2r_2}, \ldots, a_{nr_n}\} > 0$. Clearly, $\lambda_1 < 1$. We then have

$$A = \lambda_1 \Pi_1 + (1 - \lambda_1)\left\{\frac{1}{(1 - \lambda_1)}(A - \lambda_1 \Pi_1)\right\}.$$

It is clear that $B = \frac{1}{1-\lambda_1}(A - \lambda_1 \Pi_1)$ is again doubly stochastic and has at least one more zero entry than A. If B is a permutation matrix, then $A = \lambda_1 \Pi_1 + (1-\lambda_1)B$ and we are through. Otherwise, we can find a permutation Π_2 as above such that $B = \mu_2 \Pi_2 + (1 - \mu_2)C$, where C is doubly stochastic. If C is a permutation matrix, then $A = \lambda_1 \Pi_1 + \lambda_2 \Pi_2 + \lambda_3 \Pi_3$, where $\lambda_2 = (1 - \lambda_1)\mu_2, \lambda_3 = (1 - \lambda_1)(1 - \mu_2)$ and $\Pi_3 = C$. Moreover, λ_1, λ_2, and λ_3 are nonnegative and add up to unity. Also notice that in any case C has at least two more zero entries than A. If C is not a permutation matrix we can proceed as before with another iteration. Thus, after at most $n^2 - n$ such iterations, our algorithm must terminate in a permutation matrix and therefore A is a convex combination of (at most $n^2 - n + 1$) permutation matrices. This completes the proof of the theorem.   ∎

It can be shown that the algorithm described in the proof of Theorem 2.1.6 will terminate in at most $n^2 - 2n + 2$ iterations [see Johnson, Dulmage and Mendelsohn (1960), Brualdi (1982)]. Thus any $A \in \Omega_n$ can be expressed as a convex combination of at most $n^2 - 2n + 2$ permutation matrices. This fact may also be deduced directly using Carathéodory's Theorem (see Exercise 1), which asserts that if $S \subset R^n$ is convex, then any point in S can be expressed as a convex combination of at most $n + 1$ points in S [see, for example, Bazaraa and Shetty (1979)].

The proof of the Birkhoff–von Neumann Theorem given above is constructive in the sense that, given a doubly stochastic matrix A, it gives a recipe for expresssing it as a convex combination of permutation matrices. The only step in the algorithm that may be difficult is locating a positive diagonal in a doubly stochastic matrix. We now give an algorithm for doing this, which is very much in the spirit of the original proof of the Birkhoff–von Neumann Theorem due to von Neumann (1953).

If the doubly stochastic matrix A is a permutation matrix, then it has precisely one positive diagonal consisting of all the positive entries in the matrix. We therefore assume that A is not a permutation matrix. Then $0 < a_{i_1 j_1} < 1$ for some $a_{i_1 j_1}$. Now the i_1-th row has one more entry $a_{i_1 j_2}$ with $0 < a_{i_1 j_2} < 1$. Considering the j_2-th column we have $0 < a_{i_2 j_2} < 1$ for some row index $i_2 \neq i_1$. Again, suppose we have $0 < a_{i_2 j_3} < 1$ for some $j_3 \neq j_2$. Continuing this process alternately between rows and columns, we obtain a finite sequence $a_{i_1 j_1}, a_{i_1 j_2}, a_{i_2 j_2}, \ldots, a_{i_k j_k}$, where all these entries lie strictly between

0 and 1 and for the first time either $i_k = i_r$ or $j_k = j_r$ for some $r < k$. Let us suppose $i_k = i_r$. Then we have a path $\mathcal{P}$, which we again denote by $\{i_1 j_1, i_1 j_2, i_2 j_2, \ldots, i_k j_{k-1}, i_k j_1\}$ for convenience. Here each row index and each column index appears exactly twice. Let

$$\delta = a_{i_s j_s} = \min\{a_{ij} : (ij) \in \mathcal{P}\}.$$

Define $C = (c_{ij})$, where

$$c_{ij} = 0 \quad \text{if } (ij) \notin \mathcal{P}$$

and

$$c_{i_s j_s} = +1, c_{i_s j_{s+1}} = -1, c_{i_{s+1} j_{s+1}} = +1, \cdots \text{etc.}$$

Here the subscripts of i, j are to be interpreted modulo k. Now the matrix $A_1 = A - \delta C$ is also doubly stochastic. If A_1 has a nonzero entry, then the corresponding entry of A is also nonzero. Moreover, A_1 has at least one more zero entry than A. If A_1 is not a permutation matrix, then we can similarly construct a matrix A_2, with at least one more zero entry than in A_1. Since the number of zero entries is increasing each time, we will eventually arrive at a permutation matrix $A_s = P^\tau$ corresponding to a permutation τ. Now, by backward induction we have $\prod a_{i\tau(i)} > 0$. Thus τ corresponds to a positive diagonal, which we were looking for.

We illustrate the algorithm with an example. Let

$$A = \begin{bmatrix} 0 & .5 & 0 & .5 & 0 \\ .8 & 0 & 0 & .2 & 0 \\ 0 & .5 & .2 & 0 & .3 \\ .2 & 0 & .3 & 0 & .5 \\ 0 & 0 & .5 & .3 & .2 \end{bmatrix}.$$

Starting with the entry a_{12} we easily find a closed path consisting of the locations $\{(12), (14), (24), (21), (41), (43), (33), (32), (12)\}$. The minimum of the entries in this path is 0.2. This minimum value of 0.2 is alternately subtracted from or added to each entry in the path, starting with a subtraction at a minimal entry, say a_{33}. Then we get

$$A_1 = \begin{bmatrix} 0 & .3 & 0 & .7 & 0 \\ 1 & 0 & 0 & 0 & 0 \\ 0 & .7 & 0 & 0 & .3 \\ 0 & 0 & .5 & 0 & .5 \\ 0 & 0 & .5 & .3 & .2 \end{bmatrix}.$$

Again, we have the closed path $\{(12), (14), (54), (53), (43), (45), (35), (32), (12)\}$ with a minimum entry of .3 at (12). Subtracting 0.3 from that and following the earlier procedure, we get

$$A_2 = \begin{bmatrix} 0 & 0 & 0 & 1 & 0 \\ 1 & 0 & 0 & 0 & 0 \\ 0 & 1 & 0 & 0 & 0 \\ 0 & 0 & .2 & 0 & .8 \\ 0 & 0 & .8 & 0 & .2 \end{bmatrix}.$$

If we use the closed path $\{(43), (45), (55), (53), (43)\}$ in A_2 as indicated above, the next step yields

$$A_3 = \begin{bmatrix} 0 & 0 & 0 & 1 & 0 \\ 1 & 0 & 0 & 0 & 0 \\ 0 & 1 & 0 & 0 & 0 \\ 0 & 0 & 0 & 0 & 1 \\ 0 & 0 & 1 & 0 & 0 \end{bmatrix}.$$

Thus $\{a_{14}, a_{21}, a_{32}, a_{45}, a_{53}\}$ is a positive diagonal.

Doubly stochastic matrices have attracted combinatorial matrix theorists for a long time. Birkhoff (1946) stated Theorem 2.1.6 as one of three remarks in algebra and published this fundamental theorem in an obscure South American journal (see Birkhoff (1946)). Independently, while studying the equivalence of a game to an optimal assignment problem, (see section 2.4), von Neumann (1953) proved the same result. A nice exposition of this topic can be found in Minc (1988).

2.2. Fully indecomposable matrices

A square matrix A is called *partly decomposable* if there exist permutation matrices Π_1 and Π_2 such that

$$\Pi_1 A \Pi_2 = \begin{bmatrix} B & 0 \\ C & D \end{bmatrix},$$

where B and D are square matrices and 0 is a zero matrix. We call a matrix A *fully indecomposable* if it is not partly decomposable. A 1×1 matrix is partly decomposable if and only if it is the zero matrix.

The property of being fully indecomposable is stronger than irreducibility. In the study of the spectral properties of a nonnegative matrix, irreducibility plays a crucial role as observed in Chapter 1. An equally important role is played by full indecomposability in connection with the combinatorial properties of a nonnegative matrix.

Theorem 2.2.1. *If A is a fully indecomposable nonnegative matrix of order n, then $A^{n-1} > 0$.*

Proof. Let $x \geq 0$, $x \neq 0$ be a vector of order n and let $x_i = 0$ for some i. We first show that Ax has strictly fewer zero components than x. Suppose to the contrary that Ax has at least as many zero coordinates as x. For some permutation matrix Q we have $Q^T x = \begin{bmatrix} 0 \\ v \end{bmatrix}$, where v is an $n - s$ dimensional positive vector. There exists a permutation matrix P such that $PAx = \begin{bmatrix} 0 \\ u \end{bmatrix}$, where u is an $n - s$ dimensional nonnegative vector. Let PAQ be written as $PAQ = \begin{bmatrix} A_1 & A_2 \\ A_3 & A_4 \end{bmatrix}$. Here A_1 and A_4 are square matrices of order s and $n - s$, respectively. We have

$$PAQQ^T x = \begin{bmatrix} A_1 & A_2 \\ A_3 & A_4 \end{bmatrix} \begin{bmatrix} 0 \\ v \end{bmatrix} = \begin{bmatrix} 0 \\ u \end{bmatrix}.$$

Thus $A_2 v = 0$. Since $v > 0$, $A_2 = 0$ and so A is partly decomposable. This contradicts our assumption on A. Thus Ax has fewer zero coordinates than x. Similarly, $A^2 x$ will have fewer zero coordinates than Ax if Ax has a zero coordinate. In particular, $A^{n-1} > 0$. ∎

If A, B are irreducible of order n then AB need not be irreducible. In contrast, we have the following result due to Lewin (1971).

Theorem 2.2.2. *If A and B are fully indecomposable nonnegative $n \times n$ matrices, then so is* AB.

Proof. For any vector $x \geq 0$, $x \neq 0$, let $N(x)$ denote the number of zero coordinates of x. Trivially $N(x) = N(Qx)$ for any permutation Q. Also, from Theorem 2.2.1, for any permutation matrices P and Q, we have

$$N(x) = N(Qx) > N(BQx) \geq N(ABQx) = N(PABQx). \qquad (2.2.1)$$

If AB were partly decomposable, then for some permutation matrices P and Q we would have

$$PABQ = \begin{bmatrix} C & 0 \\ D & E \end{bmatrix}.$$

If $x = \begin{bmatrix} 0 \\ u \end{bmatrix}$ and $u > 0$, we get

$$PABQx = \begin{bmatrix} C & 0 \\ D & E \end{bmatrix} \begin{bmatrix} 0 \\ u \end{bmatrix} = \begin{bmatrix} 0 \\ v \end{bmatrix} \quad \text{for some } v \geq 0.$$

Thus $N(PABQx) \geq N(x)$, which contradicts (2.2.1). ∎

Matrices A and B of the same order are said to have the same *pattern* if $a_{ij} = 0$ if and only if $b_{ij} = 0$. An $n \times n$ matrix $A = (a_{ij})$ has *doubly stochastic pattern* if there exists a doubly stochastic matrix with the same pattern as A.

Theorem 2.2.3. *Let A be a nonnegative, nonzero n × n matrix. Then A has doubly stochastic pattern if and only if any positive entry of A is contained in a positive diagonal.*

Proof. We first prove the "only if" part. We may assume that A itself is doubly stochastic. By the Birkhoff–von Neumann Theorem we can write A as a convex combination

$$A = \lambda_1 \Pi_1 + \lambda_2 \Pi_2 + \cdots + \lambda_m \Pi_m,$$

where $\Pi_1, \Pi_2, \ldots, \Pi_m$ are permutation matrices. We assume that $\lambda_1, \lambda_2, \ldots, \lambda_m$ are positive. Let $a_{ij} > 0$. Then there must exist $k \in \{1, 2, \ldots, m\}$ such that the (i, j)-entry of Π_k is 1. Since $A \geq \Pi_k$, the corresponding diagonal of A is a positive diagonal and it contains a_{ij}.

We now prove the "if" part. Since A is a nonzero matrix, the hypothesis implies that per $A > 0$. Consider the matrix $B = (b_{ij})$, defined as

$$b_{ij} = \frac{a_{ij} \operatorname{per} A(i, j)}{\operatorname{per} A}, \quad i, j = 1, 2, \ldots, n.$$

Then, expanding the permanent along any row or column we see that B is doubly stochastic. Further, if $a_{ij} > 0$, then a_{ij} is contained in a positive diagonal and therefore per $A(i, j) > 0$. Thus B has the same pattern as A, and the proof is complete. ∎

Two nonzero entries of a matrix $A = (a_{ij})$ are said to be *chainable* if there is a path from one to the other, as a rook (castle) on an $n \times n$ chess board would move, with the intermediate turns of the rook at nonzero entries of A. A matrix is chainable if every pair of nonzero entries is chainable.

In the following 5×5 matrix A,

$$\begin{bmatrix} 0 & 1 & 3 & 0 & 0 \\ 5 & 0 & 0 & 2 & 4 \\ 6 & 0 & 3 & 1 & 7 \\ 0 & 2 & 1 & 4 & 5 \\ 3 & 0 & 0 & 0 & 0 \end{bmatrix}.$$

the nonzero entry a_{25} can be reached from a_{12} by rook moves with the nonzero path (12), (13), (33), (34), (24), and (25).

We will denote the cardinality of a set S (i.e., the number of elements in the set S) by card (S).

Theorem 2.2.4. *A nonnegative $n \times n$ matrix A is fully indecomposable if and only if it has a doubly stochastic pattern and is chainable.*

Proof. Let A be fully indecomposable. We first claim that any positive entry of A is contained in a positive diagonal. Otherwise, there would exist $a_{ij} > 0$ such that per $A(i, j) = 0$. By the Frobenius-König Theorem, the matrix $A(i, j)$ has an $r \times s$ zero submatrix with $r + s = n$. But then A would be partly decomposable, which is a contradiction. Thus the claim is proved. It follows by Theorem 2.2.3 that A has doubly stochastic pattern. Let B be a doubly stochastic matrix with the same pattern as A.

To prove chainability, let $a_{pq} > 0$ and let $a_{p_1 q_1}, a_{p_2 q_2}, \ldots, a_{p_k q_k}$ be the set of all positive entries that are chainable with a_{pq}. Let I denote distinct indices in $\{p_1, p_2, \ldots, p_k\}$ and let J denote distinct indices in $\{q_1, q_2, \ldots, q_k\}$. If $a_{ij} > 0$ and $i \in I$, $j \in J$, then $a_{ij} = a_{p_r q_r}$ for some r. Thus the matrix $A(I, J)$ with $i \in I$, $j \in J$ is chainable. If $I = \{1, 2, \ldots, n\}$, then, given any $a_{rs} > 0$, $r = p_t$ for some t and thus a_{pq} has a rook path to a_{rs} via $a_{p_t q_t}$. A similar proof holds if $J = \{1, 2, \ldots, n\}$. Thus if I or $J = \{1, 2, \ldots, n\}$, then A is chainable. Further, if I, J are proper subsets of $\{1, 2, \ldots, n\}$, then $A(I, J') = 0$ and $A(I', J) = 0$, where I', J' are the complements of I and J in $\{1, 2, \ldots, n\}$. Thus $B(I, J') = 0$ and $B(I', J) = 0$. Since B is doubly stochastic, the rows in I add up to card (I) = sum of the entries of $B(I, J)$. Similarly, the columns in J add up to card (J) = sum of the entries of $B(I, J)$. Thus $A(I, J)$ is a square matrix. Hence

$$A = \begin{bmatrix} A(I, J) & A(I, J') \\ A(I', J) & A(I', J') \end{bmatrix} = \begin{bmatrix} A(I, J) & 0 \\ 0 & A(I', J') \end{bmatrix}$$

is partly decomposable, which is a contradiction. Therefore, when A is fully indecomposable, A is chainable. Conversely, if A is partly decomposable and has doubly stochastic pattern, then we can show, using an argument similar to the one before that A can be written as a 2×2 block matrix with off-diagonal blocks equal to zero. However, then A is not chainable, and this completes the proof. ∎

2.3. König's Theorem and rank

Let A be an $m \times n$ matrix. By a *line* of A we mean either a row or a column of A. Thus A has $m + n$ lines. The Frobenius-König Theorem asserts that if A is a nonnegative $n \times n$ matrix, then per $A = 0$ if and only if A has a zero submatrix of $n + 1$ lines.

The *term rank* of an $m \times n$ matrix is defined as the maximal number of nonzero entries, no two of which are on the same line. For example, the term rank of the following matrix is 3:

$$\begin{bmatrix} 0 & 0 & 3^* & 0 & 0 \\ 3^* & 1 & 2 & 0 & 4 \\ 0 & 0 & 5 & 0 & 0 \\ 1 & 0 & 1 & 1 & 2^* \\ 0 & 0 & 6 & 0 & 0 \end{bmatrix}.$$

The starred entries represent a set of 3 entries, no two of which are on the same line, and it can be seen that 3 is the maximal number with this property.

König's Theorem (also known as the König-Egervary Theorem) asserts that the term rank of an $m \times n$ matrix A equals the minimal number of lines required to cover (i.e., contain) all nonzero entries in the matrix A. As an example, in the 5×5 matrix given above, rows 2 and 4 and column 3 cover all nonzero entries in the matrix.

König's Theorem implies the Frobenius-König Theorem. This can be seen as follows: Let A be a nonnegative $n \times n$ matrix and suppose per $A = 0$. Then the term rank of A is less than n. By König's Theorem there exist r rows and s columns that contain all nonzero entries in A, where $r + s < n$. The submatrix formed by the remaining $n - r$ rows and $n - s$ columns is then the zero matrix with $(n - r) + (n - s) \geq n + 1$ lines. The converse implication in the Frobenius-König Theorem is proved in a similar manner.

We next prove a general result involving matrix rank that will turn out to be stronger than König's Theorem. First we develop some preliminaries.

An $m \times n$ matrix B is said to be of *zero type* if rank $B = $ rank $B(i, j), i = 1, 2, \ldots, m; j = 1, 2, \ldots, n$. Thus a matrix is of zero type if and only if every row (column) is linearly dependent on the remaining rows (columns). By convention, a matrix whose row or column set is empty has rank zero. Thus a row or column vector is of zero type if and only if it is the zero vector.

Clearly, a zero matrix is of zero type. The matrix

$$\begin{bmatrix} 1 & 2 & 4 \\ 2 & 1 & 6 \\ 1 & -1 & 2 \end{bmatrix}$$

is also of zero type since rank $B = $ rank $B(i, j) = 2$ for any i, j.

If B is an $m \times n$ matrix, then $B(i, \cdot)$ denotes the submatrix obtained by deleting the i-th row. Similarly, $B(\cdot, j)$ denotes the submatrix obtained by deleting the j-th column.

Lemma 2.3.1. *Let B be an $m \times n$ matrix. Then B is of zero type if and only if*

$$rank\ B = rank(i, \cdot) = rank(\cdot, j), \quad i = 1, 2, \ldots, m; j = 1, 2, \ldots, n.$$

Proof. The result is obvious if $m = 1$ or $n = 1$. So we assume that $m, n \geq 2$. First suppose that B is of zero type. Then

$$\text{rank } B \geq \text{rank}(i, \cdot) \geq \text{rank}(i, j) = \text{rank } B$$

and hence rank $B = \text{rank } B(i, \cdot), i = 1, 2, \ldots, m$. Similarly, we can show that rank $B = \text{rank } B(\cdot, j), j = 1, 2, \ldots, n$.

To prove the converse, suppose B satisfies the hypothesis of the lemma. Since rank $B = \text{rank } B(1, \cdot)$, the first row of B is linearly dependent on the remaining rows. Then the first row of $B(\cdot, 1)$ is linearly dependent on the remaining rows of $B(\cdot, 1)$ and hence rank $B(\cdot, 1) = \text{rank } B(1, 1)$. Since rank $B = \text{rank } B(\cdot, 1)$, it follows that rank $B = \text{rank } B(1, 1)$. We can similarly show that rank $B = \text{rank} B(i, j), i = 1, 2, \ldots, m; j = 1, 2, \ldots, n$. ∎

Theorem 2.3.2. *Let A be an $m \times n$ matrix with rank $A < \min\{m, n\}$. Then there exists an $r \times s$ submatrix B such that B is of zero type and*

$$r + s - \text{rank } B = m + n - \text{rank } A.$$

Proof. We use induction on the number of lines in A. Thus suppose that the result is true for matrices with fewer than $m + n$ lines. If A is of zero type, then set $B = A$ and the result is proved. So we assume that A is not of zero type. Then by Lemma 2.3.1, either rank $A > \text{rank } A(i, \cdot)$ for some i or rank $A > \text{rank } A(\cdot, j)$ for some j. Suppose rank $A > \text{rank } A(i, \cdot)$ for some i. Then rank $A(i, \cdot) = \text{rank } A - 1$. Since rank $A < \min\{m, n\}$, we have rank $A(i, \cdot) < \min\{m - 1, n\}$. By the induction assumption, $A(i, \cdot)$ admits an $r \times s$ submatrix B of zero type such that

$$r + s - \text{rank } B = m + n - 1 - \text{rank } A(i, \cdot).$$

Since rank $A(i, \cdot) = \text{rank } A - 1$, B satisfies

$$r + s - \text{rank } B = m + n - \text{rank } A,$$

and the result is proved. The proof is similar if rank $A > \text{rank } A(\cdot, j)$ for some j. ∎

An examination of the proofs of Lemma 2.3.1 and Theorem 2.3.2 reveals that these results hold for matrices over an arbitrary field. This observation will be relevant in the remainder of this section.

We now wish to prove that Theorem 2.3.2 implies König's Theorem. Toward this end we introduce the following definition: An $m \times n$ matrix is said to be a *generic matrix* if its nonzero entries are distinct indeterminates. A generic matrix can be regarded as a matrix over the field generated by its nonzero entries and hence its rank is well defined. We can also use the determinantal definition of rank to easily show that the rank of a generic matrix equals its term rank. The next result follows from this observation.

Lemma 2.3.3. Let A be an $m \times n$ generic matix. Then A is of zero type if and only if it is the zero matrix.

Proof. Let A be of zero type and suppose A is not the zero matrix. We assume, without loss of generality, that $a_{11} \neq 0$. Then the term rank of A must exceed that of $A(1, 1)$. However, because the term rank equals the rank for a generic matrix, A cannot be of zero type. This is a contradiction and the proof is complete. ∎

Theorem 2.3.4 (König). Let A be an $m \times n$ matrix with term rank t. Then t is the minimal number of lines required to cover all nonzero entries in A.

Proof. Since A has t nonzero entries, no two of which are on the same line, it is clear that at least t lines are required to cover all nonzero entries in A. We must show that t lines in fact suffice. This is obvious if $t = m$ (or $t = n$) since all m rows (or n columns) cover the nonzero entries in A. We therefore assume that $t < \min\{m, n\}$. Let $\tilde{A}$ be the generic matrix obtained by replacing each nonzero entry of A by a distinct indeterminate. Then rank $\tilde{A} = t$. By Theorem 2.3.2, there exists an $r \times s$ submatrix B of $\tilde{A}$ such that B is of zero type and $r + s - \text{rank } B = m + n - t$. By Lemma 2.3.3, B is the zero matrix and hence rank $B = 0$. Therefore, $r + s = m + n - t$. Note that the lines of $\tilde{A}$ that do not intersect B are $(m - r) + (n - s) = t$ in number and that they cover all nonzero entries in $\tilde{A}$. Since A and $\tilde{A}$ have the same pattern, the proof is complete. ∎

Term ranks were first introduced by König (1936). Theorem 2.3.2 is based on Bapat (1994) and was motivated by a result due to Hartfiel and Loewy (1984) and by the work of Murota (1987, 1993) on *mixed* matrices. For some well-presented accounts of various results related to König's Theorem, see Schneider (1977), Brualdi and Ryser (1991), and Lovász and Plummer (1986).

2.4. The optimal assignment problem

Suppose there are n men and n women who attend a college for a whole year. Each man dates one of the women and each woman dates one of the men on any given day. The happiness for the pair "i-th man–j-th woman" dating over the entire year is h_{ij}. Assuming linearity of happiness, what is the dating policy that maximizes the total happiness for all the students?

Let a_{ij} be the proportion of the year the i-th man dates the j-th woman for $i, j = 1, 2, \ldots, n$. By our convenient linearity assumption, the total happiness will be $\sum_i \sum_j a_{ij} h_{ij}$. Thus we want to solve the following problem:

Problem I. Maximize $\langle H, A \rangle = \sum_i \sum_j a_{ij} h_{ij}$ subject to

$$\sum_j a_{ij} = 1 \quad i = 1, 2, \ldots, n$$

$$\sum_i a_{ij} = 1 \quad j = 1, 2, \ldots, n$$

$$a_{ij} \geq 0 \quad \text{for all } i, j.$$

Let $H = (h_{ij})$ and $A = (a_{ij})$. The problem then is to maximize the inner product $\langle H, A \rangle$ over $A \in \Omega_n$. By the Birkhoff–von Neumann Theorem, any $A \in \Omega_n$ is of the form $A = \sum_j \lambda_j \Pi_j$, where $\lambda_j \geq 0$, $\sum_j \lambda_j = 1$ and the Π_js are permutation matrices. Thus $\langle H, A \rangle = \sum_j \lambda_j \langle H, \Pi_j \rangle$ and the problem reduces to maximizing $\sum_j \lambda_j \langle H, \Pi_j \rangle$ subject to $\lambda_j \geq 0$, $\sum_j \lambda_j = 1$. This latter problem is clearly equivalent to finding the diagonal of H such that the corresponding sum of the entries on the diagonal is maximum. Suppose the maximum is attained at the diagonal corresponding to the permutation σ. Then the optimal dating policy requires that the i-th man dates the $\sigma(i)$-th woman for the whole year. (To summarize, in a linear world, monogamy is the optimal policy!)

A less romantic but a more realistic formulation of the dating problem is the optimal assignment problem.

Suppose there are n persons and n jobs. Let h_{ij} be the net output when the i-th person does the j-th job. We also might think of h_{ij} as the *worth* of the i-th person doing the j-th job. We wish to find an assignment that gives each individual exactly one job and each job to exactly one individual such that the total output (or worth) is maximum. Such an assignment (or the permutation that determines it) is called an *optimal assignment*. If τ is an optimal assignment in the matrix H, then we call $\sum_i h_{i\tau(i)}$ the maximum diagonal sum of H. Let $\Pi = (\pi_{ij})$, where

$$\pi_{ij} = 1 \quad \text{if the } i\text{-th person is assigned the } j\text{-th job}$$
$$= 0 \quad \text{otherwise.}$$

If $H = (h_{ij})$, the problem is to maximize the inner product $\langle H, \Pi \rangle$ over all permutation matrices. If we decide to search all the permutation matrices, the problem becomes too unwieldy even for small n. For example, with 10 persons and 10 jobs we have $10! = 36,28,800$ permutation matrices to try. Thus it makes sense to look for more efficient ways of solving the problem. Retracing the arguments in the previous problem, we have

$$\max_{\Pi}\langle H, \Pi \rangle = \max_{A \in \Omega_n}\langle H, A \rangle.$$

Computing $\max_{A \in \Omega_n}\langle H, A \rangle$ is a linear programming problem, formulated earlier as Problem I. The corresponding dual problem is given by the following:

Problem II. $\min(\sum_{i=1}^n u_i + \sum_{j=1}^n v_j)$ subject to

$$u_i + v_j \geq h_{ij}, \quad i, j = 1, 2, \ldots, n.$$

Because the primal problem clearly has an optimal solution, by the Duality Theorem there exists an optimal solution to the dual problem. Let $u_i^o, v_i^o, i = 1, 2, \ldots, n$ be an optimal solution to the dual problem. Let $\Pi^o = (\pi_{ij}^o)$ be an optimal solution to the primal one, where Π^o is a permutation matrix. By complementary slackness, when $\pi_{k\ell}^o = 1$ for some k, ℓ, then $u_k^o + v_\ell^o = h_{k\ell}$. In case the h_{ij}s are integers, we will see that an optimal solution to the dual problem can be found in integers.

This combination of the primal and the dual problems can be used to construct an algorithm for solving the optimal assignment problem. This is described next. We describe the procedure when the h_{ij}s are integers. We say that a matrix is a $(0, 1)$-matrix if each entry is either 0 or 1.

Let $H = (h_{ij})$ be an $n \times n$ matrix with integer entries. For a permutation τ, $\sum_{i=1}^n h_{i\tau(i)}$ is called the *diagonal sum* corresponding to τ. We want to find τ such that $\sum_i h_{i\tau(i)}$ is maximum. Let us start with $u_i = \max_j h_{ij}, i = 1, 2, \ldots, n$ and $v_j = 0, j = 1, 2, \ldots, n$. Then $u_i + v_j \geq h_{ij}$ for all i, j and therefore $u_i, v_i, i = 1, 2, \ldots, n$ are feasible for Problem II. For any permutation σ,

$$\sum_i (u_i + v_{\sigma(i)}) \geq \sum_i h_{i\sigma(i)}.$$

Thus the maximum diagonal sum of H is at most $\sum_i (u_i + v_i)$. Construct an $n \times n$ $(0, 1)$-matrix as follows: Let $a_{ij} = 1$ if $u_i + v_j = h_{ij}$ and $a_{ij} = 0$ otherwise. If A has a positive diagonal corresponding to the permutation τ, then τ is the optimal assignment in H and the problem is solved. So suppose that A does not have a positive diagonal. Then by the Frobenius-König Theorem, A contains a zero submatrix of order $r \times s$ with $r + s = n + 1$. Let I and J be subsets of cardinality r and $n - s$, respectively, such that the zero submatrix is constituted by rows in I and columns not in J.

Let

$$u_i^* = u_i - 1 \quad \text{if } i \in I,$$
$$= u_i \qquad \text{if } i \notin I,$$
$$v_j^* = v_j + 1 \quad \text{if } j \in J,$$
$$= v_j \qquad \text{if } j \notin J.$$

To verify that $u_i^*, v_i^*, i = 1, 2, \ldots, n$ are again feasible for Problem II, let us just consider the case $i \in I$, $j \notin J$. We know that $u_i + v_j > h_{ij}$. Thus $u_i^* + v_j^* = u_i - 1 + v_j \geq h_{ij}$ (we use the fact that the h_{ij}s are integers). For the other cases, the inequalities are trivially satisfied. Since $r + s = n + 1$, the set I has more elements than the set J and hence $\sum_i (u_i^* + v_i^*) < \sum_i (u_i + v_i)$. Thus the objective function in Problem II (which is to be minimized) has decreased. We repeat this procedure until the $(0, 1)$-matrix constructed in the process has a positive diagonal. Note that we have also proved that the optimal solution to the dual problem can also be found in integers.

If H is an arbitrary matrix, then the procedure is essentially the same. Instead of adding and subtracting 1 to construct u_i^*, v_i^* we now add and subtract $\delta > 0$, which is chosen as large as possible, keeping u_i^*, v_i^* feasible for Problem II.

The algorithm described above is essentially the Hungarian method first developed by Kuhn. However, we have omitted a description of the key step of locating an $r \times s$ zero submatrix with $r + s = n + 1$. This involves subtle combinatorial reasoning, and Lovász and Plummer (1986) present a rigorous treatment. The algorithm can also be formulated in the language of network flows [see Chvátal (1983)].

The optimal assignment problem is equivalent to a problem of hide-and-seek introduced by von Neumann (1953). Before describing the problem we prove the following:

Lemma 2.4.1. *Let A be a nonnegative $n \times n$ matrix with row totals and column totals not exceeding unity. Then there exists a doubly stochastic $n \times n$ matrix D such that $D \geq A$.*

Proof. If A is doubly stochastic, then there is nothing to prove. Otherwise, there exists a row or a column with total less than unity. Without loss of generality, suppose row r has total $u_r = \sum_j a_{rj} < 1$. Since the sum of all the entries in A must be less than n, there exists a column, say s, with column total $c_s = \sum_i a_{is} < 1$. Let $\delta = \min(1 - u_r, 1 - c_s)$. Adding δ to a_{rs} and keeping other entries unchanged gives a new matrix $A^{(1)}$ such that $A \leq A^{(1)}$. Also, $A^{(1)}$ is nonnegative with row totals and column totals at most unity. Furthermore, there is a row total or a column total that is less than 1 in A but equal to 1 in

$A^{(1)}$. If $A^{(1)}$ is doubly stochastic we stop; otherwise, the above procedure can be repeated until we get a doubly stochastic matrix $D \geq A$.   ∎

Now consider the following game. Player II secretly selects an entry a_{ij} of the positive matrix $A = (a_{ij})$. Player I tries to guess either the row index i or the column index j of the entry selected by Player II. He receives the payoff a_{ij} if his guess is correct; otherwise he receives nothing.

We can view this as a two-person zero-sum game (i.e., a matrix game) with $2n$ pure strategies (the n row indices and the n column indices) for Player I and n^2 pure strategies for Player II. Let $X = (x_{ij})$ be an optimal mixed strategy for Player II and let v be the value of the game. Using $X = (x_{ij})$, we find that the expected income to Player I is $\sum_j a_{ij} x_{ij} \leq v$ if he chooses i. Similarly, $\sum_i a_{ij} x_{ij} \leq v$ for all j. Let $y_{ij} = a_{ij} x_{ij}/v$. If $Y = (y_{ij})$, then by Lemma 2.4.1, there exists $Z \in \Omega_n$ such that $Y \leq Z$. If $Z = (z_{ij})$, define u_{ij} by $z_{ij} = a_{ij} u_{ij}/v$. Then

$$\sum_j a_{ij} u_{ij} = v \quad \text{and} \quad \sum_i a_{ij} u_{ij} = v \quad \text{for all } i, j.$$

Since $a_{ij} > 0$ for all i, j and since $v > 0$, we have $x_{ij} \leq u_{ij}$ for all i, j and $\sum_i \sum_j u_{ij} \geq 1$.

We now claim that in fact $\sum_i \sum_j u_{ij} = 1$. Otherwise, the matrix $W = (w_{ij})$, with $w_{ij} = u_{ij}/\sum_i \sum_j u_{ij}$, is a mixed strategy for Player II such that

$$\sum_j a_{ij} w_{ij} = \sum_j a_{ij} u_{ij} \bigg/ \sum_i \sum_j u_{ij} < v \quad \text{for all } i$$

and

$$\sum_i a_{ij} w_{ij} = \sum_i a_{ij} u_{ij} \bigg/ \sum_i \sum_j u_{ij} < v \quad \text{for all } j.$$

This contradicts the optimality of $X = (x_{ij})$. Thus $U = (u_{ij})$ is optimal for Player II. Also, since $X \leq U$, we have $X = U$. By the Birkhoff–von Neumann Theorem,

$$z_{ij} = a_{ij} x_{ij}/v = \sum_r \lambda_r z_{ij}^{(r)},$$

where $\lambda_r > 0$ and $(z_{ij}^{(r)})$ are permutation matrices. Let $z_{ij}^{(r)} = a_{ij} x_{ij}^{(r)}/v$. It is clear that

$$\sum_j a_{ij} x_{ij}^{(r)} = v \quad \text{and} \quad \sum_i a_{ij} x_{ij}^{(r)} = v \quad \text{for all } i, j. \qquad (2.4.1)$$

Since $x_{ij} = \sum_r \lambda_r x_{ij}^{(r)}$, we claim that $\sum_i \sum_j x_{ij}^{(r)} = 1$ for all r. Suppose $\sum_i \sum_j x_{ij}^{(r)} < 1$ for some r. Then $\sum_i \sum_j x_{ij}^{(s)} > 1$ for some s. Let

$$w_{ij} = x_{ij}^{(s)} \Big/ \sum_i \sum_j x_{ij}^{(s)}.$$

Then $W = (w_{ij})$ is a mixed strategy for Player II. From (2.4.1) we find

$$\sum_j a_{ij} w_{ij} < v \quad \text{for all } i \quad \text{and} \quad \sum_i a_{ij} w_{ij} < v \quad \text{for all } j.$$

This contradicts the fact that v is the value of the game. Thus we have optimal strategies of the form

$$x_{ij}^{(r)} = \frac{v}{a_{ij}} z_{ij}^{(r)} \quad i, j = 1, 2, \ldots, n, \tag{2.4.2}$$

where $(z_{ij}^{(r)})$ is a permutation matrix corresponding to, say, a permutation σ. From (2.4.2) we see that

$$v \sum_i \frac{1}{a_{i\sigma(i)}} = \sum_i \sum_j x_{ij}^{(r)} = 1. \tag{2.4.3}$$

Let $P = (p_{ij}^\tau)$ be the permutation matrix corresponding to an arbitrary permutation τ. If we define a matrix $\Psi = (\psi_{ij})$ by $p_{ij}^\tau = a_{ij} \frac{\psi_{ij}}{v}$, then

$$\sum_i \sum_j \psi_{ij} = \sum_i \frac{v}{a_{i\tau(i)}}.$$

We claim that $\theta = \sum_i \sum_j \psi_{ij} \le 1$; otherwise, $\frac{1}{\theta} \Psi$ as a mixed strategy for Player II will give

$$\sum_j a_{ij} \frac{1}{\theta} \psi_{ij} = \frac{v}{\theta} \sum_j p_{ij}^\tau < v.$$

Similarly, we can show that $\frac{v}{\theta} \sum_i p_{ij}^\tau < v$, which contradicts that v is the value. Thus $\theta \le 1$. Observe that θ is $\sum_i \frac{v}{a_{i\tau(i)}}$ and therefore, from (2.4.3),

$$\frac{1}{v} = \max_\tau \sum_i \frac{1}{a_{i\tau(i)}}.$$

Thus $\frac{1}{v}$ corresponds to the optimal value of the assignment problem with $h_{ij} = \frac{1}{a_{ij}}$. Because the solution of the optimal assignment problem is invariant under the transformation $h_{ij} \to h_{ij} + u_i + v_j$, we can always assume $a_{ij} > 0$ for all i, j in our problem.

2.5. A probabilistic algorithm

Let A be an $n \times n$ (0, 1)-matrix. Suppose we want to find out whether A contains a positive diagonal (i.e., whether per $A > 0$). As observed in connection with the optimal assignment problem, it is not easy to check each permutation and see if the corresponding diagonal is positive. In such instances randomization can be a very useful device. We describe this technique briefly with reference to the problem at hand. Just as we did in the proof of Theorem 2.3.4, let $\tilde{A}$ be the matrix obtained by replacing each 1 in A by the variable x_{ij}. Then $|\tilde{A}|$ is a polynomial in the variables x_{ij}. Clearly, per $A = 0$ if and only if $|\tilde{A}|$ is identically zero. Thus our problem is essentially that of finding out whether a given polynomial is identically zero. Suppose we evaluate the polynomial at some values of the variables chosen independently of each other and at random. If the polynomial is not identically zero, then intuitively there is a high probability that this evaluation will result in a nonzero number. In that case we can conclude, with a high degree of confidence, that the polynomial is not identically zero. Returning to our original problem, evaluating the polynomial involves only the calculation of $|\tilde{A}|$ after substituting values chosen independently and at random for the x_{ij}. Since calculating the determinant is computationally an easy task, we have found an efficient algorithm for deciding whether per $A > 0$. For further details see Exercises 10 and 11.

In the remainder of this section we describe a probabilistic algorithm for locating a positive diagonal in a (0, 1)-matrix, when we know a priori that such a diagonal exists. The algorithm is a special instance of an algorithm due to Mulmuley, Vazirani, and Vazirani (1987) for finding a perfect matching in a graph.

We need some preliminaries. Let $S = \{x_1, x_2, \ldots, x_n\}$ be a finite set and let $T \subset S$. Suppose there is a weight w_i, a real number, associated with $x_i, i = 1, 2, \ldots, n$. Then the weight of T is defined as $\sum_{x_i \in T} w_i$.

Lemma 2.5.1. *Let $S = \{x_1, x_2, \ldots, x_n\}$ be a finite set and let $F = \{S_1, S_2, \ldots, S_k\}$ be a family of nonempty subsets of S. Let elements of S be assigned integer weights chosen at random, uniformly and independently from $\{1, 2, \ldots, 2n\}$. Then the probability that there is a unique set in F with minimum weight is at least $1/2$.*

Proof. Fix the weights of all elements in S except x_i. Define the *threshold* of x_i to be the real number α_i such that if $w_i \leq \alpha_i$, then x_i is contained in some minimum weight subset in F and if $w_i > \alpha_i$, then x_i is contained in no minimum weight subset. (It is possible that $\alpha_i = \infty$.)

If $w_i < \alpha_i$, then x_i must be in every minimum weight subset. We say that x_i is ambiguous if $w_i = \alpha_i$, because in this case there is a minimum weight subset that contains x_i and one that does not.

Now observe that the threshold α_i was defined without reference to the weight w_i of x_i. Thus the random variable α_i is distributed independently of w_i. Since w_i is distributed uniformly over $\{1, 2, \ldots, 2n\}$,

$$Prob\{x_i \text{ is ambiguous, i.e., } w_i = \alpha_i\} \le \frac{1}{2n}.$$

Since S contains n elements,

$$Prob\{\text{There exists an ambiguous element}\} \le \frac{1}{2n} \times n = \frac{1}{2}.$$

Thus, with probability at least $1/2$, no element is ambiguous. In this case each element is either in every minimum weight subset or in none and therefore the minimum weight subset is unique. That completes the proof. ∎

Lemma 2.5.2. *Let W be an $n \times n$ matrix such that each entry of W is a non-negative integer. Suppose $\min_\sigma \sum_i w_{i\sigma(i)}$ is attained at the unique permutation τ and let the minimum value be θ. Let B be the $n \times n$ matrix defined as $b_{ij} = 0$ if $w_{ij} = 0$ and $b_{ij} = 2^{w_{ij}}$ otherwise. Then (i, j) lies on the diagonal τ (i.e., $j = \tau(i)$) if and only if*

$$\frac{2^{w_{ij}} |B(i, j)|}{2^\theta} \qquad (2.5.1)$$

is odd.

Proof. Fix (i, j) such that w_{ij} is nonzero. Let $\mathcal{P} = \{\sigma : \sigma(i) = j\}$. The numerator in (2.5.1) can be expressed, up to a sign, as the sum of those diagonal products of B corresponding to $\sigma \in \mathcal{P}$. Let us write

$$(-1)^{i+j} |B(i, j)| 2^{w_{ij}} = \sum_{\sigma:\sigma\in\mathcal{P}} \epsilon(\sigma) \prod_{i=1}^{n} b_{i\sigma(i)}. \qquad (2.5.2)$$

Each term in the summation in (2.5.2) is, up to a sign, either zero or 2^α for some $\alpha \ge \theta$. Further, in view of the uniqueness of τ, there exists precisely one term 2^θ in the sum if and only if $\tau \in \mathcal{P}$. Thus the expression in (2.5.2) is of the form $2^\theta (1 + u)$ for an even integer u if $\tau \in \mathcal{P}$ and $2^\theta v$ for some even integer v if $\tau \notin \mathcal{P}$. It follows that the ratio (2.5.1) is odd if and only if $\tau \in \mathcal{P}$, and the proof is complete. ∎

Let A be an $n \times n$ $(0, 1)$-matrix such that $\text{per } A > 0$. An algorithm for locating a positive diagonal consists of the following steps:

(1) Let the number of nonzero entries in A be k. For each $a_{ij} = 1$, choose an integer w_{ij} at random from $\{1, 2, \ldots, 2k\}$. Define the $n \times n$ matrix B by $b_{ij} = 0$ if $a_{ij} = 0$ and $b_{ij} = 2^{w_{ij}}$ otherwise.

(2) Compute $|B|$ and let θ be the highest power of 2 such that 2^θ divides $|B|$. [Observe that θ is precisely the minimum diagonal sum in $W = (w_{ij})$.]

(3) Compute the adjoint of B. The (j, i)-th entry of the adjoint will be $(-1)^{i+j} |B(i, j)|$.

(4) For each (i, j) compute $\frac{2^{w_{ij}} |B(i,j)|}{2^\theta}$. Select (i, j) if this number is odd.

By Lemma 2.5.1, with probability at least $1/2$, there is a unique τ such that the corresponding diagonal sum of W is minimum. In that case the hypotheses of Lemma 2.5.2 are satisfied and, therefore, the (i, j)s selected in Step 4 correspond to a positive diagonal of A.

If we repeat the algorithm m times, then the probability that it will fail every time is at most $(1/2)^m$; therefore, we will successfully locate a positive diagonal with probability at least $1 - (1/2)^m$. For example, if we repeat the algorithm 5 times, then we will locate a positive diagonal with a probability of at least 96%.

The main computational work in the algorithm involves calculating the determinant and the cofactors of the integer matrix B. Thus any reasonable algorithm that calculates the exact inverse of an integer matrix is sufficient for this purpose. Step 4 of the algorithm can be implemented in parallel on several processors, thus increasing the efficiency.

2.6. Diagonal products

The following inequality is well known; see, for example, Hardy, Littlewood, and Polya (1952). We make the convention that $0^0 = 1$ and that $0 \log 0 = 0$.

Lemma 2.6.1 (The arithmetic mean-geometric mean inequality). *Let $x, \alpha \in R^n$ be nonnegative vectors and suppose $\sum_{i=1}^n \alpha_i = 1$. Then*

$$\sum_{i=1}^n \alpha_i x_i \geq \prod_{i=1}^n x_i^{\alpha_i}.$$

Further, equality holds if and only if all the x_is corresponding to positive α_is are equal.

We now prove an inequality that turns out to be a powerful tool for proving inequalities for nonnegative matrices. Many applications of this inequality will be illustrated in Chapter 3.

Lemma 2.6.2 (The Information Inequality). *Let $x, y \in R^n$ be nonnegative* *vectors satisfying* $\sum_i x_i = \sum_i y_i > 0$. *Then*

$$\prod_i x_i^{x_i} \geq \prod_i y_i^{x_i}. \tag{2.6.1}$$

Equality holds in (2.6.1) if and only if $x = y$.

Proof. Since $\sum_i x_i = \sum_i y_i > 0$, we can assume without loss of generality that $\sum_i x_i = 1$. Let $K = \{i : x_i > 0\}$. If $x_i = 0$, then $y_i^{x_i} = x_i^{x_i} = 1$ and therefore

$$\prod_{i \notin K} x_i^{x_i} \geq \prod_{i \notin K} y_i^{x_i}. \tag{2.6.2}$$

By the arithmetic mean-geometric mean inequality (Lemma 2.6.1), we have

$$\prod_{i \in K} \left(\frac{y_i}{x_i}\right)^{x_i} \leq \sum_{i \in K} x_i \cdot \left(\frac{y_i}{x_i}\right) = 1. \tag{2.6.3}$$

Now (2.6.1) follows from (2.6.2) and (2.6.3). Suppose equality occurs in (2.6.1). Then equality must occur in (2.6.3) as well. Again, by Lemma 2.6.1, there exists a real number θ such that $y_i = \theta x_i$, $i \in K$. Since we must have $\sum_{i \in K} x_i \theta = 1$, then $\theta = 1$. It follows that $y_i = 0$ whenever $i \notin K$ and hence $x = y$. That completes the proof. ∎

The entries of any $n \times n$ nonnegative matrix are greatly constrained by the diagonal products. In the next result we show that if we have an upper bound on the diagonal products of a nonnegative matrix then the entries are also bounded in a certain sense.

Theorem 2.6.3. *Let A be a nonnegative $n \times n$ matrix with per $A > 0$ and let $c > 0$. Then the following conditions are equivalent.*

(i) $\prod_i a_{i\sigma(i)} \leq c$ *for all permutations σ of $\{1, 2, \ldots, n\}$.*
(ii) *There exist x_i, $y_i > 0$ with $\prod_i x_i y_i \leq c$ such that $a_{ij} \leq x_i y_j$ for all i, j.*

Proof. Suppose (ii) holds. Then for any permutation σ, $\prod_i a_{i\sigma(i)} \leq \prod_i x_i y_{\sigma(i)} = \prod_i x_i y_i \leq c$. Hence (i) is true. Now suppose (i) holds. Define $B = (b_{ij})$ as follows:

$$b_{ij} = \begin{cases} \log a_{ij} & \text{if } a_{ij} > 0 \\ -N & \text{if } a_{ij} = 0 \end{cases},$$

where N is a large positive number satisfying the following condition: If σ, τ are permutations of $1, 2, \ldots, n$ and $\prod_i a_{i\sigma(i)} > 0$, $\prod_i a_{i\tau(i)} = 0$, then $\sum_i b_{i\sigma(i)} >$

$\sum_i b_{i\tau(i)}$. Since there are only finitely many permutations, such a choice of N is possible. By the Birkhoff–von Neumann Theorem

$$\max_\sigma \sum_i b_{i\sigma(i)} = \max_{P \in \Omega_n} \sum_i \sum_j b_{ij} p_{ij}.$$

By the Duality Theorem there must be an optimal solution to the dual problem

$$\min\left(\sum_i u_i + \sum_j v_j\right)$$

subject to $u_i + v_j \geq b_{ij}$ for all i, j.

Let $\{\bar{u}_i, \bar{v}_i\}, i = 1, 2, \ldots, n$ be an optimal solution with $\sum_i \bar{u}_i + \sum_i \bar{v}_i = \theta$. Since A has at least one positive diagonal, $\max_\sigma \prod_i a_{i\sigma(i)} = \exp \theta \leq c$. Let $x_i = \exp \bar{u}_i$, $y_i = \exp \bar{v}_i$, $i = 1, 2, \ldots, n$. We have, for $a_{ij} > 0$,

$$b_{ij} = \log a_{ij} \leq \bar{u}_i + \bar{v}_j.$$

Thus $a_{ij} \leq x_i y_j$ if $a_{ij} > 0$ for all i, j. Trivially,

$$a_{ij} \leq x_i y_j \quad \text{if } a_{ij} = 0.$$

Further,

$$\prod_i x_i y_i = \exp\left(\sum \bar{u}_i + \sum \bar{v}_i\right) = \exp \theta \leq c.$$

That completes the proof.   ∎

Theorem 2.6.4. *Let A be a nonnegative matrix with doubly stochastic pattern. Let every diagonal product take either of the two values 0 or k for some constant k. Then there exists a matrix C of rank 1 such that the positive entries of A coincide with the corresponding entries of C.*

Proof. Suppose $\prod_i a_{i\sigma(i)} = 0$ or k for each permutation σ. Since A has doubly stochastic pattern, by Lemma 2.1.5, there exists some permutation τ such that $\prod_i a_{i\tau(i)} = k > 0$. By Theorem 2.6.3, there exist x_i, $y_i > 0$ with $\prod_i x_i y_i \leq k$ such that $a_{ij} \leq x_i y_j$ for all i, j. If $a_{ij} > 0$, we know from Theorem 2.2.3 that $\sigma(i) = j$ and $\prod_i a_{i\sigma(i)} = k$ for some permutation σ. We also know that $k = \prod_i a_{i\sigma(i)} \leq \prod_i x_i y_i \leq k$. That is, $a_{ij} = x_i y_j$ if $a_{ij} > 0$. Let $C = (c_{ij}) = (x_i y_j)$. Then C is a matrix of rank 1. Further, C satisfies the condition $a_{ij} = c_{ij}$ if $a_{ij} > 0$.   ∎

Theorem 2.6.5. *Let A and B be distinct $n \times n$ doubly stochastic matrices. Then for some permutation σ, $\prod_i a_{i\sigma(i)} > \prod_i b_{i\sigma(i)}$.*

Proof. We may assume that $b_{ij} = 0 \Rightarrow a_{ij} = 0$, for otherwise, if $a_{ij} > 0$ and $b_{ij} = 0$ for some i, j, then a positive diagonal through a_{ij} (which must exist by Theorem 2.2.3) will do. We have the convention $\log \frac{0}{0} = 1$. For any doubly stochastic matrix $X = (x_{ij})$, let $\phi(X) = \sum_i \sum_j x_{ij} \log \frac{a_{ij}}{b_{ij}}$. The linear function $\phi(X)$ with domain Ω_n attains its maximum at an extreme point of Ω_n. However, by the Information Inequality (Theorem 2.6.2), $\phi(A) = \sum_i \sum_j a_{ij} \log \frac{a_{ij}}{b_{ij}} > 0$. Thus there exists a permutation σ such that the corresponding permutation matrix $P^\sigma = (p_{ij}^\sigma)$ satisfies

$$\phi(P^\sigma) = \sum_i \sum_j p_{ij}^\sigma \log \frac{a_{ij}}{b_{ij}} = \log \left[\prod_i \frac{a_{i\sigma(i)}}{b_{i\sigma(i)}} \right] > 0.$$

It follows that $\prod_i a_{i\sigma(i)} > \prod_i b_{i\sigma(i)}$, and the proof is complete. ∎

Corollary 2.6.6. *Let A and B be doubly stochastic matrices of order n such that, for some constant k,*

$$\prod_i a_{i\sigma(i)} = k \prod_i b_{i\sigma(i)} \quad \text{for all permutations } \sigma.$$

Then A = B.

Proof. Suppose A and B are distinct. Then by Theorem 2.6.5, $k < 1$. Since $\prod_i b_{i\sigma(i)} = \frac{1}{k} \prod_i a_{i\sigma(i)}$ for all σ, another application of Theorem 2.6.5 gives $k > 1$, which is a contradiction. ∎

The results in this section are based on Bapat (1981) and Bapat and Raghavan (1980).

2.7. A self map of doubly stochastic matrices

Let A be a nonnegative $n \times n$ matrix with per $A > 0$. Define the matrix $f(A) = (f_{ij}(A))$ as

$$f_{ij}(A) = \frac{a_{ij}\text{per } A(i, j)}{\text{per } A}, \quad i, j = 1, 2, \ldots, n.$$

As observed in the proof of Theorem 2.2.3, $f(A) \in \Omega_n$. The map f and its restriction to Ω_n have interesting properties. Some of these properties will be discussed in this section.

Lemma 2.7.1. *If $A \in \Omega_n$ then A and $f(A)$ have the same pattern.*

Proof. If $A \in \Omega_n$, then per $A > 0$ and hence $f(A)$ is well defined. If $a_{ij} = 0$, then obviously $f_{ij}(A) = 0$. If $a_{ij} > 0$, then by Theorem 2.2.3, there exists σ

such that $\sigma(i) = j$ and $\prod_i a_{i\sigma(i)} > 0$. Thus per $A(i, j) > 0$ and per $A > 0$. This shows that $f_{ij}(A) > 0$. Hence the result is confirmed. ∎

We now introduce some definitions. Let X, Y be subsets of R^n and suppose $g, h : X \to Y$ are continuous maps. We say that g is *homotopic* to h if there exists a continuous map $\Phi : [0, 1] \times X \to Y$ such that for all $x \in X$, $\Phi(0, x) = g(x)$ and $\Phi(1, x) = h(x)$. The map Φ is called a *homotopy bridge*. Geometrically speaking, if g is homotopic to h, then h is a continuous deformation of g. Let $D^n = \{x \in R^n : \sum_i x_i^2 \leq 1\}$ and let ∂D^n denote the topological boundary of D^n. That is, $\partial D^n = \{x \in R^n : \sum_i x_i^2 = 1\}$. Before proceeding we shall need the following version of a topological theorem of Kronecker [see Ortega and Rheinboldt (1970), p. 161, for a proof].

Theorem 2.7.2 (Kronecker Index Theorem). *Let $g : D^n \to D^n$ be a continuous map that leaves ∂D^n invariant, i.e., g maps ∂D^n into ∂D^n. If the restriction of g to ∂D^n is homotopic to the identity map, then g maps D^n onto itself.*

We now have the following:

Lemma 2.7.3. *Let $K = \{x \in R^n : x \geq 0, Cx = b\}$, where C is an $m \times n$ matrix and $b \in R^n$. Suppose K is nonempty and bounded. Let $g : K \to K$ be a continuous map such that for any $x \in K$, $x_i = 0$ if and only if $g_i(x) = 0$. Then g maps K onto K.*

Proof. We assume that for any $i \in \{1, 2, \ldots, n\}$ there exists $x, y \in K$ such that $x_i \neq y_i$, for otherwise, we may ignore the i-th coordinate and work with R^{n-1}. Let ∂K denote the relative boundary of K, i.e., $x \in \partial K$ if and only if for any open ball $B \in R^n$ containing x, the set $B \cap \{x : Cx = b\}$ has points in K as well as points not in K. We show that ∂K is precisely the set of points in K with at least one zero coordinate.

Suppose $x \in \partial K$. If K has just one point, the result is trivial. Otherwise, let x^0 be a point in the relative interior of K. Consider points of the type $x^0 + \lambda(x - x^0)$. For $\lambda \in [0, 1]$, such points are in K. If for some $\lambda' > 1, x' = x^0 + \lambda'(x - x^0)$ is in K, then x belongs to the line segment joining x' and x^0 and therefore x is a relative interior point, which is a contradiction. Thus if $\lambda > 1$, then $x^0 + \lambda(x - x^0) \notin K$. Since $C(x + \lambda(x - x^0)) = b$, $x^0 + \lambda(x - x^0)$ has a negative coordinate for all $\lambda > 1$. Thus x has a zero coordinate.

Conversely, suppose $x \in K$ and $x_i = 0$ for some i. If the columns of C are independent, then $K = \{x\}$. Otherwise, $Cy = 0$ for some nonzero $y \in R^n$. Furthermore, we may assume that $y_i > 0$, for if $y_i = 0$ whenever $Ay = 0$, then for any two points $x, z \in K$, $x_i = z_i$ and this would contradict the assumption made at the beginning of the proof. Let B be an open ball in R^n containing x. For any small $\epsilon > 0$, the vector $x - \epsilon y$ is in $B \cap \{x : Cx = b\}$ and has its i-th coordinate negative. Thus $x \in \partial K$.

Now for $t \in [0, 1]$, define $\Phi : [0, 1] \times K \to K$ as $\Phi(t, x) = (1 - t)g(x) + tx$. Then for each t, $\Phi(t, x)$ maps ∂K into itself. Also, $\Phi(0, x) = g(x)$ and $\Phi(1, x) = x$ for all $x \in \partial K$. Thus the map $g : \partial K \to \partial K$ is homotopic to the identity map. It follows by Theorem 2.7.2 that g maps K onto K. ∎

Corollary 2.7.4. *Let $K_n \subset \Omega_n$ be the set of all $n \times n$ doubly stochastic matrices whose entries satisfy a certain given set of equality constraints. Suppose $f(A) \in K_n$ whenever $A \in K_n$, where f, as usual, is the map defined at the beginning of this section. Then f maps K_n onto K_n. In particular, each of the following sets is mapped onto itself by f:*

(i) Ω_n.

(ii) The set of symmetric matrices in Ω_n.

(iii) The set of circulants in Ω_n [i.e., matrices $A = (a_{ij})$] satisfying

$$a_{ij} = \begin{cases} b_{j-i+1}, & \text{if } i \leq j \\ b_{j-i+1+n} & \text{if } i > j \end{cases},$$

where $b_1, b_2, \ldots, b_n$ are constants.

(iv) The set of matrices in Ω_n with the first row equal to the second row.

Proof. As observed in Lemma 2.7.1, if $a_{ij} = 0$, then $f_{ij}(A) = 0$. Therefore, the result follows from Theorem 2.7.3. In examples (*i*)–(*iv*), the sets are all defined by imposing certain equality constraints on the entries of the matrix. Further, the fact that f maps the given set into itself is easily verified. That completes the proof. ∎

Theorem 2.7.5. *The map $f : \Omega_n \to \Omega_n$ is one-to-one.*

Proof. Let $f(A) = f(B)$ for two matrices A, B in Ω_n. Since per A, per $B > 0$, the matrix $C = (c_{ij})$, where $c_{ij} = (\frac{\text{per } A}{\text{per } B})^{1/n} \cdot b_{ij}$ is well defined. By assumption,

$$a_{ij} \text{ per } A(i, j)/\text{per } A = b_{ij} \text{ per } B(i, j)/\text{per } B \quad \text{for } i, j = 1, 2, \ldots, n.$$

Therefore, $a_{ij} \text{ per } A(i, j) = c_{ij} \text{ per } C(i, j)$ for $i, j = 1, 2, \ldots, n$.

Let

$$\alpha_\sigma = \prod_i a_{i\sigma(i)}, \qquad \gamma_\sigma = \prod_i c_{i\sigma(i)}, \qquad \sigma \in S_n.$$

We have

$$\prod_{\sigma} \alpha_{\sigma}^{\alpha_{\sigma}} = \prod_{\sigma}\left(\prod_{i} a_{i\sigma(i)}\right)^{\alpha_{\sigma}}$$

$$= \prod_{\sigma}\prod_{i} a_{i\sigma(i)}^{\alpha_{\sigma}}$$

$$= \prod_{i}\prod_{\sigma} a_{i\sigma(i)}^{\alpha_{\sigma}}. \tag{2.7.1}$$

Let $S_{ij} = \{\sigma : \sigma(i) = j\}$. Then

$$\prod_{i}\prod_{\sigma} a_{i\sigma(i)}^{\alpha_{\sigma}} = \prod_{i}\prod_{j}\prod_{\sigma \in S_{ij}} a_{i\sigma(i)}^{\alpha_{\sigma}}$$

$$= \prod_{i}\prod_{j}(a_{ij})^{\sum_{\sigma \in S_{ij}}\alpha_{\sigma}}$$

$$= \prod_{i}\prod_{j} a_{ij}^{a_{ij}\,\mathrm{per}\,A(i,j)}$$

$$= \prod_{i}\prod_{j} a_{ij}^{c_{ij}\,\mathrm{per}\,C(i,j)}$$

$$= \prod_{\sigma}\left(\prod_{i} a_{i\sigma(i)}\right)^{\gamma_{\sigma}}$$

$$= \prod_{\sigma} \alpha_{\sigma}^{\gamma_{\sigma}}.$$

Thus from (2.7.1) we get

$$\prod_{\sigma} \alpha_{\sigma}^{\alpha_{\sigma}} = \prod_{\sigma} \alpha_{\sigma}^{\gamma_{\sigma}}. \tag{2.7.2}$$

Similarly, we can prove

$$\prod_{\sigma} \gamma_{\sigma}^{\alpha_{\sigma}} = \prod_{\sigma} \gamma_{\sigma}^{\gamma_{\sigma}}. \tag{2.7.3}$$

Combining (2.7.2) with (2.7.3) we get

$$\prod_{\sigma}\left(\frac{\alpha_{\sigma}}{\gamma_{\sigma}}\right)^{\alpha_{\sigma}}\left(\frac{\gamma_{\sigma}}{\alpha_{\sigma}}\right)^{\gamma_{\sigma}} = 1.$$

Note that $\sum_{\sigma}\alpha_{\sigma} = \sum_{\sigma}\gamma_{\sigma} = \mathrm{per}\,A$. We have, by Lemma 2.6.2, $\alpha_{\sigma} = \gamma_{\sigma}$ for all σ, i.e., $\prod_{i} a_{i\sigma(i)} = \prod_{i} c_{i\sigma(i)} = \frac{\mathrm{per}\,A}{\mathrm{per}\,B}\prod_{i} b_{i\sigma(i)}$ for all σ. However, by Corollary 2.6.6, this is possible only when $A = B$. That completes the proof. ∎

We say that matrices A and B have proportional diagonal products if for some nonzero constant k,

$$\prod_i a_{i\sigma(i)} = k \prod_i b_{i\sigma(i)} \quad \text{for all permutations } \sigma.$$

Theorem 2.7.6. *If A is an $n \times n$ nonnegative matrix with per $A > 0$, then there exists a unique doubly stochastic matrix B such that A and B have proportional diagonal products.*

Proof. Since per $A > 0$, $f(A)$ is well defined and, by Corollary 2.7.4, we have $f(A) = f(B)$ for a doubly stochastic matrix B. A proof similar to that of Theorem 2.7.5 can be employed to show that A and B have proportional diagonal products. It follows by Corollary 2.6.6 that B is unique. ∎

To illustrate Theorem 2.7.6, consider the matrices

$$A = \begin{bmatrix} 1 & 2 & 4 \\ 0 & 2 & 0 \\ 0 & 0 & 4 \end{bmatrix} \quad \text{and} \quad B = \begin{bmatrix} 1 & 0 & 0 \\ 0 & 1 & 0 \\ 0 & 0 & 1 \end{bmatrix}.$$

Note that per $A > 0$, although A does not have doubly stochastic pattern. Matrix B is doubly stochastic and A, and B have proportional diagonal products. If in Theorem 2.7.6 we require A to have doubly stochastic pattern, then we can prove a stronger statement than given in that result. This is our next result and is an example of a scaling theorem for nonnegative matrices. We will consider more general scaling problems and their applications in Chapter 6.

Theorem 2.7.7. *Let A be a nonnegative $n \times n$ matrix with doubly stochastic pattern. Then there exist diagonal matrices D_1 and D_2 with positive diagonal entries such that $D_1 A D_2$ is doubly stochastic.*

Proof. Since A has doubly stochastic pattern, per $A > 0$. Therefore, by Theorem 2.7.6, there exists a doubly stochastic matrix B such that A and B have proportional diagonal products. Thus

$$\prod_i a_{i\sigma(i)} = k \prod_i b_{i\sigma(i)}$$

for all σ, where k is a constant. In particular, it follows by Theorem 2.2.3 that $a_{ij} = 0$ if and only if $b_{ij} = 0$. Define $Z = (z_{ij})$ by $z_{ij} = b_{ij}/a_{ij}$ if $a_{ij} > 0$ and 0 otherwise. Then, by Theorem 2.6.4, there exist constants $x_1, x_2, \ldots, x_n; y_1, y_2, \ldots, y_n$ such that $z_{ij} = x_i y_j$ if $z_{ij} > 0$. Let D_1 and D_2 be diagonal matrices with $x_1, x_2, \ldots, x_n$ and $y_1, y_2, \ldots, y_n$ along the diagonals, respectively. Then $B = D_1 A D_2$, and the proof is complete. ∎

The map f defined in this section has been studied independently by several authors–Baum and Eagon (1967), Brégman (1973), and Rothaus (1974) to mention a few. Brégman (1973) used properties of the map to settle a conjecture of Minc (1963) that asserted that if A is a 0–1 $n \times n$ matrix with row sums $r_1, \ldots, r_n$, then per $A \le \prod_{i=1}^{n} r_i!^{\frac{1}{r_i}}$. Schrijver (1978a) then gave a more elementary proof of the same conjecture. Theorem 2.7.7 is due to Sinkhorn and Knopp (1967) and was first proved for a positive matrix by Sinkhorn (1964). For an extension of Sinkhorn's Theorem on strictly positive matrices to positive operators in infinite dimensions, see Hobby and Pike (1965).

2.8. van der Waerden conjecture and its solution

It was conjectured by van der Waerden (1926) that the permanent of any $n \times n$ doubly stochastic matrix is at least $\frac{n!}{n^n}$ and that, the minimum is attained uniquely at the matrix $J_n = (\frac{1}{n})$. The conjecture was confirmed by Egorychev (1981) and independently by Falikman (1981). In this section we present a proof of the conjecture. The proof consists of two crucial steps. The first step depends on the Duality Theorem; the second uses the Alexandroff Inequality.

A matrix $A \in \Omega_n$ is called a *permanent minimizer* if per $B \ge$ per A for any $B \in \Omega_n$.

Theorem 2.8.1. *Every permanent minimizer is fully indecomposable.*

Proof. Suppose $A \in \Omega_n$ is a permanent minimizer and suppose A is partly decomposable. Then

$$G = PAQ = \begin{bmatrix} B & 0 \\ C & D \end{bmatrix},$$

where P and Q are permutation matrices and B and D are square matrices. The row sums of B are all equal to 1 and the column sums are all at most 1. It follows that the column sums of B must all be equal to 1 and hence $C = 0$. Thus per $A = $ per $PAQ = $ per B per D. Suppose B is $r \times r$. Since per $A > 0$, the last row in B has an element $b_{rk} > 0$. We can interchange the columns r and k of G if necessary and assume $b_{rr} > 0$. This clearly leaves the permanent of A unchanged. Similarly, we can assume (by permuting the rows of A) that $d_{11} > 0$. Now, for small $\theta > 0$ if we subtract θ from b_{rr} and d_{11} and add θ to G at $(r, r + 1)$ and $(r + 1, r)$ positions, we get a new matrix $G(\theta)$, which is doubly stochastic. If $G(0) = G$, $G_{r+1,r}(0) = G_{r+1,r}$ etc., then

$$\text{per } G(\theta) = \text{per } G - \theta(\text{per } D \text{ per } B_{rr} + \text{per } D_{11} \text{ per } B$$

$$-\text{per } G_{r,r+1} - \text{per } G_{r+1,r}) + \theta^2(\cdot). \tag{2.8.1}$$

Observe that per $G_{r,r+1} = 0$, for, in computing per $G_{r,r+1}$, we basically replace the r-th row of G by the unit row vector $e_{r+1} = (0, 0, \ldots 0, 1, 0, \ldots, 0)$, where 1 occurs in position $r + 1$. Thus, in the new matrix, the last row in B becomes zero and thus per $G_{r,r+1} = 0$. Similarly, per $G_{r+1,r} = 0$. Thus for $\theta > 0$ sufficiently small, we have from (2.8.1) that per $G(\theta) <$ per G. This contradicts the minimality of the permanent function at A and completes the proof. ∎

Theorem 2.8.2. *Let $A = (a_{ij})$ be a permanent minimizer in Ω_n. Then per $A(i, j) \geq$ per A for all i, j. Further, when $a_{ij} > 0$, per $A(i, j) =$ per A.*

Proof. Let us consider the linear programming problem

$$\min \sum_i \sum_j x_{ij} \, \text{per} \, A(i, j)$$

subject to $X = (x_{ij}) \in \Omega_n$. By the Duality Theorem we know that the dual problem

$$\max \left(\sum_i u_i + \sum_j v_j \right)$$

subject to

$$u_i + v_j \leq \text{per} \, A(i, j) \quad \text{for all } i, j \tag{2.8.2}$$

has an optimal solution. Further, at any optimal solution $X = (x_{ij})$ of the minimum problem, if $x_{ij} > 0$, then by complementarity, $u_i + v_j = $ per $A(i, j)$ for any optimal solution $u_1, \ldots, u_n, v_1, \ldots, v_n$ of the dual.

We prove that the minimum for the primal occurs at $X = A = (a_{ij})$. To see this, consider any $Y \in \Omega_n$. We have $f_Y(\theta) = \text{per}\,((1 - \theta)A + \theta Y) \geq$ per A for $0 \leq \theta \leq 1$.

Let $C_1, C_2, \ldots, C_n$ be the columns of A and let $Y_1, Y_2, \ldots, Y_n$ be the columns of Y. We have

$$f_Y(\theta) = \text{per}(C_1 + \theta(Y_1 - C_1), C_2 + \theta(Y_2 - C_2), \ldots, C_n + \theta(Y_n - C_n))$$

$$= \text{per}(C_1, C_2, \ldots, C_n) + \theta \sum_i \text{per}(C_1, \ldots, C_{i-1}, Y_i - C_i, \ldots, C_n)$$

$$+ \text{ terms of higher order in } \theta.$$

Thus

$$\frac{d}{d\theta} f_Y(\theta) = \sum_i \text{per}(C_1, C_2, \ldots, C_{i-1}, Y_i - C_i, \ldots, C_n) + \text{a linear term in } \theta$$

$$= -\sum_i \text{per}(C_1, \ldots, C_i, \ldots, C_n) + \sum_i \text{per}(C_1, \ldots, C_{i-1}, Y_i, \ldots, C_n)$$

$$+ \text{a linear term in } \theta.$$

Thus

$$f_Y'(0) = \left(\frac{d}{d\theta} f_Y(\theta)|\theta = 0\right) = -n \operatorname{per} A + \sum_i \sum_j y_{ij} \operatorname{per} A(i, j)$$

and $f_Y'(0) \geq 0$ for all Y. Also $f_Y'(0) = 0$ at $Y = A$. Thus

$$\min_{Y \in \Omega_n} \sum_i \sum_j y_{ij} \operatorname{per} A(i, j) = n \operatorname{per} A \qquad (2.8.3)$$

is attained at $Y = A$.

By Theorem 2.8.1 the matrix A is fully indecomposable and we can assume that the main diagonal of A is positive. Moreover, by Theorem 2.2.2, AA^T and $A^T A$ are fully indecomposable and hence irreducible. Moreover, AA^T and $A^T A$ are clearly doubly stochastic. From (2.8.2), (2.8.3), and complementarity it follows that

$$u_i + \sum_j a_{ij} v_j = \sum_j a_{ij} \operatorname{per} A(i, j) = \operatorname{per} A, \quad i = 1, 2, \ldots, n.$$

That is,

$$u + Av = \operatorname{per} A \cdot \mathbf{1},$$

where $\mathbf{1}$ denotes the column with each entry equal to 1. Similarly,

$$A^T u + v = \operatorname{per} A \cdot \mathbf{1}.$$

Thus $A^T A v = v$ and $A A^T u = u$. Clearly, $u = \alpha \cdot \mathbf{1}$ and $v = \beta \cdot \mathbf{1}$ for some constants α and β. Now $\sum_i u_i + \sum_j v_j = n(\alpha + \beta) = n \operatorname{per} A$. Thus $\alpha + \beta = \operatorname{per} A$ and we have $\operatorname{per} A(i, j) \geq \operatorname{per} A$ for all i, j with equality when $a_{ij} > 0$. That completes the proof of the theorem.   ■

We now prove an inequality of Alexandroff. The original inequality applies to a more general function called the mixed discriminant, which we will consider in Chapter 5. The present version is sufficient for our purpose at the moment. We first prove the following preliminary result:

Lemma 2.8.3. *Let $C_1, \ldots, C_{n-1}$ be positive vectors in R^n, $n \geq 2$. Let $C_n \in R^n$ and suppose that C_{n-1} and C_n are linearly independent. Then the equation*

$$per(C_1, \ldots, C_{n-2}, C_n + yC_{n-1}, C_n + yC_{n-1}) = 0$$

has two distinct real roots, say $\alpha_1 < \alpha_2$. Furthermore, for any positive vector $z \in R^n$, the root of the linear equation

$$per(C_1, \ldots, C_{n-2}, C_n + yC_{n-1}, z) = 0$$

is in the interval (α_1, α_2).

Proof. Let $n = 2$ and let $A = (C_1, C_2)$. Then clearly

$$\text{per}(C_2 + yC_1, C_2 + yC_1) = \text{per}\begin{bmatrix} a_{12} + ya_{11} & a_{12} + ya_{11} \\ a_{22} + ya_{21} & a_{22} + ya_{21} \end{bmatrix} = 0$$

has roots $-\frac{a_{12}}{a_{11}}, -\frac{a_{22}}{a_{21}}$.

Since $(a_{11}, a_{21})^T$ and $(a_{12}, a_{22})^T$ are linearly independent, the roots are distinct. It is easily verified that for any positive vector (z_1, z_2), the linear function

$$\phi(y) = \text{per}\begin{bmatrix} a_{12} + ya_{11} & z_1 \\ a_{22} + ya_{21} & z_2 \end{bmatrix}$$

of y has opposite signs at $y = -\frac{a_{12}}{a_{11}}, y = -\frac{a_{22}}{a_{21}}$. Hence the root of $\phi(y) = 0$ must be strictly between $-\frac{a_{12}}{a_{11}}$ and $-\frac{a_{22}}{a_{21}}$.

Let $n \geq 3$. We proceed by induction on n. Suppose the lemma is true for $n - 1$. Let

$$C_y = (C_1, \ldots, C_{n-2}, C_n + yC_{n-1}, C_n + yC_{n-1}).$$

Then, expanding along the first row,

$$\text{per } C_y = \left(\sum_{j=1}^{n-2} C_{j1} \text{ per } C_y(1, j) \right) + 2(C_{n1} + yC_{n-1,1})\text{per } C_y(1, n), \qquad (2.8.4)$$

where C_{j1} denotes the first coordinate of C_j.

Let y_0 be the root of per $C_y(1, n) = 0$. By the induction assumption, y_0 lies strictly between the two roots of per $C_y(1, j) = 0$, $j = 1, 2, \ldots, n - 2$.

Since per $C_y(1, j) \to \infty$ as $y \to \pm\infty$, $j = 1, 2, \ldots, n - 2$, we conclude that

$$\text{per } C_{y_0}(1, j) < 0, \quad j = 1, 2, \ldots, n - 2.$$

Hence, from (2.8.4), per $C_{y_0} < 0$. Thus per C_y, a quadratic in y, must have two distinct real roots, say $\alpha_1 < \alpha_2$. Note that in the argument above we have also shown that the root of per $C_y(1, n) = 0$ lies in (α_1, α_2). We can similarly show that the root of per $C_y(i, n) = 0$ lies in (α_1, α_2), $i = 1, 2, \ldots, n$. In particular, it follows that

$$\text{per } C_{\alpha_1}(i, n) < 0, \qquad \text{per } C_{\alpha_2}(i, n) > 0, \quad i = 1, 2, \ldots, n. \qquad (2.8.5)$$

Now let $z \in R^n$ be a positive vector. Then

$$\text{per}(C_1, \ldots, C_{n-2}, C_n + yC_{n-1}, z) = \sum_{i=1}^{n} z_i \text{ per } C_y(i, n). \qquad (2.8.6)$$

It follows from (2.8.5) and (2.8.6) that

$$\text{per}(C_1, \ldots, C_{n-2}, C_n + \alpha_1 C_{n-1}, z) < 0,$$

$$\text{per}(C_1, \ldots, C_{n-2}, C_n + \alpha_2 C_{n-1}, z) > 0,$$

and therefore the root of

$$\text{per}(C_1, \ldots, C_{n-2}, C_n + yC_{n-1}, z) = 0$$

is in (α_1, α_2).   ∎

Theorem 2.8.4 (Alexandroff Inequality). *Let $C_1, \ldots, C_{n-1}$ be positive vectors in R^n, $n \geq 2$. Then for any $C \in R^n$,*

$$\{\text{per}(C_1, \ldots, C_{n-1}, C)\}^2$$
$$\geq \text{per}(C_1, \ldots, C_{n-2}, C_{n-1}, C_{n-1})\text{per}(C_1, \ldots, C_{n-2}, C, C).$$

Furthermore, equality occurs in the inequality if and only if C_{n-1}, C are linearly dependent.

Proof. If C_{n-1} and C are linearly dependent, then clearly equality holds in the inequality. So suppose C_{n-1} and C are linearly independent. By expanding

$$\text{per}(C_1, \ldots, C_{n-2}, C + yC_{n-1}, C + yC_{n-1}) \tag{2.8.7}$$

with respect to the last two columns we get a quadratic in y whose discriminant is

$$4\{\text{per}(C_1, \ldots, C_{n-1}, C)\}^2$$
$$- 4\text{per}(C_1, \ldots, C_{n-2}, C_{n-1}, C_{n-1})\text{per}(C_1, \ldots, C_{n-2}, C, C).$$

By Lemma (2.8.3), the quadratic in (2.8.7) has two distinct real roots and hence its mixed discriminant must be positive. That completes the proof.   ∎

Theorem 2.8.5. *If $A \in \Omega_n$ is a permanent minimizer then $\text{per } A(i, j) = \text{per } A$ for all i, j.*

Proof. Let $A = (C_1, C_2, \ldots, C_n)$. If $a_{rs} > 0$, then $\text{per } A(r, s) = \text{per } A$ by Theorem 2.8.2. So suppose $a_{rs} = 0$. Then there exists $a_{rt} > 0$. By Theorem 2.8.4 we have

$$(\text{per } A)^2$$
$$\geq \text{per}(C_1, C_2, \ldots, C_s, \ldots, C_s, \ldots, C_n)\text{per}(C_1, \ldots, C_t, \ldots, C_t, \ldots, C_n)$$
$$= \sum_{k=1}^{n} a_{ks} \text{ per } A(k, t) \sum_{k=1}^{n} a_{kt} \text{ per } A(k, s). \tag{2.8.8}$$

If $\text{per } A(r, s) > \text{per } A$, then since $\text{per } A(i, j) \geq \text{per } A$ for all i, j, by Theorem 2.8.2, the right-hand side in Equation (2.8.8) is strictly greater than $(\text{per } A)^2$, which is a contradiction.   ∎

To prove the validity of van der Waerden's conjecture we need one more result.

Lemma 2.8.6. *If $A = (C_1, C_2, \ldots, C_n) \in \Omega_n$ is a permanent minimizer, then so is*

$$A^* = \left(C_1, C_2, \ldots, \frac{C_i + C_j}{2}, C_{i+1}, \ldots, \frac{C_i + C_j}{2}, C_{j+1}, \ldots, C_n \right).$$

Proof. We have

$$\operatorname{per} A^* = \frac{1}{2}\operatorname{per} A + \frac{1}{4}\operatorname{per}(C_1, \ldots, C_i, \ldots, C_i, \ldots, C_n)$$

$$+ \frac{1}{4}\operatorname{per}(C_1, \ldots, C_j, \ldots, C_j, \ldots, C_n)$$

$$= \frac{1}{2}\operatorname{per} A + \frac{1}{4}\sum_k a_{ki} \operatorname{per} A(k, j) + \frac{1}{4}\sum_k a_{kj} \operatorname{per} A(k, i) = \operatorname{per} A,$$

by Theorem 2.8.5, which gives the desired result. ∎

Theorem 2.8.7 (van der Waerden–Egorychev–Falikman Inequality). *If $A \in \Omega_n$, then $\operatorname{per} A \geq \frac{n!}{n^n}$. Equality occurs only for $A = J_n = (\frac{1}{n})$.*

Proof. Since Ω_n is a compact set and the permanent of a matrix is continuous, there exists a permanent minimizer A in Ω_n. By Theorem 2.8.1, in every row of A there are at least two positive elements. Let C_n be the last column of A. By applying Lemma 2.8.6 repeatedly, we get an $A^* > 0$ whose first $n - 1$ columns are positive, whose last column is C_n, and such that A^* is also a permanent minimizer. Let $A^* = (C_1^*, C_2^*, \ldots, C_{n-1}^*, C_n)$. Applying Theorems 2.8.4 and 2.8.5 to $\operatorname{per}(C_1^*, C_2^*, \ldots, C_{n-1}^*, C_n)$, we see that C_n is a scalar multiple of C_{n-1}^* and, similarly, we conclude that $C_n = \lambda_i C_i^*, \lambda_i > 0, i = 1, 2, \ldots, n-1$. Since $C_n + \sum_{i=1}^{n-1} C_i^* = 1$, we get $C_n = \frac{1}{n}\mathbf{1}$. We can similarly show that every column of A is $\frac{1}{n}\mathbf{1}$ and therefore $A = J_n$. That completes the proof. ∎

A significant portion of the results on permanents available today were discovered in the process of trying to solve the van der Waerden conjecture. The first characterization of a permanent minimizer, given in Theorem 2.8.2, was obtained by Marcus and Newman (1959). London (1971) gave a proof of the same characterization using the Duality Theorem. Egorychev (1981) and, independently, Falikman (1981) proved the van der Waerden conjecture. The Alexandroff Inequality appears in Alexandroff (1938), where it was proved for a more general function called the *mixed discriminant*, which arises as a measure of the *mixed volume* of several convex bodies; see Burago and Zalgaller (1988). This general inequality will be proved in Chapter 5.

2.9. Cooperative games with side payments

Optimal assignment problems occur naturally in markets for the sale or posses-
sion of indivisible commodities such as a house, a car, or a boat. The participants
in this market can be thought of as players in a game, each seeking cooperation. In this section we first give a very preliminary development of cooperative
game theory. We then analyze some games involving two-sided markets using
the solution concept known as the core. The prices that emerge as solutions to
the game can be viewed as the dual solutions of an optimal assignment problem.

An n-person cooperative game in *characteristic function* form with side
payments consists of a finite set of players $N = \{1, 2, \ldots, n\}$. Subsets of N
are called *coalitions*. A real-valued set function $v(S)$ is defined for all $S \subset$
$\{1, 2, \ldots, n\}$, where $v(\emptyset) = 0$. The amount $v(S)$ is to be interpreted as the
maximum a coalition S can collect on its own. For simplicity, we assume linear
utility and identical but insatiable taste among the participants. The following
example can be converted into such a game.

Example 2.1. Mowing the lawn as an operation requires frequent starting and
stopping of the lawn mower, a skill that Raghavan possesses but not his wife
Usha nor their daughter Deepa. Raghavan can, on his own, mow only half the
lawn. With a helping hand from his wife he could finish mowing the entire lawn.
Raghavan and his daughter together can finish three quarters of the lawn. The
three together can finish mowing the entire lawn as well. (This last assumption
is in fact not very realistic. It has been observed that when all three get together,
they would rather watch a Tamil movie on video rather than mow the lawn). The
problem is the following: If a watermelon, which all three family members enjoy
equally, is on the table, what could be a reasonable share for the participants of
this game as a reward for their skills?

We summarize the above game with the following characteristic function.
Let R, U, and D represent Raghavan, his wife Usha, and their daughter Deepa,
respectively. We have $N = \{R, U, D\}$, the player set. Here

$$v(R) = \frac{1}{2}, \ v(U) = v(D) = v(U, D) = 0, \ v(R, D) = \frac{3}{4},$$

$$v(U, R) = v(R, U, D) = 1.$$

Suppose it is suggested that R gets $\frac{7}{8}$ of the watermelon and U and D each get
$\frac{1}{16}$ of the watermelon. Then certainly it is possible for U and R to join together,
and with shares $\frac{29}{32}$ and $\frac{3}{32}$ of the melon they can keep D away, for nothing. Thus
the distribution $(\frac{7}{8}, \frac{1}{16}, \frac{1}{16})$ is not "stable." However, if it turns out that $(\frac{3}{4}, \frac{1}{4}, 0)$
was originally suggested for R, U, and D, respectively, then it can be checked
that no recontracting will help to improve the individual shares.

Formally, we have the following definition. Given an n-person game with
player set $N = \{1, 2, \ldots, n\}$ and the characteristic function $v(S)$, the core is

defined by

$$C = \left\{ (x_1, x_2, \ldots, x_n) : \sum_{i \in S} x_i \geq v(S) \quad \text{for all} \right.$$

$$\left. S \subset \{1, 2, \ldots, n\}, \sum_i x_i = v(N) \right\}.$$

In general, the core could be empty. For example, the three-person game with a majority coalition winning and the rest losing, defined by $N = \{1, 2, 3\}$, $v(1, 2) = v(2, 3) = v(1, 3) = v(1, 2, 3) = 1$ and $v(1) = v(2) = v(3) = 0$, has empty core, since the inequalities $x_1 \geq 0$, $x_2 \geq 0$, $x_3 \geq 0$, $x_1 + x_2 \geq 1$, $x_2 + x_3 \geq 1$, $x_1 + x_2 \geq 1$, and $x_1 + x_2 + x_3 = 1$ are inconsistent.

The following important theorem gives a necessary and sufficient condition for a game to possess core elements.

Theorem 2.9.1 (Bondareva-Shapley). *An n-person game in characteristic function form has a nonempty core if and only if for any family $\mathcal{L}$ of nonempty subsets of N, and for any $\delta_S > 0$, $S \subset \{1, 2, \ldots, n\}$, $S \in \mathcal{L}$, satisfying*

$$\sum_{S \ni i, S \in \mathcal{L}} \delta_S = 1, \quad i = 1, 2, \ldots, n, \tag{2.9.1}$$

we have

$$\sum_{S \in \mathcal{L}} \delta_S v(S) \leq v(N). \tag{2.9.2}$$

Proof. The core is nonempty if and only if the minimum of $\sum_i x_i$ subject to $\sum_{i \in S} x_i \geq v(S)$, $S \subset \{1, 2, \ldots, n\}$ has an optimal solution with optimal value $v(N)$. By the Duality Theorem, this is so if and only if the dual problem

$$\max \sum_S \delta_S v(S)$$

subject to

$$\sum_S \delta_S \chi_S(i) = 1 \quad \text{for } i = 1, 2, \ldots, n$$

has an optimal solution. Here

$$\chi_S(i) = 1 \quad \text{if } S \ni i$$
$$= 0 \quad \text{if } S \not\ni i$$

and the optimal value is $v(N)$. If $\mathcal{L}$ denotes those S with $\delta_S > 0$ we see that

$$1 = \sum_S \delta_S \chi_S(i) = \sum_{S \ni i, S \in \mathcal{L}} \delta_S \quad \text{and} \quad \sum_{S \in \mathcal{L}} \delta_S v(S) \leq v(N).$$

That completes the proof. ∎

Conditions (2.9.1) and (2.9.2) can be given the following interpretation: Players might be active at different levels in different coalitions where they are present. Let δ_S be the level of participation in coalition S by player i. Thus $\sum_{S \ni i} \delta_S = 1$ says that player i splits the unit level of participation among various coalitions in which the player is present. The second inequality says that players are better off forming the grand coalition.

Verifying the existence of core elements is somewhat easy for certain classes of games. We now describe two such classes, assignment games and permutation games. The following is an example of an assignment game.

Real estate game. There are n house owners in a locality trying to sell their houses. There are n newcomers to the area and each one wants to own just one house. House owner i thinks that house i is worth c_i, $i = 1, 2, \ldots, n$. The potential buyer j thinks that house i is worth only h_{ij}. Suppose that a price p_i is announced on house i, $i = 1, 2, \ldots, n$. Buyer j's gain is $h_{ij} - p_i$, if positive. Seller i's gain is $p_i - c_i$. The gain for the coalition i, j is $h_{ij} - p_i + p_i - c_i$. Thus a transaction between i and j could take place only when $h_{ij} - c_i > 0$.

The problem is to determine who will buy which house and at what price. Before we suggest any prices, we convert our problem to a game in characteristic function form.

Let B denote the set of buyers and let C be the set of sellers. Let $a_{ij} = v(i, j) = \max(h_{ij} - c_i, 0)$ if $i \in C$, $j \in B$. We can define

$$v(S) = 0 \quad \text{if } S \subset B \quad \text{or} \quad S \subset C$$

$$= \max \sum_{\alpha=1}^{k} a_{i_\alpha j_\alpha}, \tag{2.9.3}$$

where $k = \min[\text{card } (S \cap B), \text{card } (S \cap C)]$, and $i_1, i_2, \ldots, i_k; j_1, j_2, \ldots, j_k$ denote distinct elements in $S \cap C$, $S \cap B$, respectively. [The amount $v(S)$ is obtained when buyers and sellers in any coalition S find the right partners among them in the sense of maximizing their total gain.]

The buyers and sellers will be looking for a harmonious split of the gains accruing out of sales. Suppose buyer j expects a net gain of at least $v_j \geq 0$ out of the transaction. Similarly, suppose seller i expects a net gain of $u_i \geq 0$ out of the sales. Thus $u_i + v_j = a_{ij}$ if coalition i, j is formed, and $u_i + v_j \geq a_{ij}$ if coalition i, j is not formed. Our question is the following: Are there nonnegative real numbers u_i, v_j satisfying

$$u_i + v_j = a_{ij} \quad \text{if } i, j \text{ coalition ultimately forms}$$

$$\geq a_{ij} \quad \text{otherwise?} \tag{2.9.4}$$

The following theorem summarizes the relationship between this game and the assignment problem.

Theorem 2.9.2. *The real estate game has a nonempty core and is precisely the set of optimal solutions to the dual of the linear programming problem*

$$maximize \sum_i \sum_j a_{ij} x_{ij}$$

subject to

$$\sum_i x_{ij} \leq 1 \quad for\ all\ j$$

$$\sum_j x_{ij} \leq 1 \quad for\ all\ i$$

$$and\ x_{ij} \geq 0 \quad for\ all\ i,\ j. \tag{2.9.5}$$

Proof. By Lemma 2.4.1 we know that any $X = (x_{ij})$ satisfying (2.9.5) is dominated entrywise by a doubly stochastic matrix Q. Since $a_{ij} \geq 0$ for all i, j, we can also assume that the maximum occurs at a doubly stochastic matrix, which the Birkhoff–von Neumann Theorem can be taken to be a permutation matrix. Thus, the maximum value exists and is given by

$$v(N) = \max_{\sigma \in \Omega_n} \sum_i a_{i\sigma(i)} = v(B \cup C).$$

The dual to our linear program is the problem

$$minimize \left(\sum_i u_i + \sum_j v_j \right)$$

subject to

$$u_i + v_j \geq a_{ij} \quad for\ all\ i,\ j$$
$$u_i,\ v_j \geq 0 \quad for\ all\ i,\ j. \tag{2.9.6}$$

By complementarity, if $X^* = (x_{ij}^*)$ is an optimal solution to the primal and $x_{ij}^* > 0$, we have $u_i + v_j = a_{ij}$. Thus when the coalition i, j forms they could split the net gain a_{ij} by amounts $u_i, v_j \geq 0$. Also, for any coalition S,

$$\sum_{i \in S \cap C} u_i + \sum_{j \in S \cap B} v_j \geq \sum_{i \in S \cap C, j \in S \cap B} a_{ij} \geq v(S),$$

from (2.9.3) and hence $v(N) = \sum_i u_i + \sum_j v_j$. Thus the core is nonempty and is precisely the set of optimal solutions to the dual problem. ∎

The problem of arriving at a set of prices in the real estate game is now quite easy. The seller thinks that his house is worth c_i. If he expects a net gain u_i, then he announces the price $p_i = q_i + c_i$. Buyer j thinks that the house is worth h_{ij} and is selling for p_i. Pursuing a transaction is not worth any effort if $h_{ij} - p_i < 0$. Among the houses that are worth his consideration, if any, the buyer adopts comparative shopping to select the best house. This amounts to maximizing $a_{ij} - u_i$ among the houses worth the effort. We know that $a_{ij} - u_i \le v_j$ and $a_{ij} - u_i = v_j$ for some j if u_i, v_j is optimal for the dual. Thus comparative shopping amounts to a price system consistent with gain expectations that are precisely core elements. In general, there is a leverage for more than one set of prices due to overall price movement in the market.

The following numerical example illustrates the problem at hand, for a real estate market with 3 buyers and 3 sellers. Suppose the base prices of the sellers (in thousands) are 18, 15, and 19, respectively. Let h_{ij} denote the j-th buyer's evaluation of the i-th house (in thousands) and suppose the matrix $H = (h_{ij})$ is given by

$$\begin{bmatrix} 23 & 26 & 20 \\ 22 & 24 & 21 \\ 21 & 22 & 17 \end{bmatrix}.$$

The matrix (a_{ij}), measured in thousands, is

$$\begin{array}{c c c c} & v_1 = 2 & v_2 = 4 & v_3 = 0 \\ u_1 = 4 & \begin{pmatrix} 5 & 8^* & 2 \\ u_2 = 6 & 7 & 9 & 6^* \\ u_3 = 0 & 2^* & 3 & 0 \end{pmatrix} \end{array}.$$

The unique assignment for the given gain expectations is starred in the above matrix. At the given core point, the prices are

$$p_1 = 18,000 + 4,000, \quad p_2 = 15,000 + 6,000, \quad \text{and } p_3 = 17,000 + 0.$$

At these prices, buyer 1 feels it best to buy house 3 with maximum gain of 4,000, the second buyer to buy house 1 with a maximum gain of 5,000, and the third buyer to buy house 2 with a maximum gain of 4,000. Since the core has other elements, there could be many other sets of prices.

2.10. Lexicographic center

Given a game (N, v) and an imputation x, let $f(x, S) = \sum_{i \in S} x_i - v(S)$. We call $f(x, S)$ the *satisfaction of the coalition* S for imputation x. For each imputation x let $\theta(x)$ be the vector of all satisfactions for all possible coalitions arranged in a nondecreasing order. Thus $\theta_k(x) \le \theta_l(x)$ whenever $k < l$. We say $x >_L y$ if and only if lexicographically the vector $\theta(x)$ dominates $\theta(y)$. Formally, if k is the least index with $\theta_k(x) \ne \theta_k(y)$, then $\theta_k(x) > \theta_k(y)$. By the *lexicographic center* of a subset D of the imputation set, we mean the unique point $x^* \in D$

that lexicographically maximizes the vector $\theta(x)$ over D. If D is any closed, bounded convex set, Schmeidler (1969) showed that the lexicographic center exists and is unique. When D is chosen as the set of all imputations, this unique imputation x^* is called the *nucleolus* of the game (N, v). When the core is nonempty the nucleolus lies in the core. Even though the determination of the nucleolus is quite difficult in general, it can be located efficiently for special subclasses of games. We will describe an algorithm to locate the nucleolus for the real estate game. We will reinterpret the game slightly differently as follows:

Stable real estate commissions. Homeowners $M = \{U_1, U_2, \ldots, U_m\}$, each possessing one house, and home buyers $N = \{V_1, V_2, \ldots, V_n\}$, each wanting to buy one house, approach a common real estate agent. Not revealing the identity of the buyers and sellers, the agent wants an up-front commission $a_{ij} \geq 0$ if he links seller U_i to buyer V_j. The sellers and buyers prefer fixed commissions $u_1, u_2, \ldots, u_m$ and $v_1, \ldots, v_n$. The agent has no objection if they meet his expectation for every possible link. He guarantees their money's worth in his effort and promises to take no commission from a seller (buyer) if he cannot find a suitable buyer (seller).

With the introduction of $M_0 := M \cup \{0\}$, $N_0 := N \cup \{0\}$, $a_{i0} := 0 \; \forall i \in M$, $a_{0j} := 0 \; \forall j \in N$, $a_{00} := 0$, $u_0 := 0$, and $v_0 := 0$, all the constraints in Equation 2.9.6 can be uniformly written as

$$f_{ij}(u, v) := u_i + v_j - a_{ij} \geq 0, \quad \forall (i, j) \in (M_0, N_0). \tag{2.10.1}$$

If $\sigma \subseteq (M, N)$ is an optimal assignment and D is the core, we get

$$(i, j) \in \sigma \text{ implies } f_{ij}(u, v) = 0, \quad \forall (u, v) \in D. \tag{2.10.2}$$

With the convention that $(0, 0) \in \sigma$ we write $(i, 0) \in \sigma$ [$(0, j) \in \sigma$] if in σ row $i \in M$ (column $j \in N$) is not assigned to any column $j \in N$ (row $i \in M$). Here σ is extended to a subset of (M_0, N_0) so that (2.10.2) also expresses that D lies in the hyperplane $u_i = 0$ (or $v_j = 0$) for any unassigned row i (column j). It is easily seen that

$$D = \{(u, v) : f_{ij}(u, v) = 0 \; \forall (i, j) \in \sigma, \; f_{ij}(u, v) \geq 0 \; \forall (i, j) \notin \sigma\}. \tag{2.10.3}$$

Here and now on $(i, j) \notin \sigma$ is written instead of $(i, j) \in (M_0, N_0) \setminus \sigma$.

Among many sets of commissions $u_0, u_1, \ldots, u_m; v_0, v_1, \ldots, v_n$ in D, for the agent, he wants to choose one that is neutral and stable. The lexicographic center is a possible option that is both neutral and stable for all pairs.

For every $(u, v) \in D$, the first $\max(m, n) + 1$ components (those coordinates $k = (i, j)$ corresponding to $(i, j) \in \sigma$) of $\theta(u, v)$ are equal to 0. Let

$$\alpha^1 := \max_{(u,v) \in D} \left(\min_{(i,j) \notin \sigma} f_{ij}(u, v) \right),$$

$$D^1 := \left\{ (u, v) \in D : \min_{(i,j) \notin \sigma} f_{ij}(u, v) = \alpha^1 \right\},$$

and
$$\sigma^1 = \{(i, j) : f_{ij}(u, v) = \text{constant on } D^1\}.$$
The set σ^1 can be regarded as an "assignment" between the equivalence classes
of the relation $\sim^1$ defined on M_0 and N_0 by

$$i_1 \sim^1 i_2 \quad \text{if and only if } u_{i_1} - u_{i_2} \text{ is constant on } D^1,$$

$$j_1 \sim^1 j_2 \quad \text{if and only if } v_{j_1} - v_{j_2} \text{ is constant on } D^1,$$

respectively.

Similarly,
$$\alpha^2 := \max_{(u,v) \in D^1} \left(\min_{(i,j) \notin \sigma^1} f_{ij}(u, v) \right),$$

$$D^2 := \left\{ (u, v) \in D^1 : \min_{(i,j) \notin \sigma^1} f_{ij}(u, v) = \alpha^2 \right\},$$

and
$$\sigma^2 := \{(i, j) \in (M_0, N_0) : f_{ij}(u, v) \text{ is constant on } D^2\}.$$

Let $i \sim^2 k$ if and only if $u_i - u_k$ is a constant on D^2. Observe that $\sigma^2 \supset \sigma^1 \supset \sigma$.
Therefore, after some $t \leq \min(m, n)$ rounds, the process terminates with

$$\sigma^t := \{(i, j) = (M_0 \times N_0) : f_{ij}(u, v) \text{ is constant on } D^t\}.$$

Consequently, a subset of D is found that is parallel to all hyperplanes defining
D. Since these hyperplanes include $u_i = 0$ for all $i \in M$ and $v_j = 0$ for all
$j \in N$, this subset must consist of a single point. It can be proved [see Solymosi
and Raghavan (1994)] that this point is precisely the lexicographic center of D.

Next we illustrate how to implement the procedure leading to the lexicographic center. Given

$$A = \begin{bmatrix} 6 & 7 & 7 \\ 0 & 5 & 6 \\ 2 & 5 & 8 \end{bmatrix},$$

where $M = \{1, 2, 3\} = N$, the unique optimal assignment for A is $\sigma = \{(1, 1), (2, 2), (3, 3)\}$, i.e., the entries in the main diagonal. Starting with all
commissions collected entirely from sellers, one could use the procedure to be
described below to locate the u worst point $(u^1, v^1) = (0, 6, 4, 6 : 0, 0, 1, 2)$
in D. Furthermore, with rows numbered 0, 1, 2, and 3 and columns numbered
0, 1, 2, and 3 we can read off (u^1, v^1) from column 0 and row 0 of the matrix

$$[f_{ij}(u^1, v^1)] = \begin{bmatrix} 0 & 0^* & 1 & 2 \\ 6 & 0 & 0^* & 1 \\ 4 & 4 & 0 & 0^* \\ 6 & 4 & 2 & 0 \end{bmatrix}.$$

Even though the coordinates for starred entries above are the next set with higher f_{ij} values in the lexicographic ranking, they are still 0. However, from now on there will be strict improvement with higher value when we follow the iteration. We want to move in a direction (s, t) inside D, with one end at the extreme solution (u^1, v^1). Let the new point be $(u^2, v^2) = (u^1, v^1) + \beta \cdot (s, t)$ for some $\beta \geq 0$. Becuase the point (u^1, v^1) constitutes the worst for all sellers in terms of commissions in D they would like their commissions reduced.

Since (u^2, v^2) is the farthest from the hyperplanes indexed by $(0, 1)$, $(1, 2)$, and $(2, 3)$ (indicated by a* in the above matrix) this translates to the requirements

$$s_0 + t_1 \geq 1, \qquad s_1 + t_2 \geq 1, \qquad s_2 + t_3 \geq 1, \qquad (2.10.4)$$

with at least one equality. Since we must remain in D we also have

$$s_i + t_i = 0, \quad i = 1, 2, 3. \qquad (2.10.5)$$

Combining (2.10.4) and (2.10.5) gives

$$t_1 - t_0 \geq 1, \qquad t_2 - t_1 \geq 1, \qquad t_3 - t_2 \geq 1, \qquad (2.10.6)$$

with at least one equality to hold. Thus the direction for improvement for sellers is $(s, t) = (0, -1, -2, -3 : 0, 1, 2, 3)$. Next we determine how far we can move along this direction inside D, starting from the initial u-worst corner of D.

To calculate the maximal distance β in the direction (s, t) we use the change of distance matrix:

$$[f_{ij}((u^1, v^1) + \beta \cdot (s, t))] = \begin{bmatrix} 0 & 0 + \beta & 1 + 2\beta & 2 + 3\beta \\ 6 - \beta & 0 & 0 + \beta & 1 + 2\beta \\ 4 - 2\beta & 4 - \beta & 0 & 0 + \beta \\ 6 - 3\beta & 4 - 2\beta & 2° - \beta & 0 \end{bmatrix}.$$

Here $f_{32} = 2 - \beta$ is the first to reach the increasing minimum level $0 + \beta$. From $2 - \beta = 0 + \beta$ we get $\beta = 1$ and $(u^2, v^2) = (0, 5, 2, 3; 0, 1, 3, 5)$. It can be shown that (u^2, v^2) is the u-worst corner (v-best corner) in D^1. The updated distance matrix is

$$[f_{ij}(u^2, v^2)] = \begin{bmatrix} 0 & 1^* & 3 & 5 \\ 5 & 0 & 1^* & 3 \\ 2 & 3 & 0 & 1^* \\ 3 & 2 & 1^* & 0 \end{bmatrix}.$$

To improve further from $(u^2, v^2) = (0, 5, 2, 3; 0, 1, 3, 5)$ we need to find a direction to move inside D^1. If the rows and columns are numbered 0,1,2,3 as before, then the starred value 1 at entries $(2, 3)$ and $(3, 2)$ implies

$$f_{23}(u, v) \geq 1, \quad f_{32}(u, v) \geq 1$$

for all $(u, v) \in D^1$. Indeed, using identities $f_{22}(u, v) \equiv 0$, $f_{33}(u, v) \equiv 0$ on D^1, we get the new identities

$$f_{23}(u, v) \equiv 1, \quad f_{32}(u, v) \equiv 1 \forall (u, v) \in D^1. \tag{2.10.7}$$

Hence, the new direction (s, t) must satisfy, $s_0 + t_1 \geq 1$, $s_1 + t_2 \geq 1$, $s_2 + t_3 = 0$, $s_3 + t_2 = 0$, with at least one of the inequalities becoming an equality. Using (2.10.2) we get $(s, t) = (0, -1, -2, -2 : 0, 1, 2, 2)$. Now to determine the new step size β for the new direction we proceed as follows:

$$[f_{ij}((u^2, v^2) + \beta \cdot (s, t))] = \begin{bmatrix} 0 & 1+\beta & 3+2\beta & 5+2\beta \\ 5-\beta & 0 & 1+\beta & 2+\beta \\ 2° - 2\beta & 3-\beta & 0 & 1 \\ 3-2\beta & 2-\beta & 1 & 0 \end{bmatrix}.$$

The decreasing distance $f_{20} = 2-2\beta$ is the first to reach the increasing second smallest distance $1+\beta$. It happens when $\beta = 1/3$. So the maximal distance into this direction is $\beta = 1/3$, and the u-worst corner of the set D^2 of points with the second smallest distance $\frac{4}{3}$ is $(u^3, v^3) = (0, 14/3, 4/3, 7/3; 0, 4/3, 11/3, 17/3)$. The updated distance matrix is

$$[f_{ij}(u^3, v^3)] = \begin{bmatrix} 0 & 4/3^* & 11/3 & 17/3 \\ 14/3 & 0 & 4/3^* & 7/3 \\ 4/3^* & 8/3 & 0 & 1 \\ 7/3 & 5/3 & 1 & 0 \end{bmatrix}.$$

Again (u^3, v^3) is the u-worst corner (v-best corner) in D^2. To move inside D^2, we look for direction (s, t). Using starred entries, (s, t) must satisfy $s_0 + t_1 \geq \frac{4}{3}$, $s_1 + t_2 \geq \frac{4}{3}$, $s_2 + t_0 \geq \frac{4}{3}$. Also, since $f_{23}(u, v) = f_{32}(u, v) \equiv 1$ on D^2, we easily find the above system of inequalities inconsistent. Thus no more movement inside is possible. We have reached the lexicographic center.

Remark. Starting with the worst set of commissions for all sellers and using Kuhn's Hungarian method (1955), the algorithm locates unique set of commissions that again favor all the buyers in the restricted new domain D of commissions. The next ingredient involves locating the unique direction (s, t) and the unique step size β to find the new set of commissions. We have not used in our example any efficient procedure to find the direction (s, t). In Solymosi and Raghavan (1994), they develop an explicit graph theoretic algorithm to find this direction [also see Maschler, Peleg and Shapley (1979), Reijnierse (1995). The decomposition of the payoff space and the lattice structure of the feasible set at each iteration are utilized in associating a directed graph. In case the graph is acyclic, finding the direction (s, t) for movement is translated to determining the longest paths to each vertex. Cycles are used to collapse vertices so that the graph has fewer vertices. The algorithm stops when the graph is reduced to just one vertex. The assignment game is the simplest type of cooperative games, which are balanced and hence have nonempty core. The real estate game was first considered by Shapley and Shubik (1972). The same problem was viewed in the context of competitive pricing of indivisible goods by Gale (1960). Theorem 2.9.1 was obtained independently by Bondareva (1962) and Shapley (1965). Pooling peoples' utility amounts to interpersonal comparisons and hence has remained alien to mainstream economists. For a version of the real estate game without side payments, see Shapley and Scarf (1974). A closely related class of games with nonempty core are the so-called permutation games due to Tijs et al. (1984).

Permutation game. A set of n customers are waiting in a line for a service that takes unit time per customer. Customers have different costs for waiting. (One should see the assortment of people in a queue for rice in a ration shop in an Indian city–women with children, workers, old men, boys, etc.). Persons in the queue may prefer to mutually exchange positions with friends in the queue to cut total waiting cost, without affecting other customers. If this is permissible, we have a cooperative game with the characteristic function

$$v(S) = \min_{\sigma_S} \sum_{i \in S} c_{i\sigma_S(i)}, \quad \text{where} \quad \sigma_S : S \to S \text{ is a permutation}, \quad (2.10.8)$$

and c_{ij} is the cost for customer i to be served at time j. We would like to know whether this game has a nonempty core, with the obvious definition of core to be the set

$$C = \left\{ (x_1, x_2, \ldots, x_n) : \sum_{i \in S} x_i \le v(S) \quad \text{for all} \right.$$

$$\left. S \subset \{1, 2, \ldots, n\} = N \text{ with equality for } S = N \right\}.$$

Theorem 2.10.1. *A permutation game has nonempty core.*

Proof. In view of Theorem 2.9.1, it is enough to prove the following: Given a collection $\mathcal{L}$ of nonempty subsets of $N = 1, 2, \ldots, n$, and given $\sigma_S > 0$ for $S \in \mathcal{L}$ with $\sum_{S \ni i} \sigma_S = 1, i = 1, 2, \ldots, n$,

$$\sum_{S \in \mathcal{L}} \sigma_S v(S) \ge v(N).$$

Consider the collection $\mathcal{K}$ of all $n \times n$ (0, 1)-matrices. Given any permutation $\sigma_S : S \to S$, we can associate with it a matrix X in $\mathcal{K}$ with

$$x_{ij} = 0 \quad \text{if } i \notin S$$

$$= 0 \quad \text{if } i \in S, \ j \neq \sigma_S(i)$$

$$= 1 \quad \text{if } i \in S, \ j = \sigma_S(i).$$

For fixed S, we denote the collection of such matrices by $P(S)$. Thus from (2.10.8), it follows that $v(S) = \sum_i \sum_j c_{ij} p_{ij}^S$ for some $(p_{ij}^S) \in P(S)$. Observe that $Q = (q_{ij})$ defined by

$$q_{ij} = \sum_{S \in \mathcal{L}} \sigma_S p_{ij}^S$$

satisfies

$$\sum_j q_{ij} = \sum_j \sum_{S \in \mathcal{L}} \sigma_S p_{ij}^S = \sum_{S \in \mathcal{L}} \sigma_S \sum_j p_{ij}^S = \sum_S \sigma_S \chi_i(S)$$

$$= \sum_{S \ni i} \sigma_S = 1, \quad i = 1, 2, \ldots, n.$$

Similarly, $\sum_i q_{ij} = 1$, $j = 1, 2, \ldots, n$. Thus Q is doubly stochastic. We know that $Q = \sum_k \alpha_k P^{(k)}$ for some permutation matrices $P^{(k)}$. Also,

$$\sum_i \sum_j c_{ij} q_{ij} = \sum_k \alpha_k \sum_i \sum_j c_{ij} p_{ij}^{(k)} \geq \sum_k \alpha_k \min_{\alpha \in \Omega_n} \sum_i c_{i\sigma(i)}$$

$$= \sum_k \alpha_k v(N) = v(N).$$

Thus

$$\sum_{S \in \mathcal{L}} \delta(S) v(S) = \sum_S \delta(S) \sum_i \sum_j c_{ij} p_{ij}^S$$

$$= \sum_i \sum_j c_{ij} q_{ij} \geq v(N).$$

That completes the proof.   ∎

The following is an example of a permutation game.

Example 2.2. A clerk, a professor, and an executive enter an auto service center, as the first, second, and third customers for tuneups. The service center has just one mechanic. On each car it takes an hour for a tuneup. Waiting costs per hour are, respectively, $15 for the clerk, $30 for the professor, and $50 for the executive. (Waiting costs are often proportional to salaries!) The mechanic is willing to do the jobs in any order if the customers agree to switch places.

The total cost for the clerk is $15, for the professor next in line it is $60 for the two-hour wait, and for the executive it is $150 for the three-hour wait. Any cooperation amounts to reversing the order to cut costs. We have

$$C(1) = 15, \quad C(2) = 60, \quad C(3) = 150, \quad C(1, 2) = 60$$

$$C(1, 3) = 95, \quad C(2, 3) = 190, \quad C(1, 2, 3) = 155.$$

Clearly, $(0, 60, 95)$ is a core allocation of costs. This situation is achieved by the clerk and the executive exchanging their positions and the executive just covering the clerk for his waiting cost of $45. Thus the clerk gets his work done and his pay cut is made up by the executive.

Remark. Though we showed that assignment games and permutation games have non-empty core, this is not the case for many closely related classes of combinatorial games like the so called sequencing games. For details see Curiel et al (1994).

2.11. Open shop scheduling

An interesting application of the Birkhoff–von Neumann Theorem can be made to the problem of open shop scheduling, due to Gonzales and Sahni (1976) [also see Gonzales (1979)]. For a related application see Brualdi (1988) and Brualdi and Csima (1992). A timetabling problem described in Bondy and Murty (1976) is also similar.

Consider the service department of an auto dealer. Cars often require several types of multiple repairs: (a) wheel alignment (b) exhaust pipe replacement, (c) engine tuneup, (d) fixing a bumper, and others. These tasks can be carried out in any order. However, alignment, tuneup, and body repair may have to be done in different departments, and this makes it impossible to carry out more than one task at a time on any car. The logistic problem of servicing these tasks in the given time with the available personnel leads to the following mathematical formulation of the problem.

Suppose a shop has m machines. Each machine performs different tasks. Let there be n jobs. (Think of them as cars to be serviced.) Each job requires certain time on each machine. Let t_{ij} be the time needed on machine i for job j. A *schedule* for a processor (machine) is defined as a sequence (l_j, s_{l_j}, f_{l_j}). Here l_j refers to the job index, s_{l_j} is the starting time of job l_j, and f_{l_j} is the finishing time of job l_j. Thus the tuples of this index are arranged in the order $s_{l_j} < f_{l_j} \leq s_{l_{(j+1)}}$. There may be more than one tuple per job. A schedule for an m-shop (i.e., a shop with m machines) is a set of m schedules, one for each machine. The *finishing time* of a schedule is defined as the latest completion time on the individual machines. This represents the time at which all jobs are completed. An optimal finishing time schedule is the one with the least finishing time among all schedules. A schedule is called *preemptive* if the machines can be used more than once on the same job at different time points for its completion. The following hypothetical example illustrates the problem.

Example 2.3. There are four cars at a service station. Each one needs at least one of the following tasks: (1) tuneup, (2) oil change, (3) tire rotation. Since the cars are of different brands and sizes they need varying times for the execution of these jobs. The matrix below summarizes the information.

$$
\begin{array}{c}
\quad\quad\quad\; C_1 \quad C_2 \quad C_3 \quad C_4 \\
\begin{array}{c}
\text{tuneup} \\
\text{oil change} \\
\text{rotate tires}
\end{array}
\left(
\begin{array}{cccc}
25 & - & 20 & - \\
15 & 15 & - & 25 \\
25 & 15 & 10 & 15
\end{array}
\right).
\end{array}
$$

For example, Car 2 needs 15 minutes for an oil change and 15 minutes for the rotation of tires. It does not need a tuneup.

Under any type of scheduling, Car 1 will take 65 minutes for completion. In the same way the machine shop for tire rotations has to be kept open for at least 65 minutes for the completion of the tire rotation jobs. Thus the maximum of the two time periods, that is of the row and column totals, is the minimum time the shop definitely has to be open. This does not necessarily mean that all the jobs can be done in that time with any arbitrary scheduling policy. For example, if all jobs are to be completed on each car before another car is taken for service on a first-come first-served basis, the shop has to be kept open for at least 165 minutes. However, some machines might remain idle while jobs of different types are being done on a car. It is contructive to find the optimal time taken for preemptive scheduling and also to determine the optimal preemptive scheduling. The next result will be used to demonstrate the interesting fact that in 65 minutes we can get all the jobs finished in the above problem under preemptive scheduling.

Lemma 2.11.1. *Let* $T = (t_{ij})$ *be an* $m \times n$ *nonnegative matrix with row sums* r_i *and column sums* c_j, *for* $1 \leq i \leq m$, $1 \leq j \leq n$. *Let* $\alpha = \max(r_i, c_j)$. *Then there exists an* $(m + n) \times (m + n)$ *matrix* $A = (a_{ij})$, *with* T *as the leading* $m \times n$ *submatrix, and with all row sums and all column sums equal to* α.

Proof. Let D be the nonnegative $(m + n) \times (m + n)$ matrix

$$
D = \begin{bmatrix} T & D_2 \\ D_3 & 0 \end{bmatrix},
$$

where D_2 and D_3 are diagonal matrices with $(D_2)_{ii} = \alpha - r_i$, $i = 1, 2, \ldots, m$, $(D_3)_{jj} = \alpha - c_j$, $j = 1, 2, \ldots, n$. As in Theorem 2.4.1 we can find a matrix A with row sums and column sums α and with $D \leq A$. Since the first m row sums of D and the first n column sums of D are α,

$$
A = \begin{bmatrix} T & D_2 \\ D_3 & A_4 \end{bmatrix}.
$$

That completes the proof.   ■

The matrix A constructed in Lemma 2.11.1 is a scalar multiple of a doubly stochastic matrix and, by the Birkhoff–von Neumann Theorem, it can be expressed as a nonnegative linear combination of permutation matrices. Furthermore, a constructive procedure for arriving at the linear combination was described in the proof of the Birkhoff–von Neumann Theorem.

Continuing with our example, let us add fictitious cars and fictitious jobs to make the matrix a 7×7 matrix:

	1	2	3	4	5	6	7
tuneup	25	–	20	–	–	–	–
oil change	15	15	–	25	–	–	–
rotate tires	25	15	10	15	–	–	–
dummy job 1	–	–	–	–	–	–	–
dummy job 2	–	–	–	–	–	–	–
dummy job 3	–	–	–	–	–	–	–
dummy job 4	–	–	–	–	–	–	–

Since we want row sums and column sums to be $\max(r_i, c_j) = 65$, we add entries to make it a 7×7 matrix with row sums and column sums 65. Fill up the first three rows to make row totals 65. Now fill up the first 4 columns to make column sums 65. We get the matrix

	1	2	3	4	5	6	7
tuneup	25	0	20	0	20	0	0
oil change	15	15	0	25	0	10	0
rotate tires	25	15	10	15	0	0	0
dummy job 1	0	0	0	0	0	0	0
dummy job 2	0	35	0	0	0	0	0
dummy job 3	0	0	35	0	0	0	0
dummy job 4	0	0	0	25	0	0	0

A matrix with row sums and column sums 65 that dominates the above matrix is, for example,

$$
\begin{bmatrix}
25 & 0 & 20 & 0 & 20 & 0 & 0 \\
15 & 15 & 0 & 25 & 0 & 10 & 0 \\
25 & 15 & 10 & 15 & 0 & 0 & 0 \\
0 & 0 & 0 & 10 & 35 & 20 & 0 \\
0 & 35 & 0 & 0 & 10 & 20 & 0 \\
0 & 0 & 35 & 0 & 0 & 15 & 15 \\
0 & 0 & 0 & 15 & 0 & 0 & 50
\end{bmatrix}.
$$

It may be verified that the above matrix can be expressed as

$$15B_1 + 10B_2 + 10B_3 + 10B_4 + 5B_5 + 5B_6 + 5B_7 + 5B_8,$$

where B_i, $i = 1, 2, \ldots, 8$ are permutation matrices corresponding to the following permutations, respectively : (5416237), (1234567), (1645237), (3125674), (3215674), (1425637), (3415267), (5146237). The following summarizes the corresponding scheduling operations:

	15 m	10 m	10 m	10 m	5 m	5 m	5 m	5 m
tuneup	$-$	C_1	C_1	C_3	C_3	C_1	C_3	$-$
oil change	C_4	C_2	$-$	C_1	C_2	C_4	C_4	C_1
rotate tires	C_1	C_3	C_4	C_2	C_1	C_2	C_1	C_4

Here we have discarded any information regarding the dummy cars and dummy jobs. The final interpretation is as follows. For the first 15 minutes change the oil and rotate the tires on cars 4 and 1, respectively; for the next 10 minutes tuneup car 1, change the oil on car 2, rotate the tires on car 3; for the next 10 minutes tuneup car 1, rotate the tires on car 4, and so on.

2.12. A fair division problem

In this section we describe an interesting application of the Frobenius-König Theorem to a problem of fair division due to Kuhn (1967).

Imagine n persons (henceforth called players) fighting over the division of an estate to get their fair share. Suppose $n = 2$ and that the estate is a finely divisible object, say a coffee plantation. The following procedure can be considered fair. A player is selected by the toss of an ordinary coin. She divides the land into two parts. The other player selects the part he likes. This procedure presupposes implicitly that the chooser can find at least one part acceptable to him and the divider is capable of dividing the estate into two portions so that either portion is acceptable to her. With three players we can proceed as follows: Cards numbered 1, 2, and 3 are thoroughly shuffled and one of them is selected. Depending on the number i selected, player i is chosen to initiate the division. He partitions the land into three sets S_1, S_2, and S_3. If for some permutation σ of the labels 1, 2, and 3, $S_{\sigma(1)}$ is acceptable to player 1, $S_{\sigma(2)}$ is acceptable to player 2, and $S_{\sigma(3)}$ is acceptable to player 3, then the problem is over. If it is not so, then at least one portion is not acceptable to the two choosers. If we want to make any progress in the problem we at least need to assume that the divider is willing to accept any of the three portions of the property and that, if both choosers reject a portion, then the complement of the property is a reasonable share to be divided among the two choosers. Thus they may start all over again with the complementary set for a fair division. The following is a direct generalization of this procedure to the n-player case.

Let S be a set to be partitioned into n subsets. Let $\mathcal{F}_1, \mathcal{F}_2, \ldots, \mathcal{F}_n$ denote collections of subsets of the set S that are acceptable to players $1, 2, \ldots, n$, respectively. Any element of $\mathcal{F}_i$ is called a *fair share* for player i. A *fair division* of S consists of a partition $S_1, S_2, \ldots, S_n$ of S and a permutation σ of $\{1, 2, \ldots, n\}$ such that $S_{\sigma(i)} \in \mathcal{F}_i$, $i = 1, 2, \ldots, n$. In this case player i receives as his fair share the set $\mathcal{F}_i$. For any nonempty coalition $M \subset N = \{1, 2, \ldots, n\}$ with m members, a set $T \subset S$ is said to be acceptable to the members if and only if there exists a partition of T into m subsets and a bijection $\theta : M \rightarrow \{1, 2, \ldots, m\}$ such that $T_{\theta(j)} \in \mathcal{F}_j$, $j \in M$. A set T is called a *fair restriction* to a coalition M if and only if for *some* partition $\mathcal{Q} = (T_{m+1}, T_{m+2}, \ldots, T_n)$ of $S \setminus T$ into $n - m$ sets, none of these sets in the partition are acceptable to any of the players $j \in M$ (i.e., $T_k \notin \mathcal{F}_j$ if $k \geq m + 1$, $j \in M$).

We stipulate the following conditions on $\mathcal{F}_i$, $i = 1, 2, \ldots, n$:

(i) Each player i can partition S into n sets such that all the sets in the particular partition lie in $\mathcal{F}_i$.

(ii) Given a partition $\prod$ of S into n sets, each player can find at least one of the sets acceptable to him.

(iii) Given a coalition M with m members and a fair restriction T, conditions similar to (i) and (ii) hold good on the set T for players in M (i.e., each player $p \in M$ can find a partition $\mathcal{B}^p$ of the set T into m sets such that any of the sets in the partition $\mathcal{B}^p$ is acceptable to player p and for any partition of T into m sets each player in M can find at least one set acceptable to him.)

In this setup we now show that there exists a *fair division scheme*. A rigorous definition of a fair division scheme would require the notion of a cooperative game in extensive form [see Kuhn (1967)]. Intuitively, a division scheme is fair if each player can assure himself a fair share by using an appropriate strategy.

The scheme runs as follows: Let one among the n players be selected at random by a chance mechanism. Without loss of generality, we take him to be player 1. By condition (i), he has a partition $\prod^1$ of S into n sets, where all the sets belong to $\mathcal{F}_1$. We now define a square matrix $A = (a_{ij})$ of order n such that

$$a_{ij} = 1 \quad \text{if} \quad S_j \in \mathcal{F}_i \text{ (that is, } S_j \text{ is acceptable to player i)}$$
$$= 0 \quad \text{if} \quad S_j \notin \mathcal{F}_i \text{ (that is, } S_j \text{ is not acceptable to player i).}$$

Assumptions (i) and (ii) say that

$$a_{11} = a_{12} = \cdots = a_{1n} = 1 \quad \text{and} \quad \sum_{j=1}^{n} a_{ij} \geq 1, \quad i = 1, 2, \ldots, n.$$

There are two cases to discuss.

Case (i): per $A > 0$.

In this case for some permutation σ, $\prod_i a_{i\sigma(i)} > 0$. Since A is a $(0, 1)$-matrix, $a_{i\sigma(i)} = 1$ for all i. Thus $S_{\sigma(i)} \in \mathcal{F}_i, i = 1, 2, \ldots, n$ and we have a fair division.

Case (ii): per $A = 0$.

By the Frobenius-König Theorem there exists an $s \times t$ zero submatrix of A such that $s + t = n + 1$. We can also assume that s is maximal, satisfying the above property. If $s = n - 1$, by reordering the columns if necessary, the matrix A can be brought to the form

$$\begin{bmatrix} 1 & 1 & \cdots & 1 \\ 0 & a_{22} & \cdots & a_{2n} \\ 0 & \cdots & \ddots & \cdots \\ 0 & a_{n2} & \cdots & a_{nn} \end{bmatrix}.$$

Since $a_{i1} = 0$ for $i \neq 1$, S_1 is not acceptable to any of the players $2, 3, \ldots, n$. If S_1 is assigned to player 1, then by assumption (iii), we can use an induction argument to find a partition of $S \setminus S_1$ into $n - 1$ subsets such that players $2, 3, \ldots, n$ get a fair share out of $S \setminus S_1$. We therefore assume that $s \leq n - 2$. Since $s + t = n + 1$, then $t \geq 3$. We can rename the players and rearrange the columns if necessary so that the submatrix formed by the last s rows and the first t columns of A is the zero matrix. Note that the first row of A must consist of all 1s. Let B be the submatrix formed by the first t columns and rows $2, 3, \ldots, t - 1$ of A.

Since s is maximal, the matrix B has no $k \times l$ zero submatrix with $k + l = t$, for in that case, $(k + s) + l = n + 1$ and we would have a $(k + s) \times l$ zero matrix chosen from the rows of B and from the last s rows of A. This contradicts the maximality of s. Now consider the $(t - 1) \times (t - 1)$ submatrix of A formed by rows $1, 2, \ldots, t - 1$ and columns $1, 2, \ldots, t - 1$. Because there is no $k \times l$ zero submatrix of this $(t - 1) \times (t - 1)$ submatrix with $k + l = t$, it has, by the Frobenius-König Theorem, a positive permanent. Thus for some permutation π of $\{1, 2, \ldots, t - 1\}$, $S_{\pi(1)}, S_{\pi(2)}, \ldots, S_{\pi(t-1)}$ is a fair division for players $1, 2, \ldots, t - 1$. Further, $(S_1 \cup S_2 \cup \cdots \cup S_{(t-1)})^c$ is a fair restriction of S to the coalition $\{t, t + 1, \ldots, n\}$. (Because the last s rows in the first t columns in the original matrix are zero, none of the sets $S_1, S_2, \ldots, S_{t-1}$ are acceptable to players $t, t + 1, \ldots, n$.) By assumption (iii), for the coalition $\{t, t + 1, \ldots, n\}$, the set $S_t \cup S_{t+1} \cup \cdots \cup S_n$ is acceptable and we can find a fair division for them by an inductive argument. Since in each attempt at least one person gets his fair share, the procedure terminates in finitely many steps. This demonstrates the existence of a fair division scheme.

Exercises

1. Show that Ω_n may be viewed as a subset of the $(n-1)^2$-dimensional Euclidean space. Now use this observation, the fact that permutation matrices are precisely the extreme points in Ω_n, and Carathéodory's Theorem to show that any $A \in \Omega_n$ can be expressed as a convex combination of at most $n^2 - 2n + 2$ permutation matrices.

2. Let A be a nonnegative $n \times n$ matrix and suppose that the permanent of any $(n-1) \times (n-1)$ submatrix of A is zero. Show that A must have a zero submatrix of order $r \times s$, where $r + s = n + 2$.

3. A *Latin rectangle* of order $m \times n$ $(m < n)$ is an $m \times n$ matrix such that each row of the matrix is a permutation of $1, 2, \ldots, n$ and that in each column all the entries are distinct. Given an $m \times n$ Latin rectangle, define the $n \times n$ matrix A as follows: a_{ij} is 1 if i does not appear in column j and is 0 otherwise. Show that A is a constant multiple of a doubly stochastic matrix. Conclude that any $m \times n$ Latin rectangle can be extended to an $(m+1) \times n$ Latin rectangle by adding a row.

4. Is it true that any symmetric doubly stochastic matrix can be expressed as a convex combination of symmetric permutation matrices?

5. If A is a matrix (or a vector), then the Frobenius norm of A is defined as $||A||_F = \{\text{trace}\,(A^T A)\}^{\frac{1}{2}}$. Let A, B be symmetric $n \times n$ matrices with eigenvalues $\alpha_i, \beta_i, i = 1, 2, \ldots, n$, respectively. Then show that there exist permutations σ, τ of $1, 2, \ldots, n$ such that

$$\sum_{i=1}^n |\alpha_i - \beta_{\sigma(i)}|^2 \leq ||A - B||_F \leq \sum_{i=1}^n |\alpha_i - \beta_{\tau(i)}|^2.$$

[The result also holds when A and B are normal matrices, and is the well-known Hoffman-Wielandt Theorem [see Hoffman and Wielandt (1953)]. The proof uses the Birkhoff–von Neumann Theorem. Elsner (1993) has introduced some additional simplifications, proving in the process a more general result due to Bhatia and Bhattacharyya (1993) for commuting m-tuples of normal matrices. For several related results on spectral variation, see Bhatia (1987).]

6. Let $\mathcal{D}$ be a set of indeterminates and let A be an $n \times n$ matrix over $R \cup \mathcal{D}$, where R is the set of real numbers, such that an element of $\mathcal{D}$ occurs at most once in A. If $|A| = 0$, then show that there exists an $r \times s$ submatrix B such that each entry of B is real, $r + s = n + p$, and rank $B \leq p - 1$.
[This result, due to Hartfiel and Loewy (1984), is a consequence of Theorem 2.3.2.]

7. Let A be an $n \times n$ matrix over an arbitrary field. Show that A is singular if and only if it has a zero type submatrix B with at least $n + 1 - \text{rank}(B)$ lines.

8. Let A be an $m \times n$ matrix. If S, T are subsets of $\{1, 2, \ldots, m\}, \{1, 2, \ldots, n\}$, respectively, then let $\lambda(S, T)$ denote the rank of the submatrix of A determined by rows indexed by S and columns indexed by T. Show that for any $S, U \subset \{1, 2, \ldots, m\}, T, V \subset \{1, 2, \ldots, n\}$,

$$\lambda(S \cap U, T \cup V) + \lambda(S \cup U, T \cap V) \leq \lambda(S, T) + \lambda(U, V).$$

[This property is known as *bisubmodularity*. See Schrijver (1978b) for a proof.]
Deduce Lemma 2.3.1.

9. A ballet based on the Indian epic Ramayana is a last minute addition to a local cultural program for children in Chicago. With very little time available for dance rehearsals, the ballet teacher is tensed up with the five children, Vani, Tara, Tina, Bhuma, and Deepa, to whom the roles of Rama, Sita, Lakshmana, Bharata, and Guha must be assigned. The following matrix $C = (c_{ij})$ gives the rehearsal time, in hours, needed for child i to act in role j.

	Rama	Sita	Lakshmana	Bharata	Guha
Vani	9	3	5	6	4
Tara	7	7	9	7	11
Tina	6	13	10	10	12
Bhuma	4	11	10	9	13
Deepa	8	10	15	17	9

Given the time constraints and the abilities of children for the various roles, solve the problem of minimizing the total rehearsal time.

10. Let $p(x_1, x_2, \ldots, x_n)$ be a polynomial of degree $d \geq 1$ with real coefficients and let S be a set of integers with cardinality $k \geq 1$. Substitute values for the variables $x_1, x_2, \ldots, x_n$, say, $a_1, a_2, \ldots, a_n$, chosen randomly and independently from the set S. Show that the probability that $p(a_1, a_2, \ldots, a_n) = 0$ is at most d/k.

[Hint: First condition with respect to all variables except one; see Schwartz (1980)].

11. Let A be an $n \times n$ (0, 1)-matrix with per $A > 0$. Replace each 1 in A by an integer chosen randomly and independently from $\{1, 2, \ldots, 2n\}$. Show that the resulting matrix will be nonsingular with probability at least 1/2. Discuss the relevance of this result in the problem of deciding whether a given (0, 1)-matrix has a positive diagonal.

12. Let A be a nonnegative $n \times n$ matrix with row sums $r_1, \ldots, r_n$. Show that per $A \leq \prod_{i=1}^{n} r_i$.

13. If A and B are nonnegative $n \times n$ matrices, show that

$$\text{per } AB \geq (\text{per } A)(\text{per } B).$$

14. Let A be a doubly stochastic $n \times n$ matrix and suppose that we want to express A as a convex combination $\sum \alpha_\sigma P^\sigma$ of permutation matrices in such a way that maximizes $\prod_\sigma \alpha_\sigma{}^{\alpha_\sigma}$. Show that this is achieved precisely when α_σ is proportional to $\prod_i b_{i\sigma(i)}$, where $f(B) = A$. (See Section 2.7 for the definition of the map f).

15. Let $0 < \alpha < 1$ and for a nonnegative $n \times n$ matrix A define

$$\text{per}_\alpha A = \sum_\sigma \prod_{i=1}^{n} a_{i\sigma(i)}^\alpha.$$

Show that $\text{per}_\alpha A$ is minimized over the set of doubly stochastic matrices at the matrix J_n. (Hint: Use Holder's Inequality and the van der Waerden–Egorychev–Falikman Inequality.)

16. Let A and B be distinct $n \times n$ doubly stochastic matrices. Show that there exists $\sigma \in S_n$ such that

$$\prod_{i=1}^{n} a_{i\sigma(i)} + \prod_{i=1}^{n} b_{i\sigma(i)} > \frac{2}{n^n}.$$

17. A committee of three experts is to evaluate four candidates who are being considered for a position. The evaluation is to be carried out through a process in which each expert is to interview each candidate separately. The (i, j)-th

entry of the matrix A,

$$A = \begin{bmatrix} 15 & 20 & 45 & 15 \\ 25 & 40 & 30 & 25 \\ 20 & 45 & 15 & 30 \end{bmatrix},$$

indicates the total amount of time (in minutes and possibly in different sessions) for which the i-th expert is to interview the j-th candidate. Find a schedule that allows all the interviews to be completed in a span of two hours.

18. An used car dealer has the following prices for negotiation:

Buick Skylark 1990 model—$4500,

Toyota Camry 1989 model—$6400,

Honda Accord 1991 model—$8300.

Four customers are willing to pay the following maximum prices for the cars:

buyer/model	Buick	Toyota	Honda
Tom	3800	5900	7200
Dick	4000	5800	7500
Harry	4200	6100	8000
Peter	3900	600	7600

Convert this into a cooperative game and find a set of prices reflecting a core element. What is the lexicographic center of your model?

3

Inequalities

Although inequalities form an integral part of practically every branch of mathematics, the theory of matrices is particularly endowed with a rich collection of interesting inequalities. In this chapter we wish to give a flavor of the subject by considering inequalities pertaining to nonnegative matrices and also to positive semidefinite matrices. In the first section we prove some elementary bounds for the Perron root. The second section deals with several applications of a basic inequality derived from the Information Inequality. The next few sections are concerned with some related topics such as inequalities of Levinger, Kingman, and Cohen as well as sum-symmetric matrices and circuit geometric means. We then discuss the Hadamard Inequality and various related results such as the Fischer Inequality, Oppenheim's Inequality, and their refinements based on Schur complements. Some basic properties of the Schur power matrix are discussed in the next section. The last three sections focus on the topics of majorization inequalities for eigenvalues, the parallel sum, and symmetric function means.

3.1. Perron root and row sums

If $A \geq 0$ is an $n \times n$ matrix, then $r(A)$ will denote the spectral radius of A. By the Perron-Frobenius Theorem, $r(A)$ is an eigenvalue of A and there are vectors $x \geq 0$, $y \geq 0$ such that

$$Ay = r(A)y, \qquad x^T A = r(A)x^T.$$

Recall that we refer to $r(A)$ as the Perron root of A, y as a right Perron vector, and x as a left Perron vector. If A is irreducible, then x, y are positive and unique up to a scalar multiple.

In this section we obtain some simple lower and upper bounds for the Perron root. An application of the bounds will be illustrated by an example.

115

Lemma 3.1.1. *Let $A \geq 0$ be an $n \times n$ matrix with row sums $r_1, \ldots, r_n$. Then*

$$\min_{1 \leq i \leq n} r_i \leq r(A) \leq \max_{1 \leq i \leq n} r_i. \tag{3.1.1}$$

Proof. First suppose that A is irreducible and let y be a right Perron vector so that $Ay = r(A)y$. We assume, after a row and column permutation, if necessary, that $r_1 \leq \cdots \leq r_n$; $y_1 \leq \cdots \leq y_n$. Then

$$r(A) = \frac{\sum_{j=1}^{n} a_{1j} y_j}{y_1} \geq r_1$$

and

$$r(A) = \frac{\sum_{j=1}^{n} a_{nj} y_j}{y_n} \leq r_n.$$

The result follows by a continuity argument when A is reducible. ∎

Lemma 3.1.2. *Let $A \geq 0$ be an $n \times n$ matrix and let $z \in R^n$, $z > 0$. Then*

$$\min_{1 \leq i \leq n} \frac{(Az)_i}{z_i} \leq r(A) \leq \max_{1 \leq i \leq n} \frac{(Az)_i}{z_i}.$$

Proof. Define $B = (b_{ij})$ with $b_{ij} = a_{ij} z_j / z_i$ for all i, j. Then $r(A) = r(B)$ and the result is proved by applying Lemma 3.1.1 to the matrix B. ∎

We now illustrate an application of Lemma 3.1.2. Let $\{f_n\}$, $n = 1, 2, \ldots$ denote the *Fibonacci sequence*, defined by $f_1 = f_2 = 1$, $f_i = f_{i-1} + f_{i-2}$, $i = 3, 4, \ldots$. It is well known and easy to prove that

$$\lim_{n \to \infty} \frac{f_n}{f_{n-1}} = \frac{1 + \sqrt{5}}{2},$$

which is the *golden ratio*. We also note that $\left(\frac{1+\sqrt{5}}{2}\right)^{-1} = \frac{\sqrt{5}-1}{2}$.

For any odd positive integer $n = 2m + 1$, let $A_n = (a_{ij}^{(n)})$ denote the $n \times n$ matrix defined as

$$a_{ij}^{(n)} = \begin{cases} 1 & \text{if } |i - j| = 1 \\ 1 & \text{if } i = j = m + 1 \\ 0 & \text{otherwise.} \end{cases}$$

For example, A_3, A_5, and A_7 are, respectively,

$$\begin{bmatrix} 0 & 1 & 0 \\ 1 & 1 & 1 \\ 0 & 1 & 0 \end{bmatrix}, \quad \begin{bmatrix} 0 & 1 & 0 & 0 & 0 \\ 1 & 0 & 1 & 0 & 0 \\ 0 & 1 & 1 & 1 & 0 \\ 0 & 0 & 1 & 0 & 1 \\ 0 & 0 & 0 & 1 & 0 \end{bmatrix}, \quad \text{and} \quad \begin{bmatrix} 0 & 1 & 0 & 0 & 0 & 0 & 0 \\ 1 & 0 & 1 & 0 & 0 & 0 & 0 \\ 0 & 1 & 0 & 1 & 0 & 0 & 0 \\ 0 & 0 & 1 & 1 & 1 & 0 & 0 \\ 0 & 0 & 0 & 1 & 0 & 1 & 0 \\ 0 & 0 & 0 & 0 & 1 & 0 & 1 \\ 0 & 0 & 0 & 0 & 0 & 1 & 0 \end{bmatrix}.$$

This leads us to the following theorem:

Theorem 3.1.3. $r(A_n) \le \sqrt{5}, n = 1, 3, 5, \dots$.

Proof. Let $n = 2m + 1$ be fixed and let $z \in R^n, z > 0$ be such that $z_i = z_{n-i+1}, i = 1, 2, \dots$. Then

$$(A_n z)_i = \begin{cases} z_2 & \text{if } i = 1, n \\ 2z_{m-1} + z_m & \text{if } i = m + 1 \\ z_{i-1} + z_{i+1} & \text{otherwise.} \end{cases}$$

Now fix a positive integer k and let $z_i = f_{k+i}, i = 1, 2, \dots, m + 1$. Then

$$\frac{(A_n z)_i}{z_i} = \begin{cases} f_{k+2}/f_{k+1} & \text{if } i = 1, n \\ 1 + 2f_{k+m-1}/f_{k+m} & \text{if } i = m + 1 \\ 2f_{k+i+1}/f_{k+i} - 1 & \text{otherwise.} \end{cases}$$

Let $k \to \infty$. Then, by Lemma 3.1.2,

$$r(A_n) \le \max \left\{ \frac{1 + \sqrt{5}}{2}, 1 + 2\left(\frac{\sqrt{5} - 1}{2}\right), 2\left(\frac{1 + \sqrt{5}}{2}\right) - 1 \right\}$$
$$= \sqrt{5}.$$

That completes the proof. It is in fact true that $\lim_{n \to \infty} r(A_n) = \sqrt{5}$, and the proof of this is left as an exercise. ∎

For other inequalities related to Lemmas 3.1.1, and 3.1.2, see Varga (1962) and Horn and Johnson (1985). Theorem 3.1.3 is based on an observation in Bapat and Sunder (1991).

3.2. Applications of the Information Inequality

We first introduce some useful notation. Let P^n denote the set of probability vectors of order n. Thus

$$P^n = \left\{ x \in R^n : x \geq 0, \sum_{i=1}^{n} x_i = 1 \right\}.$$

Let $A \geq 0$ be an $m \times n$ nonzero matrix and let $x \in P^m$, $y \in P^n$. It clearly follows that $x^T A y > 0$. We denote by $\tilde{x}$ (respectively, $\tilde{y}$), the normalized row-sum (respectively, column-sum) vector of the matrix $(a_{ij} x_i y_j)$. Formally,

$$\tilde{x}_i = \frac{\sum_{j=1}^{n} a_{ij} x_i y_j}{x^T A y}, \quad i = 1, 2, \ldots, m \tag{3.2.1}$$

$$\tilde{y}_j = \frac{\sum_{i=1}^{m} a_{ij} x_i y_j}{x^T A y}, \quad j = 1, 2, \ldots, n. \tag{3.2.2}$$

Note that the definition of $\tilde{x}$, $\tilde{y}$ depends on the matrix A. However, for simplicity, this dependence is not made explicit in the notation. It is easily seen that $\tilde{x} \in P^m$, $\tilde{y} \in P^n$ for any $x \in P^m$, $y \in P^n$. Let $\phi_A : P^m \times P^n \to P^m \times P^n$ be the map defined by $\phi_A(x, y) = (\tilde{x}, \tilde{y})$. The next result is obvious in view of the definition of $\tilde{x}$, $\tilde{y}$.

Lemma 3.2.1. *Let $A \geq 0$ be an $m \times n$ matrix, and let $(x, y) \in P^m \times P^n$. Then $x_i = 0 \Rightarrow \tilde{x}_i = 0$ and $y_j = 0 \Rightarrow \tilde{y}_j = 0$.*

We can now use the Kronecker Index Theorem (Theorem 2.7.2) as in Chapter 2 to conclude that for any positive $m \times n$ matrix A, ϕ_A maps $P^m \times P^n$ onto itself. One interpretation of the onto property of ϕ_A is that any positive $m \times n$ matrix A can be "scaled" to achieve specified row and column sums. More precisely, if A is a positive $m \times n$ matrix, then there exist diagonal matrices D_1, D_2 with positive entries on the diagonal such that $D_1 A D_2$ has prescribed row and column sums. Compare this with Theorem 2.7.7, which showed that any nonnegative matrix with doubly stochastic pattern can be scaled to obtain a doubly stochastic matrix. More general results of this type will be discussed in Chapter 6.

We now prove a basic inequality that will serve as a useful tool for proving several inequalities. The proof uses the Information Inequality (Lemma 2.6.2). Recall the convention that for any real z, $(z/0)^0 = 1$.

Theorem 3.2.2. *Let $A \geq 0$, $B \geq 0$ be $m \times n$ matrices and let x, y, u, and v be nonnegative vectors such that $x^T A y > 0$. Let $\tilde{x}$, $\tilde{y}$ be as defined in (3.2.1) and*

(3.2.2). Then

$$\frac{u^T B v}{x^T A y} \geq \prod_{i=1}^{m} \left(\frac{u_i}{x_i}\right)^{\tilde{x}_i} \prod_{j=1}^{n} \left(\frac{v_j}{y_j}\right)^{\tilde{y}_j} \prod_{i=1}^{m} \prod_{j=1}^{n} \left(\frac{b_{ij}}{a_{ij}}\right)^{\frac{a_{ij} x_i y_j}{x^T A y}}. \tag{3.2.3}$$

Furthermore, equality holds in (3.2.3) if and only if there exists $\lambda \geq 0$ such that $b_{ij} u_i v_j = \lambda a_{ij} x_i y_j$ for all i, j.

Proof. First suppose that $u^T B v = 0$. Since $x^T A y > 0$, there exist $k \in \{1, 2, \ldots, m\}, \ell \in \{1, 2, \ldots, n\}$ such that $x_k > 0$, $y_\ell > 0$, and $a_{k\ell} > 0$. This implies that $\tilde{x}_k > 0$, $\tilde{y}_\ell > 0$. Since $u^T B v = 0$, then one of the numbers u_k, v_ℓ, or $b_{k\ell}$ must be zero. If $b_{k\ell} = 0$, then

$$\prod_{i=1}^{m} \prod_{j=1}^{n} \left(\frac{b_{ij}}{a_{ij}}\right)^{a_{ij} x_i y_j} = 0$$

and (3.2.3) is true. Similarly, if $u_k = 0$, then $(u_k/x_k)^{\tilde{x}_k} = 0$, whereas if $v_\ell = 0$, then $(v_\ell/y_\ell)^{\tilde{y}_\ell} = 0$ and (3.2.3) holds. So we may assume that $u^T B v > 0$. Let

$$K = \{(i, j) : a_{ij} x_i y_j > 0\}.$$

An application of the Information Inequality yields

$$\prod_{i=1}^{m} \prod_{j=1}^{n} \left\{\frac{b_{ij} u_i v_j}{u^T B v}\right\}^{\frac{a_{ij} x_i y_j}{x^T A y}} \leq \prod_{i=1}^{m} \prod_{j=1}^{n} \left\{\frac{a_{ij} x_i y_j}{x^T A y}\right\}^{\frac{a_{ij} x_i y_j}{x^T A y}}$$

and (3.2.3) follows after some simple algebraic manipulation using the definition of $\tilde{x}_i$, $\tilde{y}_j$.

If there exists $\lambda \geq 0$ such that $b_{ij} u_i v_j = \lambda a_{ij} x_i y_j$ for all i, j, then clearly equality holds in (3.2.3). Conversely, suppose equality holds in (3.2.3). If $u^T B v = 0$, then we may take $\lambda = 0$. If $u^T B v > 0$, then the result follows from the assertion concerning equality in Theorem 2.6.2. ∎

We now use Theorem 3.2.2 to obtain a fairly general inequality for the Perron root, which will then be used to derive some consequences.

Theorem 3.2.3. *Let $A \geq 0$, $B \geq 0$ be $n \times n$ matrices, where A is irreducible, and let x and y be the left and the right Perron vectors of A, satisfying $x^T y = 1$. Then for any vectors $u \geq 0$, $v \geq 0$,*

$$u^T B v \geq r(A) \prod_{i=1}^{n} \left(\frac{u_i v_i}{x_i y_i}\right)^{x_i y_i} \prod_{i,j=1}^{n} \left(\frac{b_{ij}}{a_{ij}}\right)^{\frac{a_{ij} x_i y_j}{r(A)}}. \tag{3.2.4}$$

In particular, if $u_i v_i = x_i y_i$, $i = 1, 2, \ldots, n$, then

$$u^T B v \geq r(A) \prod_{i,j=1}^{n} \left(\frac{b_{ij}}{a_{ij}} \right)^{\frac{a_{ij} x_i y_j}{r(A)}}. \tag{3.2.5}$$

Proof. Since $Ay = r(A)y$, $x^T A = r(A)x^T$, and $x^T y = 1$, it follows that $x^T A y = r(A)$ and that

$$\tilde{x}_i = \frac{\sum_{j=1}^{n} a_{ij} x_i y_j}{x^T A y} = x_i y_i, \quad i = 1, 2, \ldots, n,$$

$$\tilde{y}_j = \frac{\sum_{i=1}^{n} a_{ij} x_i y_j}{x^T A y} = x_j y_j, \quad j = 1, 2, \ldots, n.$$

The result then follows from Theorem 3.2.2. ∎

Suppose $A \geq 0$, $B \geq 0$ are $n \times n$ matrices such that $B \geq A$ entrywise. Then it is easy to show that $r(B) \geq r(A)$. (See Exercise 13 in Chapter 1.) In the next result we show that even without a restriction of the type $B \geq A$, it is possible to compare the Perron roots of A and B in terms of the ratios b_{ij}/a_{ij}.

Theorem 3.2.4. *Let $A \geq 0$, $B \geq 0$ be irreducible $n \times n$ matrices and let x and y be the left and right Perron vectors of A such that $x^T y = 1$. Then*

$$\frac{r(B)}{r(A)} \geq \prod_{i,j=1}^{n} \left(\frac{b_{ij}}{a_{ij}} \right)^{\frac{a_{ij} x_i y_j}{r(A)}}. \tag{3.2.6}$$

Furthermore, equality holds in (3.2.6) if and only if there exist $\lambda > 0$, $\alpha_i > 0$, $i = 1, 2, \ldots, n$ such that

$$b_{ij} = \lambda a_{ij} \frac{\alpha_i}{\alpha_j}, \quad i, j = 1, 2, \ldots, n.$$

Proof. Let v be a right Perron vector of B and set

$$u_i = \frac{x_i y_i}{v_i}, \quad i = 1, 2, \ldots, n.$$

Then $u^T B v = r(B)$ and (3.2.6) follows from (3.2.5). Also, equality holds in (3.2.6) if and only if, for some $\lambda > 0$,

$$b_{ij} u_i v_j = \lambda a_{ij} x_i y_j, \quad i, j = 1, 2, \ldots, n.$$

Thus

$$b_{ij} = \lambda a_{ij} \frac{v_i}{v_j} \cdot \frac{y_j}{y_i}, \quad i, j = 1, 2, \ldots, n.$$

If we set $\alpha_i = v_i/y_i, i = 1, 2, \ldots, n$, then $b_{ij} = \lambda a_{ij}\alpha_i/\alpha_j$ for all i, j and the proof is complete. ∎

Suppose $A \geq 0$ is irreducible with left and right Perron vectors x, y satisfying $x^T y = 1$. Then clearly $x^T A y = r(A)$. It is a rather remarkable and important property that if we "flip" the Perron vectors, we get an expression at least as big as $r(A)$; more specifically, it is true that $y^T A x \geq r(A)$. This result is contained in the following.

Theorem 3.2.5. *Let $A \geq 0$ be an irreducible $n \times n$ matrix and let x, y be the left and the right Perron vectors of A satisfying $x^T y = 1$. Then for any vectors $u > 0, v > 0$ satisfying $u_i v_i = x_i y_i, i = 1, 2, \ldots, n$ it is true that $u^T A v \geq r(A)$. In particular $y^T A x \geq r(A)$.*

Proof. The result follows by setting $B = A, u = y$, and $v = x$ in (3.2.5). ∎

The map defined in (3.2.1) and (3.2.2) is important in mathematical genetics. An application is given in Exercise 1, which has been interpreted as the discrete form of the fundamental theorem of natural selection; see Karlin (1984) and Bapat (1986a). Theorems 3.2.3 and 3.2.4 appear in Bapat (1987a, 1989a) [also see Fiedler et al (1985), Horn and Johnson (1991), and Section 5.7].

3.3. Inequalities of Levinger and Kingman

The next few results are concerned with an inequality due to Levinger, which essentially says that for any $A \geq 0$, the Perron root, considered as a function along the line segment joining A and A^T, is concave. We first prove a more general result.

Theorem 3.3.1. *Let $A \geq 0, B \geq 0$ be irreducible $n \times n$ matrices such that they have a common right Perron vector y and a common left Perron vector x. Then for any $\tau \in [0, 1]$,*

$$r(\tau A + (1 - \tau)B^T) \geq \tau r(A) + (1 - \tau)r(B). \tag{3.3.1}$$

Furthermore, if $0 < \tau < 1$, then equality holds in (3.3.1) if and only if x, y are linearly dependent.

Proof. We assume, after normalizing if necessary, that $x^T y = 1$. The inequality (3.3.1) is obvious if $\tau = 0, 1$. So suppose $0 < \tau < 1$. Let μ be a right Perron vector of $\tau A + (1-\tau)B^T$ and define the vector λ as $\lambda_i = x_i y_i / \mu_i$, $i = 1, 2, \ldots, n$. An application of (3.2.5) gives the following inequalities

$$\sum_{i,j=1}^{n} a_{ij}\lambda_i\mu_j \geq r(A) \qquad (3.3.2)$$

$$\sum_{i,j=1}^{n} b_{ij}\mu_i\lambda_j \geq r(B), \qquad (3.3.3)$$

which yields

$$\sum_{i,j=1}^{n} (\tau a_{ij} + (1-\tau)b_{ji})\lambda_i\mu_j \geq \tau r(A) + (1-\tau)r(B). \qquad (3.3.4)$$

Since μ is a right Perron vector of $\tau A + (1-\tau)B^T$, (3.3.1) follows from (3.3.4). Equality occurs in (3.3.1) if it occurs in both (3.3.2) and (3.3.3). Referring back to the equality assertion in Theorem 3.2.2 and interpreting it for this special case we find that this occurs if and only if, for some positive α, β,

$$a_{ij}\lambda_i\mu_j = \alpha a_{ij}x_i y_j, \qquad b_{ij}\mu_i\lambda_j = \beta b_{ij}x_i y_j, \qquad i, j = 1, 2, \ldots, n.$$

Since $\lambda_i = x_i y_i / \mu_i$, $i = 1, 2, \ldots, n$, we have

$$a_{ij}\frac{y_i}{y_j} = \alpha a_{ij}\frac{\mu_i}{\mu_j}, \qquad b_{ij}\frac{x_j}{x_i} = \beta b_{ij}\frac{\mu_j}{\mu_i}, \qquad i, j = 1, 2, \ldots, n. \qquad (3.3.5)$$

Since A is irreducible, for any i, j there exist $i = i_1, i_2, \ldots, i_k = j$ such that

$$a_{i_1 i_2}a_{i_2 i_3} \cdots a_{i_{k-1} i_k}\frac{y_i}{y_j} = \alpha^{k-1}a_{i_1 i_2}a_{i_2 i_3} \cdots a_{i_{k-1} i_k}\frac{\mu_i}{\mu_j}$$

and hence

$$\frac{y_i}{y_j} = \alpha^{k-1}\frac{\mu_i}{\mu_j},$$

where k, of course, depends on i, j. Using this fact for all i, j, it is easy to conclude that $\alpha = 1$ and that y and μ are linearly dependent. Similarly, x and μ are linearly dependent and, hence, so are x and y. Conversely, if x and y are linearly dependent, then clearly equality holds in (3.3.1) and the proof is complete.   ∎

Corollary 3.3.2. *If $A \geq 0$ is an irreducible $n \times n$ matrix, then for any $\tau \in [0, 1]$,*

$$r(\tau A + (1-\tau)A^T) \geq r(A). \qquad (3.3.6)$$

Furthermore, if $0 < \tau < 1$, then equality holds in (3.3.6) if and only if any right Perron vector of A is also a left Perron vector.

Proof. The result follows by setting $B = A$ in Theorem 3.3.1.   ■

We now prove Levinger's Inequality.

Theorem 3.3.3. *Let $A \geq 0$ be an irreducible $n \times n$ matrix. Then the function*
$$\phi(t) = r(tA + (1 - t)A^T)$$
is either constant in $[0, 1]$ or it is increasing in $(0, \frac{1}{2})$ and decreasing in $(\frac{1}{2}, 1)$. Furthermore, it is constant in $[0, 1]$ if and only if A and A^T have a common right Perron vector.

Proof. Let $0 < t_1 < t_2 < \frac{1}{2}$. Then
$$\phi(t_2) = r(t_2 A + (1 - t_2)A^T)$$
$$= r\{\alpha(t_1 A + (1 - t_1)A^T) + (1 - \alpha)(t_1 A^T + (1 - t_1)A)\},$$
where
$$\alpha = \frac{1 - t_1 - t_2}{1 - 2t_1}.$$
Since $t_1 A + (1 - t_1)A^T$ is irreducible, it follows by Corollary 3.3.2 that $\phi(t_2) \geq \phi(t_1)$. If $\phi(t_2) = \phi(t_1)$, then again by Corollary 3.3.2, A and A^T have a common right Perron vector and then
$$r(tA + (1 - t)A^T) = r(A), \quad 0 \leq t \leq 1,$$
so that ϕ is constant in $[0, 1]$. Thus, either ϕ is constant in $[0, 1]$ or it is (strictly) increasing in $(0, \frac{1}{2})$. The interval $(\frac{1}{2}, 1)$ can be handled similarly, and the proof is complete.   ■

We now proceed to prove an inequality due to Kingman. We first recall that the Schur product of $A = (a_{ij})$, $B = (b_{ij})$ is defined as $A \circ B = (a_{ij}b_{ij})$. If $A \geq 0$ and $\alpha \geq 0$, then $A^{(\alpha)}$ is defined to be the matrix (a_{ij}^α).

Theorem 3.3.4. *Let $A \geq 0$, $B \geq 0$ be $n \times n$ matrices and let $\alpha \in [0, 1]$. Then*
$$r(A^{(\alpha)} \circ B^{(1-\alpha)}) \leq r(A)^\alpha r(B)^{1-\alpha}.$$

Proof. We assume that A and B are irreducible and the general case will then follow by a continuity argument. Let
$$C = A^{(\alpha)} \circ B^{(1-\alpha)}$$
and let u and v be right Perron vectors of A and B, respectively. Define
$$z = u^{(\alpha)} \circ v^{(1-\alpha)}.$$

Then, by Holder's inequality,

$$\sum_{j=1}^{n} c_{ij} z_j = \sum_{j=1}^{n} (a_{ij} u_j)^{\alpha} (b_{ij} v_j)^{1-\alpha}$$

$$\leq \left(\sum_{j=1}^{n} a_{ij} u_j \right)^{\alpha} \left(\sum_{j=1}^{n} b_{ij} v_j \right)^{1-\alpha}$$

$$= r(A)^{\alpha} r(B)^{1-\alpha} z_i, \quad i = 1, 2, \ldots, n.$$

Consequently, by Lemma 3.1.2,

$$r(A)^{\alpha} r(B)^{1-\alpha} \geq \frac{\sum_{j=1}^{n} c_{ij} z_j}{z_i}, \quad i = 1, 2, \ldots, n$$

$$\geq \max_{1 \leq i \leq n} \frac{\sum_{j=1}^{n} c_{ij} z_j}{z_i}$$

$$\geq r(C),$$

and the proof is complete.   ∎

Theorem 3.3.3 was announced in Levinger (1970). See Marek (1984) for a generalization of the result to cone preserving maps. Theorem 3.3.4 is from Kingman (1961). Karlin and Ost (1985) and Elsner and Johnson (1989) provide several related results and applications.

3.4. Sum-symmetric matrices

We begin with some definitions. A nonzero $n \times n$ matrix $A \geq 0$ is said to be *sum-symmetric* if any row sum of A is equal to the corresponding column sum, i.e., if

$$\sum_{k=1}^{n} a_{ik} = \sum_{k=1}^{n} a_{ki}, \quad i = 1, 2, \ldots, n.$$

Clearly, any doubly stochastic matrix provides a simple example of a sum-symmetric matrix. In this section we study the combinatorial structure of sum-symmetric matrices and obtain certain inequalities.

An $n \times n$ (0, 1)-matrix A is said to be a *circuit matrix* if there are distinct integers $i_1, i_2, \ldots, i_k$ in $\{1, 2, \ldots, n\}$ such that A has 1s at positions

$$(i_1, i_2), (i_2, i_3), \ldots, (i_{k-1}, i_k), (i_k, i_1)$$

and zeros elsewhere.

Here are some 3×3 circuit matrices:

$$\begin{bmatrix} 0 & 0 & 0 \\ 0 & 1 & 0 \\ 0 & 0 & 0 \end{bmatrix}, \quad \begin{bmatrix} 0 & 0 & 1 \\ 0 & 0 & 0 \\ 1 & 0 & 0 \end{bmatrix}, \quad \begin{bmatrix} 0 & 1 & 0 \\ 0 & 0 & 1 \\ 1 & 0 & 0 \end{bmatrix}.$$

It is obvious that a circuit matrix is sum-symmetric.

An $n \times n$ matrix $A \geq 0$ is said to be *completely reducible* if it is irreducible or if there exists a permutation matrix P such that

$$P^T A P = \begin{bmatrix} A_{11} & 0 & \cdots & 0 \\ 0 & A_{22} & \cdots & 0 \\ \vdots & \vdots & \ddots & \vdots \\ 0 & 0 & \cdots & A_{kk} \end{bmatrix},$$

where the A_{ii} are square, irreducible matrices, $i = 1, 2, \ldots, k$.

Lemma 3.4.1. *If A is sum-symmetric, then it is completely reducible.*

Proof. If A is irreducible, there is nothing to prove. So suppose that A is reducible, in which case there exists a permutation matrix P such that

$$P^T A P = \begin{bmatrix} B & 0 \\ C & D \end{bmatrix},$$

where B and D are square matrices. Suppose B is $r \times r$. Since A is sum-symmetric, the sum of the elements in the first r rows of A equals the sum of the elements in the first r columns and hence $C = 0$. Now B and D must both be sum-symmetric, and the proof can be completed by an induction argument. ∎

In the combinatorial theory of doubly stochastic matrices, the permutation matrices play an important role. A similar role is played by the circuit matrices in the case of sum-symmetric matrices.

Lemma 3.4.2. *Let A be an $n \times n$ sum-symmetric matrix. Then there exists an $n \times n$ circuit matrix B such that for any i, j, $b_{ij} = 1 \Rightarrow a_{ij} > 0$.*

Proof. The result is obvious for $n = 1$. If $n = 2$ and if a_{11} or a_{22} is positive, the result is true. If $a_{11} = a_{22} = 0$, then, since A is sum-symmetric, $a_{12} > 0$, $a_{21} > 0$ and we may take

$$B = \begin{bmatrix} 0 & 1 \\ 1 & 0 \end{bmatrix}.$$

Suppose the result is true for matrices of order less than or equal to n and let A be $n \times n$. If A is irreducible, then $G(A)$ is strongly connected (see Section 1.1). In the terminology of Markov chains, we would say that any state has access to any other state. In particular, state 1 has access to itself and thus there must exist distinct positive integers $i_1, \ldots, i_k$ such that the (i_t, i_{t+1}) entry of A is positive for each t, where $t = 1, 2, \ldots, k$, where $i_{k+1} = i_1$. Thus B can be defined as the circuit matrix with 1s precisely at positions $(i_t, i_{t+1}), t = 1, 2, \ldots, k$. If A is reducible, then by Lemma 3.4.1, A is completely reducible and hence it has a principal submatrix that is irreducible. Now the proof is completed by induction. ∎

The proof of the next result may be compared with that of the Birkhoff–von Neumann Theorem (Theorem 2.1.6).

Lemma 3.4.3. *Let A be an $n \times n$ sum-symmetric matrix. Then A can be expressed as a nonnegative linear combination of circuit matrices.*

Proof. By Lemma 3.4.2 there exists a circuit matrix B such that $a_{ij} > 0$ whenever $b_{ij} = 1$. Let

$$\alpha = \min_{i,j: b_{ij}=1} a_{ij}$$

and let $C = A - \alpha B$. Then C is sum-symmetric, $a_{ij} > 0$ whenever $c_{ij} > 0$, and C has at least one more zero entry than A. The same argument can be repeated with C instead of A and hence the result is proved. ∎

Example 3.1. Suppose A is the sum-symmetric matrix given by

$$A = \begin{bmatrix} 1 & 2 & 1 \\ 1 & 0 & 3 \\ 2 & 2 & 1 \end{bmatrix}.$$

Then we can write

$$A = B_1 + B_2 + 2B_3 + B_4 + B_5,$$

where

$$B_1 = \begin{bmatrix} 1 & 0 & 0 \\ 0 & 0 & 0 \\ 0 & 0 & 0 \end{bmatrix}, \quad B_2 = \begin{bmatrix} 0 & 0 & 0 \\ 0 & 0 & 0 \\ 0 & 0 & 1 \end{bmatrix}, \quad B_3 = \begin{bmatrix} 0 & 1 & 0 \\ 0 & 0 & 1 \\ 1 & 0 & 0 \end{bmatrix}$$

$$B_4 = \begin{bmatrix} 0 & 0 & 0 \\ 0 & 0 & 1 \\ 0 & 1 & 0 \end{bmatrix}, \quad B_5 = \begin{bmatrix} 0 & 0 & 1 \\ 1 & 0 & 0 \\ 0 & 1 & 0 \end{bmatrix}.$$

In Sections 2.7 and 3.2 we discussed the problem of scaling a nonnegative matrix to a doubly stochastic matrix or to a matrix with prescribed row and column sums. It turns out that a similar theory can be developed if we consider scalings of the type DAD^{-1}, which are sum-symmetric, where D is a diagonal matrix with positive diagonal entries. We now obtain a scaling result of this type. We first need some preliminaries. We denote by R_+^n the set of positive vectors in R^n.

Theorem 3.4.4. *Let $A \geq 0$ be an irreducible matrix and let*

$$f(x) = \sum_{i,j=1}^{n} a_{ij} \frac{x_i}{x_j}, \quad x \in R_+^n.$$

Then f attains a minimum over R_+^n.

Proof. Let

$$\mathcal{U} = \left\{ x \in R_+^n : \sum_{i=1}^{n} x_i = 1 \right\}.$$

Let e be the vector in R^n with all entries equal to 1 and let α be the least positive entry of A. If $n = 1$, the result is trivial, so suppose $n > 1$. Let

$$\delta = \frac{1}{n} \left(\frac{\alpha}{f(e)} \right)^n$$

and let

$$\mathcal{U}_\delta = \left\{ x \in \mathcal{U}; \min_{1 \leq i \leq n} x_i \geq \delta \right\}.$$

Let $x \in \mathcal{U}$. There exist $p, k \in \{1, 2, \ldots, n\}$ such that $x_p \geq \frac{1}{n}$ and $x_k = \min_{1 \leq i \leq n} x_i$. In particular, $x_p \geq x_k$. Since A is irreducible, there exist distinct integers $i_0, i_1, \ldots, i_{q-1}, i_q$, where $i_0 = p, i_q = k$ such that $a_{i_t i_{t+1}} > 0, t = 0, 1, \ldots, q - 1$.

By the arithmetic mean-geometric mean inequality,

$$\sum_{t=0}^{q-1} a_{i_t i_{t+1}} \frac{x_{i_t}}{x_{i_{t+1}}} \geq q \left\{ \prod_{t=0}^{q-1} a_{i_t i_{t+1}} \frac{x_{i_t}}{x_{i_{t+1}}} \right\}^{1/q}$$

$$\geq q\alpha \left(\frac{x_p}{x_k} \right)^{1/q}$$

$$\geq q\alpha \left(\frac{x_p}{x_k} \right)^{1/n}$$

$$\geq q\alpha \left(\frac{1}{nx_k} \right)^{1/n}. \tag{3.4.1}$$

It follows from (3.4.1) that

$$f(x) \geq \frac{1}{q} f(x) \geq \alpha \left(\frac{1}{nx_k} \right)^{1/n}. \tag{3.4.2}$$

Now let $x \in \mathcal{U}$, $x \notin \mathcal{U}_\delta$. Then

$$\min_{1 \leq i \leq n} x_i < \delta,$$

and from (3.4.2),

$$f(x) \geq \alpha \left(\frac{1}{n\delta} \right)^{1/n} = f(e) = f\left(\frac{1}{n} e \right). \tag{3.4.3}$$

Since A is irreducible, it has no zero row and hence $f(e) \geq n\alpha$. Therefore, $\delta \leq \frac{1}{n}$ and hence $\frac{1}{n} e \in \mathcal{U}_\delta$. This observation and (3.4.3) imply that

$$\inf_{x \in \mathcal{U}_\delta} f(x) = \inf_{x \in \mathcal{U}} f(x). \tag{3.4.4}$$

Also, because $f(\lambda x) = f(x)$ for any $\lambda > 0$, we have

$$\inf_{x \in \mathcal{U}} f(x) = \inf_{x \in R_+^n} f(x). \tag{3.4.5}$$

Since f is continuous and $\mathcal{U}_\delta$ is compact, the result follows from (3.4.4) and (3.4.5).   ■

Theorem 3.4.5. *Let $A \geq 0$ be an $n \times n$ matrix. Then the following conditions are equivalent:*

(i) A is completely reducible.
(ii) $f(x) = \sum_{i,j=1}^n a_{ij} \frac{x_i}{x_j}$ attains a minimum over R_+^n.
(iii) There exists $z \in R_+^n$ such that $(a_{ij} \frac{z_i}{z_j})$ is sum-symmetric.

Proof. (i) $\Rightarrow$ (ii): By Theorem 3.4.4, if A is irreducible, then f attains a minimum over R_+^n. The result follows in view of the definition of complete reducibility.

(ii) $\Rightarrow$ (iii): Suppose f attains a minimum over R_+^n at z. Since f is differentiable over R_+^n, the gradient of f much vanish at z, i.e.,

$$\frac{\partial f}{\partial x_k}(z) = 0, \quad k = 1, 2, \ldots, n.$$

A simple calculation shows that

$$\frac{\partial f}{\partial x_k} = \sum_{\substack{j=1 \\ j \neq k}}^n \frac{a_{kj}}{x_j} - \sum_{\substack{i=1 \\ i \neq k}}^n a_{ik} \frac{x_i}{x_k^2}, \quad k = 1, 2, \ldots, n$$

and hence

$$\sum_{j=1}^{n} a_{kj} \frac{z_k}{z_j} = \sum_{i=1}^{n} a_{ik} \frac{z_i}{z_k}, \quad k = 1, \ldots, n.$$

Thus $(a_{ij} \frac{z_i}{z_j})$ is sum-symmetric.

$(iii) \Rightarrow (i)$: It follows from Lemma 3.4.1 that $(a_{ij} \frac{z_i}{z_j})$, and hence A, must be completely reducible. ∎

We now obtain certain inequalities involving sum-symmetric matrices.

Lemma 3.4.6. *Let $A \geq 0$ be an $n \times n$ matrix and suppose $x \in R_+^n$ such that $(a_{ij} \frac{x_i}{x_j})$ is sum-symmetric. Then for any $u \in R_+^n$,*

$$\sum_{i,j=1}^{n} a_{ij} \frac{u_i}{u_j} \geq \sum_{i,j=1}^{n} a_{ij} \frac{x_i}{x_j}.$$

Proof. Make the following substitutions in Theorem 3.2.2: $B = A$, $y_i = \frac{1}{x_i}$, $v_i = \frac{1}{u_i}$, $i = 1, 2, \ldots, n$. Since $(a_{ij} \frac{x_i}{x_j})$ is sum-symmetric, it follows that $\tilde{x}_i = \tilde{y}_i$, $i = 1, 2, \ldots, n$, and the result follows from (3.2.2). ∎

Corollary 3.4.7. *If A is an $n \times n$ sum-symmetric matrix, then for any $u \in R_+^n$,*

$$\sum_{i,j=1}^{n} a_{ij} \frac{u_i}{u_j} \geq \sum_{i,j=1}^{n} a_{ij}.$$

Proof. The result is obtained by setting $x_i = 1$, $i = 1, 2, \ldots, n$ in Lemma 3.4.6. ∎

If $A \geq 0$ is an $n \times n$ completely reducible matrix, then from (i) and (ii) of Theorem 3.4.5, there exists a diagonal matrix D with positive diagonal elements $x_1, \ldots, x_n$ such that DAD^{-1} is sum-symmetric. We now consider the question of the uniqueness of the scaling factors $x_1, \ldots, x_n$.

Lemma 3.4.8. *Let $A \geq 0$ be an irreducible $n \times n$ matrix. Then there exist positive numbers $z_1, \ldots, z_n$, which are unique up to a scalar multiple, such that $(a_{ij} \frac{z_i}{z_j})$ is sum-symmetric.*

Proof. The existence of $z_1, \ldots, z_n$ follows from Theorem 3.4.5. Suppose there are positive numbers $u_1, \ldots, u_n$ such that $(a_{ij} \frac{u_i}{u_j})$ is sum-symmetric. By

Lemma 3.4.6,

$$\sum_{i,j=1}^{n} a_{ij}\frac{z_i}{z_j} \geq \sum_{i,j=1}^{n} a_{ij}\frac{u_i}{u_j} \geq \sum_{i,j=1}^{n} a_{ij}\frac{z_i}{z_j}.$$

Equality must hold in the above inequalities and as in Theorems 3.2.2, and 3.3.1 it follows that there exists $\lambda > 0$ such that

$$a_{ij}\frac{z_i}{z_j} = \lambda a_{ij}\frac{u_i}{u_j}, \quad i, j = 1, 2, \ldots, n.$$

Hence

$$\frac{z_i}{z_j} = \frac{u_i}{u_j} \quad \text{if} \quad a_{ij} > 0. \tag{3.4.6}$$

Fix $k, \ell \in \{1, 2, \ldots, n\}, k \neq \ell$. Since A is irreducible there exist distinct integers $i_0 = k, i_1, \ldots, i_q = \ell$ such that for $t = 0, 1, \ldots, q - 1$, the (i_t, i_{t+1}) entry of A is positive. Thus from (3.4.6),

$$\prod_{t=0}^{q-1} \frac{z_{i_t}}{z_{i_{t+1}}} = \prod_{t=0}^{q-1} \frac{u_{i_t}}{u_{i_{t+1}}},$$

which simplifies to $z_k/z_\ell = u_k/u_\ell$. Thus (3.4.6) is satisfied for all i, j. Letting $\mu = z_1/u_1$ so that we have $z_i = \mu u_i, i = 1, 2, \ldots, n$ completes the proof. $\blacksquare$

If A is completely reducible, then by Theorem 3.4.5 there exist positive numbers $z_1, \ldots, z_n$ such that $(a_{ij}\frac{z_i}{z_j})$ is sum-symmetric. If A is reducible then we can reduce A (after a permutation similarity) to a direct sum of irreducible matrices. By Lemma 3.4.8 we can assert that the $z_i s$, which correspond to an irreducible principal block of A, are unique up to a scalar multiple.

The material in this section is mainly based on the material in Eaves et al (1985), although some of the proofs are different.

3.5. Circuit geometric means

Let $A \geq 0$ be an $n \times n$ matrix and let $i_1, \ldots, i_k$ be distinct integers in $\{1, 2, \ldots, n\}$. The product

$$a_{i_1 i_2} a_{i_2 i_3} \cdots a_{i_{k-1} i_k} a_{i_k i_1}$$

is called a *circuit product* of A of length k and the k-th root of the product is called a *circuit geometric mean* of A. The maximum and the minimum of all circuit geometric means of A are denoted by $g_{\max}(A)$ and $g_{\min}(A)$, respectively. Equivalently, a circuit product of A is a product of the form $\prod_{i,j=1}^{n} a_{ij}^{b_{ij}}$, where B is a circuit matrix.

For any positive integer n, we define C_n as the class of all $n \times n$ sum-symmetric matrices A such that $\sum_{i,j=1}^{n} a_{ij} = 1$. Then C_n is clearly a compact, convex set.

Lemma 3.5.1. *The extreme points of C_n are precisely matrices of the form $\frac{1}{k}Q$, where k is a positive integer and Q is a circuit matrix with k ones.*

Proof. Suppose $A = \frac{1}{k}Q$, where k is a positive integer and Q is a circuit matrix with precisely k ones. If $A = \frac{1}{2}(B + C)$, where $B, C \in C_n$, then it must be true that $q_{ij} = 0 \Rightarrow b_{ij} = c_{ij} = 0$. However, since B and C are also sum-symmetric it follows that $B = C = A$, and hence A is an extreme point of C_n. The converse follows from Lemma 3.4.3, and the proof is complete.   ∎

Lemma 3.5.2. *If $C \in C_n$, then for any $n \times n$ matrix $A \geq 0$,*

$$g_{\min}(A) \leq \prod_{i,j=1}^{n} a_{ij}^{c_{ij}} \leq g_{\max}(A).$$

Proof. First suppose that $A > 0$. The general case can then be obtained by a continuity argument. Let

$$f(C) = \sum_{i,j=1}^{n} c_{ij} \log a_{ij}$$

be a real-valued function defined on C_n. Then by Lemma 3.5.1, f attains its maximum and minimum on C_n at matrices of the form $\frac{1}{k}Q$, where k is a positive integer and Q is a circuit matrix with k ones. Since

$$\prod_{i,j=1}^{n} a_{ij}^{\frac{1}{k}q_{ij}}$$

is a circuit geometric mean of A, the result follows.   ∎

Lemma 3.5.3. *If $A \geq 0$ is an $n \times n$ matrix, then $g_{\max}(A) \leq r(A)$.*

Proof. Let k be a positive integer. The (i, i)-th entry of A^k is given by

$$\sum a_{i_1 i_2} a_{i_2 i_3} \cdots a_{i_{k-1} i_k} a_{i_k i_1},$$

where the sum is over all tuples $\{i_1, \ldots, i_k\}$ of integers from $\{1, 2, \ldots, n\}$ such that $i_1 = i$. In particular, it follows that the maximum circuit product of A of length k is less than or equal to the largest diagonal entry of A^k. Since the

Perron root of a nonnegative matrix is greater than or equal to any of its diagonal entries, we have

$$[r(A^k)]^{1/k} \geq g_{\max}(A).$$

However, since $r(A^k) = r(A)^k$, the result is proved. ∎

Lemma 3.5.4. *If $B \geq 0, C \geq 0$ are $n \times n$ matrices, then*

$$r(B \circ C) \leq r(B)g_{\max}(C) \leq r(B)r(C).$$

Proof. We give a proof when $B > 0, C > 0$; the general case will follow by a continuity argument. Let $A = B \circ C$. Let x and y be the left and the right Perron vectors of A such that $x^T y = 1$. By Theorem 3.2.4,

$$r(B \circ C) \leq r(B) \prod_{i,j=1}^{n} c_{ij}^{\frac{a_{ij}x_iy_j}{r(A)}}. \tag{3.5.1}$$

Clearly, the matrix $(\frac{a_{ij}x_iy_j}{r(A)})$ is in C_n and hence, by Lemma 3.5.2 and Equation (3.5.1),

$$r(B \circ C) \leq r(B)g_{\max}(C).$$

The second inequality of the lemma follows by Lemma 3.5.3, and the proof is complete. ∎

Corollary 3.5.5. *If $A \geq 0, B \geq 0$ are $n \times n$ matrices, then $r(A \circ B) \leq r(A)r(B)$.*

If $A \geq 0$ and $s > 0$, recall that the matrix $A^{(s)}$ is defined as $A^{(s)} = (a_{ij}^s)$.

Theorem 3.5.6. *Let $A \geq 0$ be an $n \times n$ matrix. Then the following assertions hold:*

(i) *If $0 < s < t$, then $[r(A^{(s)})]^{1/s} \geq [r(A^{(t)})]^{1/t}$.*
(ii) $\lim_{s \to \infty}[r(A^{(s)})]^{1/s} = g_{\max}(A).$

Proof. (i). Let $0 < s < t$. Then $A^{(t)} = A^{(s)} \circ A^{(t-s)}$. From Lemma 3.5.4 we conclude that

$$r(A^{(t)}) \leq r(A^{(s)})g_{\max}(A^{(t-s)}).$$

Thus

$$[r(A^{(t)})]^{1/t} \leq [r(A^{(s)})]^{1/s}[r(A^{(s)})]^{\frac{1}{t}-\frac{1}{s}}\{g_{\max}(A^{(t-s)})\}^{1/t}.$$

The result will follow if we show that

$$g_{\max}(A^{(t-s)}) \leq [r(A^{(s)})]^{\frac{t-s}{s}}.$$

Since

$$r(A^{(s)}) \geq g_{\max}(A^{(s)}) = \{g_{\max}(A^{(t-s)})\}^{\frac{s}{t-s}},$$

the proof is complete.

(ii). Let E_n be the $n \times n$ matrix with all entries equal to 1. Then by Lemma 3.5.4,

$$\begin{aligned}
r(A^{(s)}) &= r(A^{(s)} \circ E_n) \\
&\leq r(E_n) g_{\max}(A^{(s)}) \\
&= n \{g_{\max}(A)\}^s.
\end{aligned}$$

Hence

$$[r(A^{(s)})]^{1/s} \leq n^{1/s} g_{\max}(A). \tag{3.5.2}$$

Also, by Lemma 3.5.4,

$$r(A^{(s)}) \geq g_{\max}(A^{(s)}) = \{g_{\max}(A)\}^s$$

and hence

$$[r(A^{(s)})]^{1/s} \geq g_{\max}(A). \tag{3.5.3}$$

It follows from (3.5.2) and (3.5.3) that

$$\lim_{s \to \infty} [r(A^{(s)})]^{1/s} = g_{\max}(A),$$

and the proof is complete.   ∎

The next few results are centered around an interesting result of Cohen (1979), which essentially says that for a matrix $A \geq 0$, the Perron root $r(A)$, considered as a function of the main diagonal entries of A, is convex.

Lemma 3.5.7. *Let $A \geq 0$ be an irreducible $n \times n$ matrix and let x, y be positive vectors such that $x^T y = 1$ and that $(a_{ij} x_i y_j)$ is sum-symmetric. Then $r(A) \geq x^T A y$.*

Proof. In Theorem 3.2.2, let $B = A$, let v be a right Perron vector of A, and let u be defined as $u_i = x_i y_i / v_i$, $i = 1, 2, \ldots, n$. Because $(a_{ij} x_i y_j)$ is sum-symmetric, $\tilde{x}_i = \tilde{y}_i$, $i = 1, 2, \ldots, n$. The result then follows from (3.2.2)   ∎

Lemma 3.5.8. *Let $A \geq 0$, $B \geq 0$ be irreducible $n \times n$ matrices and let x and y be the left and the right Perron vectors of $A + B$ such that $x^T y = 1$. If $(a_{ij} x_i y_j)$ is sum-symmetric, then $r(A + B) \leq r(A) + r(B)$.*

Proof. Note that $((a_{ij} + b_{ij})x_i y_j)$ must be sum-symmetric. Since $(a_{ij}x_i y_j)$ is sum-symmetric, so is $(b_{ij}x_i y_j)$. Now by Lemma 3.5.7, $r(A) \geq x^T Ay, r(B) \geq x^T By$, and hence

$$r(A) + r(B) \geq x^T Ay + x^T By$$
$$= x^T (A + B)y$$
$$= r(A + B).$$

That completes the proof. ∎

We now prove Cohen's result.

Theorem 3.5.9. *Let $A \geq 0$ be an irreducible $n \times n$ matrix. Let ϕ be defined on the set of nonnegative diagonal matrices D as $\phi(D) = r(A + D)$. Then ϕ is a convex function.*

Proof. Let $D \geq 0, E \geq 0$ be $n \times n$ diagonal matrices and let $\alpha \in [0, 1]$. Let

$$C = \alpha(A + D) + (1 - \alpha)(A + E),$$

and let x and y be the left and the right Perron vectors of C such that $x^T y = 1$. Then $(c_{ij}x_i y_j)$ is sum-symmetric. Since C and $\alpha(A+D)$ differ possibly in their diagonal elements only, it follows that $(\alpha(a_{ij}+d_{ij})x_i y_j)$ is also sum-symmetric. Hence by Lemma 3.5.8 we have

$$r(C) \leq \alpha r(A + D) + (1 - \alpha)r(A + E).$$

Thus ϕ is convex and the proof is complete. ∎

For Theorem 3.5.6, see Karlin and Ost (1985) and Friedland (1986). Theorem 3.5.9 appears in Cohen (1979) [also see Cohen (1981)]. For alternative proofs of the result, see Deutsch and Neumann (1984), Friedland (1981), and Elsner (1984). Sum-symmetric matrices were also considered by Afriat (1974).

3.6. The Hadamard Inequality

In the remaining sections of this chapter we will be concerned with positive definite and positive semidefinite matrices. We first introduce some definitions and notation. Recall that an $n \times n$ real matrix A is said to be positive definite if it is symmetric and $x^T Ax > 0$ for any $x \in R^n$, $x \neq 0$. The matrix A is *positive semidefinite* if it is symmetric and $x^T Ax \geq 0$ for any $x \in R^n$. The notation $A \succeq B$ means that A, B, and $A - B$ are all positive semidefinite. The ordering $\succeq$ is known as the *Löwner ordering*.

If A is an $n \times n$ matrix and if $S \subset \{1, 2, \ldots, n\}$, then $A[S]$ denotes the principal submatrix of A formed by taking rows and columns of A with indices in S. We denote by $A(S)$ the matrix $A[S']$, where S' is the complement of S in $\{1, 2, \ldots, n\}$. The determinant of A will be denoted by $|A|$, as usual. We adopt the convention that if $S = \phi$, then $|A[S]| = 1$.

If A is a symmetric $n \times n$ matrix, then the eigenvalues of A, which are necessarily real, are denoted by $\lambda_1(A) \geq \cdots \geq \lambda_n(A)$.

It may be remarked that although we consider only real symmetric matrices, the results hold for complex Hermitian matrices as well. The statements as well as the proofs of the results require only trivial modifications to achieve this.

We first prove some elementary results that will be used in this section.

Lemma 3.6.1. *Let A be a symmetric $n \times n$ matrix. Then*

$$\lambda_1(A) = \max_{x \neq 0} \frac{x^T A x}{x^T x}, \quad \lambda_n(A) = \min_{x \neq 0} \frac{x^T A x}{x^T x}.$$

Proof. By the Spectral Theorem we can write $A = P^T D P$, where P is orthogonal and D is the diagonal matrix with $\lambda_1(A), \ldots, \lambda_n(A)$ along the diagonal. We have

$$\begin{aligned}
\max_{x \neq 0} \frac{x^T A x}{x^T x} &= \max_{x \neq 0} \frac{x^T P^T D P x}{x^T x} \\
&= \max_{y \neq 0} \frac{y^T D y}{y^T y} \\
&= \max_{y \neq 0} \frac{\lambda_1(A) y_1^2 + \cdots + \lambda_n(A) y_n^2}{y_1^2 + \cdots + y_n^2} \\
&= \lambda_1(A).
\end{aligned}$$

The expression for $\lambda_n(A)$ can be obtained similarly, and the proof is complete. ∎

Several elementary properties of positive definite and positive semidefinite matrices are summarized next.

Lemma 3.6.2. *The following assertions hold:*

(i) *If A is positive semidefinite, then A is positive definite if and only if A is nonsingular.*

(ii) *If A is positive semidefinite, then for any matrix C, $C^T A C$ is positive semidefinite. Furthermore, if A is positive definite and if C is nonsingular, then $C^T A C$ is positive definite.*

(iii) If $A \succeq B$, then $|A| \geq |B|$.

(iv) If B is positive semidefinite, then there exists a positive semidefinite matrix C such that $C^2 = B$.

(v) If A, B are positive semidefinite, then AB has only nonnegative eigenvalues.

Proof. (*i*). Clearly, if A is positive definite, then A is nonsingular. Conversely, if A is positive semidefinite, then $\lambda_i(A) \geq 0, i = 1, 2, \ldots, n$. However, if A is nonsingular, then in fact $\lambda_i(A) > 0, i = 1, 2, \ldots, n$ and hence A is positive definite.

(*ii*). For any $x, x^T C^T A C x = y^T A y \geq 0$, where $y = Cx$. Thus $C^T A C$ is positive semidefinite. If A is positive definite and C is nonsingular, then $C^T A C$ is nonsingular and hence $C^T A C$ is positive definite by (*i*).

(*iii*). The result is trivial if B is singular so we assume that B is nonsingular. By the Spectral Theorem, $B = Q^T D Q$, where Q is orthogonal and D has $\lambda_i(B), i = 1, 2, \ldots, n$ along the diagonal. Let $D^{1/2}$ be the matrix with $\sqrt{\lambda_i(B)}, i = 1, 2, \ldots, n$ along the diagonal and let $D^{-1/2}$ be the inverse of $D^{1/2}$. Since $A \succeq B$ is positive semidefinite, by (*ii*), $D^{-1/2} Q A Q^T D^{-1/2} \succeq I$. By Lemma 3.6.1 all the eigenvalues of $D^{-1/2} Q A Q^T D^{-1/2}$ must be greater than or equal to 1 and hence

$$|D^{-1/2} Q A Q^T D^{-1/2}| \geq 1.$$

It follows that $|A| \geq |B|$ by the multiplicative property of the determinant.

(*iv*). Using the notation in (*iii*) we may take $C = Q^T D^{1/2} Q$. Then $C^2 = B$. We remark that C can be shown to be unique, although we omit a proof of this fact. We refer to C as the square root of B and denote it by $B^{1/2}$.

(*v*). We use the fact that if X, Y are $n \times n$ matrices then XY and YX have the same eigenvalues. Thus AB and $B^{1/2} A B^{1/2}$ have the same eigenvalues. However, by (*ii*), $B^{1/2} A B^{1/2}$ is positive semidefinite, and the result follows.   ■

We will use the results contained in Lemmas 3.6.1 and 3.6.2 frequently without making an explicit reference to them each time. We now prove the Hadamard Inequality, which gets its name from the paper by Hadamard (1893).

Theorem 3.6.3 (The Hadamard Inequality). *Let A be an $n \times n$ positive semidefinite matrix. Then $|A| \leq a_{11} \cdots a_{nn}$.*

Proof. If A is singular, then $|A| = 0$, whereas since A is positive semidefinite, $a_{ii} \geq 0, i = 1, 2, \ldots, n$ and the theorem is proved. So suppose A is nonsingular so that A is positive definite. Then $a_{ii} > 0, i = 1, 2, \ldots, n$. Let D be the $n \times n$ diagonal matrix with its i-th diagonal entry equal to $a_{ii}^{-1/2}, i = 1, 2, \ldots, n$ and let $B = DAD$. Then B is positive definite, $b_{ii} = 1, i = 1, 2, \ldots, n$, and $|B| = |A|(a_{11} \cdots a_{nn})^{-1}$.

By the arithmetic mean-geometric mean inequality,

$$\sum_{i=1}^{n} \lambda_i(B) \geq n \left(\prod_{i=1}^{n} \lambda_i(B) \right)^{1/n} = n|B|^{1/n}. \tag{3.6.1}$$

However, $\sum_{i=1}^{n} \lambda_i(B) = \text{trace}(B) = n$ and hence, from (3.6.1), $|B| \leq 1$. This is equivalent to $|A| \leq a_{11} \cdots a_{nn}$, and the proof is complete. ∎

The importance of the Hadamard inequality can be briefly summarized as follows:

It provides an upper bound for the determinant of a positive definite matrix that is extremely simple, conceptually as well as computationally. In many applications this bound, although somewhat crude, has been found to be adequate for the purpose at hand.

The inequality is self-refining, that is to say, it can be used to get other inequalities, which are apparently more general. One such example is the Fischer inequality, which we will prove.

The inequality can be generalized and extended in several directions, each such generalization or extension in itself leading to a new family of results. In this and subsequent sections we will consider the following generalizations:

(a) Fiedler's Inequality: If A is positive semidefinite, then $A \circ A^{-1} \succeq I$.
(b) Oppenheim's Inequality: If A, B are $n \times n$ positive semidefinite matrices, then $|A \circ B| \geq |B|a_{11} \cdots a_{nn}$.
(c) Schur's Inequality: If A is an $n \times n$ positive semidefinite matrix and if G is a subgroup of S_n, the permutation group of degree n, then $|A| \leq \sum_{\sigma \in G} \prod_{i=1}^{n} a_{i\sigma(i)}$.
(d) Schur's Majorization Theorem: If A is a symmetric $n \times n$ matrix, then the eigenvalues of A majorize its diagonal elements. (The concept of majorization is defined in Section 3.9.)

The following result is known as the Fischer Inequality. Although it appears stronger than the Hadamard Inequality, it can actually be derived using the Hadamard Inequality.

Theorem 3.6.4. *Let A be an $n \times n$ positive semidefinite matrix, which is partitioned as*

$$A = \begin{bmatrix} A_{11} & A_{12} \\ A_{21} & A_{22} \end{bmatrix}, \tag{3.6.2}$$

where A_{11} and A_{22} are square. Then

$$|A| \leq |A_{11}||A_{22}| \leq a_{11} \cdots a_{nn}.$$

Proof. Let P and Q be orthogonal matrices such that $PA_{11}P^T$ and $QA_{22}Q^T$ are both diagonal matrices and let

$$R = \begin{bmatrix} P & 0 \\ 0 & Q \end{bmatrix}.$$

Then

$$
\begin{aligned}
RAR^T &= \begin{bmatrix} P & 0 \\ 0 & Q \end{bmatrix} \begin{bmatrix} A_{11} & A_{12} \\ A_{21} & A_{22} \end{bmatrix} \begin{bmatrix} P^T & 0 \\ 0 & Q^T \end{bmatrix} \\
&= \begin{bmatrix} PA_{11}P^T & PA_{12}Q^T \\ QA_{21}P^T & QA_{22}Q^T \end{bmatrix}.
\end{aligned}
\tag{3.6.3}
$$

The product of the diagonal entries of RAR^T is the same as the product of the diagonal entries of $PA_{11}P^T$ and $QA_{22}Q^T$. But since the latter two matrices are diagonal, it actually equals $|PA_{11}P^T||QA_{22}Q^T|$. Therefore, using the Hadamard Inequality and (3.6.3), we have

$$|RAR^T| \leq |PA_{11}P^T||QA_{22}Q^T|$$

and hence $|A| \leq |A_{11}||A_{22}|$. The second inequality of the theorem also follows by applying the Hadamard Inequality to A_{11} and A_{22}. ∎

Let A be an $n \times n$ matrix partitioned as

$$A = \begin{bmatrix} A_{11} & A_{12} \\ A_{21} & A_{22} \end{bmatrix},$$

where A_{11} is nonsingular. The *Schur complement* of A_{11} in A, denoted simply by $\tilde{A}_{11}$ for convenience, is defined as

$$\tilde{A}_{11} = A_{22} - A_{21}A_{11}^{-1}A_{12}.$$

Similarly, the Schur complement of A_{22} in A is $\tilde{A}_{22} = A_{11} - A_{12}A_{22}^{-1}A_{21}$ if A_{22} is nonsingular.

Let A be a square matrix and consider the system of linear equations $Ax = y$. Suppose these equations are expressed in partitioned form as

$$\begin{bmatrix} A_{11} & A_{12} \\ A_{21} & A_{22} \end{bmatrix} \begin{bmatrix} x^{(1)} \\ x^{(2)} \end{bmatrix} = \begin{bmatrix} y^{(1)} \\ y^{(2)} \end{bmatrix}. \tag{3.6.4}$$

Then (3.6.4) is equivalent to

$$A_{11}x^{(1)} + A_{12}x^{(2)} = y^{(1)}, \qquad A_{21}x^{(1)} + A_{22}x^{(2)} = y^{(2)}. \qquad (3.6.5)$$

Suppose A_{11} is nonsingular. If we solve the first equation in (3.6.5) for $x^{(1)}$ we get

$$x^{(1)} = A_{11}^{-1}y^{(1)} - A_{11}^{-1}A_{12}x^{(2)}. \qquad (3.6.6)$$

Substituting (3.6.6) in the second equation in (3.6.5) we have

$$(A_{22} - A_{21}A_{11}^{-1}A_{12})x^{(2)} = y^{(2)} - A_{21}A_{11}^{-1}y^{(1)}.$$

Thus the Schur complement naturally arises as the coefficient matrix in the reduced equations.

Lemma 3.6.5. *Let A be an $n \times n$ matrix partitioned as in (3.6.2), and suppose A_{11} is nonsingular. Then*

(i) $|A| = |A_{11}||\tilde{A}_{11}|.$
(ii) If A is positive semidefinite, then $\tilde{A}_{11}$ is positive semidefinite.
(iii) If A is positive semidefinite, then

$$\begin{bmatrix} A_{11} & A_{12} \\ A_{21} & A_{22} - \tilde{A}_{11} \end{bmatrix}$$

is positive semidefinite.

Proof. (*i*). Consider A partitioned as in (3.6.2). Premultiply the top block by $A_{12}A_{11}^{-1}$ and subtract the resulting matrix from the lower block. Since this operation leaves the determinant unchanged, we have

$$|A| = \begin{vmatrix} A_{11} & A_{12} \\ 0 & A_{22} - A_{21}A_{11}^{-1}A_{12} \end{vmatrix} = |A_{11}||\tilde{A}_{11}|.$$

(*ii*). It is easily verified that if we set

$$B = \begin{bmatrix} I & 0 \\ -A_{21}A_{11}^{-1} & I \end{bmatrix},$$

then

$$BAB^T = \begin{bmatrix} A_{11} & 0 \\ 0 & \tilde{A}_{11} \end{bmatrix}.$$

Since A is positive semidefinite, so is BAB^T and hence $\tilde{A}_{11}$, which is a principal submatrix of BAB^T, is also positive semidefinite. Note that if A is positive definite, then $\tilde{A}_{11}$ is positive definite.

(*iii*). We have

$$\begin{bmatrix} A_{11} & A_{12} \\ A_{21} & A_{22} - \tilde{A}_{11} \end{bmatrix} = \begin{bmatrix} A_{11} & A_{12} \\ A_{21} & A_{21}A_{11}^{-1}A_{12} \end{bmatrix}$$

$$= \begin{bmatrix} A_{11}^{1/2} \\ A_{21}A_{11}^{-1/2} \end{bmatrix} \begin{bmatrix} A_{11}^{1/2}, & A_{11}^{-1/2}A_{12} \end{bmatrix},$$

which is positive semidefinite.   ∎

The following simple facts about the inverse of a partitioned matrix are often useful. Part (iii) of the result is known as *Jacobi's formula*. The proof is easy and hence is omitted.

Lemma 3.6.6. *Let A be a nonsingular n × n matrix, let $B = A^{-1}$, and suppose A and B are conformally partitioned as*

$$A = \begin{bmatrix} A_{11} & A_{12} \\ A_{21} & A_{22} \end{bmatrix}, \qquad B = \begin{bmatrix} B_{11} & B_{12} \\ B_{21} & B_{22} \end{bmatrix}.$$

Then

(*i*) $A_{11}^{-1} = \tilde{B}_{22}$, *if A_{11} is nonsingular.*
(*ii*) $A_{22}^{-1} = \tilde{B}_{11}$, *if A_{22} is nonsingular.*
(*iii*) $|A_{11}| = |B_{22}|/|B|$, $|A_{22}| = |B_{11}|/|B|$.

If A is an $n \times n$ positive semidefinite matrix and if $S \subset \{1, 2, \ldots, n\}$, then by Theorem 3.6.4,

$$|A| \le |A[S]||A(S)|. \tag{3.6.7}$$

The next result is a natural generalization of (3.6.7) and is often known as the Hadamard-Fischer Inequality.

Theorem 3.6.7. *Let A be an n × n positive semidefinite matrix and let S, T ⊂ $\{1, 2, \ldots, n\}$. Then*

$$|A[S \cup T]||A[S \cap T]| \le |A[S]||A[T]|. \tag{3.6.8}$$

Proof. We may assume, without loss of generality, that $S \cup T = \{1, 2, \ldots, n\}$, for otherwise, we can reformulate the problem in terms of $A[S \cup T]$ and its principal minors. After making this assumption, (3.6.8) reduces to

$$|A||A[S \cap T]| \le |A[S]||A[T]|. \tag{3.6.9}$$

If A is singular, then (3.6.9) is obvious, so suppose A is positive definite. Let $B = A^{-1}$. By Jacobi's formula,

$$|A[S]| = \frac{|B(S)|}{|B|}, \qquad |A[T]| = \frac{|B(T)|}{|B|}, \qquad |A[S \cap T]| = \frac{|B(S \cap T)|}{|B|}.$$

Hence (3.6.9) reduces to

$$|B(S \cap T)| \leq |B(S)||B(T)| \tag{3.6.10}$$

and (3.6.10) is equivalent to

$$|B[S' \cup T']| \leq |B[S']||B[T']|. \tag{3.6.11}$$

Since $S' \cap T' = \phi$, (3.6.9) follows from Theorem 3.6.4 and Equation (3.6.11). That completes the proof. ∎

3.7. Inequalities of Fiedler and Oppenheim

If A and B are $n \times n$ matrices, then as defined in Section 3.3, their *Schur product*, denoted by $A \circ B$, is the $n \times n$ matrix $(a_{ij} b_{ij})$. The term "Hadamard product" is also commonly used in the literature for the "Schur product." The term Schur product is justified in view of the following result, first proved by Schur.

Lemma 3.7.1. *If A and B are $n \times n$ positive semidefinite matrices, then $A \circ B$ is positive semidefinite.*

Proof. If $B = 0$, there is nothing to prove. Now suppose that B is of rank one, so that $B = (x_i x_j)$ for some vector $x \in R^n$. Then

$$A \circ B = (a_{ij} b_{ij})$$
$$= (a_{ij} x_i x_j)$$
$$= DAD,$$

where D is the diagonal matrix with $x_1, \ldots, x_n$ on its diagonal. Since A is positive semidefinite, so is DAD. In general, by the Spectral Theorem, we can write

$$B = B_1 + \cdots + B_r,$$

where B_i is positive semidefinite, $i = 1, 2, \ldots, r$, and each B_i is of rank one. Then $A \circ B_i$ is positive semidefinite, $i = 1, 2, \ldots, r$, and

$$A \circ B = A \circ B_1 + \cdots + A \circ B_r.$$

Therefore, $A \circ B$ is positive semidefinite, and the proof is complete. ∎

Let A be an $n \times n$ positive definite matrix. The i-th diagonal entry of A^{-1} is given by $|A(i)|/|A|$, $i = 1, 2, \ldots, n$. By Theorem 3.6.4,

$$|A| \le a_{ii}|A(i)|, \quad i = 1, 2, \ldots, n,$$

and so the diagonal entries of $A \circ A^{-1}$ are all greater than or equal to 1. It turns out that a stronger result is in fact true, namely, if A is positive definite, then $A \circ A^{-1} \succeq I$. We refer to this result as Fiedler's Inequality, although it was independently observed by other authors. The following result gives a stronger version of Fiedler's Inequality.

Theorem 3.7.2. *Let A be an $n \times n$ positive definite matrix partitioned as in (3.6.2). Then*

$$A \circ A^{-1} \succeq \begin{bmatrix} A_{11} \circ A_{11}^{-1} & 0 \\ 0 & \tilde{A}_{11} \circ \tilde{A}_{11}^{-1} \end{bmatrix}.$$

Proof. Let $B = A^{-1}$ and suppose B is also conformally partitioned as A. By (*iii*) of Lemma 3.6.5,

$$\begin{bmatrix} A_{11} & A_{12} \\ A_{21} & A_{22} - \tilde{A}_{11} \end{bmatrix}, \quad \begin{bmatrix} B_{11} - \tilde{B}_{22} & B_{12} \\ B_{21} & B_{22} \end{bmatrix}$$

are both positive semidefinite and hence, by Lemma 3.7.1, so is their Schur product. After some simplification, this gives

$$A \circ B \succeq \begin{bmatrix} A_{11} \circ \tilde{B}_{22} & 0 \\ 0 & \tilde{A}_{11} \circ B_{22}. \end{bmatrix}.$$

By Lemma 3.6.6, $\tilde{B}_{22} = A_{11}^{-1}$, $B_{22} = \tilde{A}_{11}^{-1}$, and the proof is complete. ∎

Fiedler's Inequality follows from Theorem 3.7.2 by induction.

Theorem 3.7.3. *Let A and B be $n \times n$ positive semidefinite matrices. Then*

$$\lambda_n(A \circ B) \ge \lambda_n(AB).$$

Proof. If A is singular, then so is AB and $\lambda_n(AB) = 0$. However, $\lambda_n(A \circ B) \ge 0$, and the result is proved. So suppose A is nonsingular. We have

$$\lambda_n(AB) = \lambda_n(A^{1/2}BA^{1/2})$$
$$= \min_{x \ne 0} \frac{x^T A^{1/2} B A^{1/2} x}{x^T x}$$
$$= \min_{y \ne 0} \frac{y^T B y}{y^T A^{-1} y}.$$

Therefore, for any y,

$$y^T B y \geq \lambda_n(AB) y^T A^{-1} y$$

and hence $B \succeq \lambda_n(AB) A^{-1}$. Hence

$$A \circ B \succeq \lambda_n(AB) A \circ A^{-1}.$$

By Theorem 3.7.2, $A \circ A^{-1} \succeq I$ and therefore, $A \circ B \succeq \lambda_n(AB)I$. It follows that $\lambda_n(A \circ B) \geq \lambda_n(AB)$, and the proof is complete. ∎

The next few results are centered around an interesting inequality due to Oppenheim.

Theorem 3.7.4. *Let A, and B be $n \times n$ positive definite matrices that are conformally partitioned as*

$$A = \begin{bmatrix} A_{11} & A_{12} \\ A_{21} & A_{22} \end{bmatrix}, \qquad B = \begin{bmatrix} B_{11} & B_{12} \\ B_{21} & B_{22} \end{bmatrix}. \tag{3.7.1}$$

Then $|A \circ B| \geq |A_{11} \circ B_{11}||A_{22} \circ \tilde{B}_{11}|$.

Proof. By Lemma 3.6.5,

$$\begin{bmatrix} B_{11} & B_{12} \\ B_{21} & B_{22} - \tilde{B}_{11} \end{bmatrix}$$

is positive definite and hence

$$A \circ \begin{bmatrix} B_{11} & B_{12} \\ B_{21} & B_{22} - \tilde{B}_{11} \end{bmatrix} = \begin{bmatrix} A_{11} \circ B_{11} & A_{12} \circ B_{12} \\ A_{21} \circ B_{21} & A_{22} \circ (B_{22} - \tilde{B}_{11}) \end{bmatrix}$$

is positive semidefinite. Therefore, by Lemma 3.6.5,

$$A_{22} \circ (B_{22} - \tilde{B}_{11}) \succeq (A_{21} \circ B_{21})(A_{11} \circ B_{11})^{-1}(A_{12} \circ B_{12}).$$

Hence

$$A_{22} \circ B_{22} - (A_{21} \circ B_{21})(A_{11} \circ B_{11})^{-1}(A_{12} \circ B_{12}) \succeq A_{22} \circ \tilde{B}_{11}$$

and therefore

$$|A_{22} \circ B_{22} - (A_{21} \circ B_{21})(A_{11} \circ B_{11})^{-1}(A_{12} \circ B_{12})| \geq |A_{22} \circ \tilde{B}_{11}|. \tag{3.7.2}$$

Now we have

$$|A \circ B| = \begin{vmatrix} A_{11} \circ B_{11} & A_{12} \circ B_{12} \\ A_{21} \circ B_{21} & A_{22} \circ B_{22} \end{vmatrix}$$

$$= |A_{11} \circ B_{11}||A_{22} \circ B_{22} - (A_{21} \circ B_{21})(A_{11} \circ B_{11})^{-1}(A_{12} \circ B_{12})|$$

$$\geq |A_{11} \circ B_{11}||A_{22} \circ \tilde{B}_{11}|$$

by (3.7.2), and the proof is complete. ∎

We now prove Oppenheim's Inequality.

Theorem 3.7.5. Let A and B be $n \times n$ positive semidefinite matrices. Then

$$|A \circ B| \geq a_{11} \cdots a_{nn} |B|.$$

Proof. We assume that A and B are positive definite and the general result will follow by a continuity argument. Suppose A and B are partitioned as in (3.7.1). By Theorem 3.7.4,

$$|A \circ B| \geq |A_{11} \circ B_{11}||A_{22} \circ \tilde{B}_{11}|. \qquad (3.7.3)$$

If $n = 1$, then the result of the theorem is trivial. Suppose the result is true for matrices of order less than n and let us proceed by induction. Then by the induction assumption,

$$|A_{11} \circ B_{11}| \geq a_{11} \cdots a_{kk}|B_{11}|, \qquad |A_{22} \circ \tilde{B}_{11}| \geq a_{k+1,k+1} \cdots a_{nn}|\tilde{B}_{11}|, \qquad (3.7.4)$$

where A_{11} is assumed to be $k \times k$. From (3.7.3) and (3.7.4) we have

$$|A \circ B| \geq a_{11} \cdots a_{nn} |B_{11}||\tilde{B}_{11}|.$$

Since by Lemma 3.6.5, $|B| = |B_{11}||\tilde{B}_{11}|$, the proof is complete. ∎

Corollary 3.7.6. Let A, B be $n \times n$ positive semidefinite matrices. Then $|A \circ B| \geq |AB|$.

Proof. By Theorem 3.7.5, $|A \circ B| \geq a_{11} \cdots a_{nn}|B|$. Since $|A| \leq a_{11} \cdots a_{nn}$ by the Hadamard Inequality, and since $|AB| = |A||B|$, the proof is complete. ∎

We remark that in Theorem 3.7.5 if we set $A = I$, then we get the Hadamard Inequality.

Theorem 3.7.2 is from Bapat and Kwong (1987). Theorem 3.7.3 appears in Fiedler (1983). The following more general result was conjectured by Johnson and Bapat (1988) and has recently been proved by Ando (1995) and by Visick (1995). *Let A, B be $n \times n$ positive semidefinite matrices. Then*

$$\prod_{i=k}^{n} \lambda_i (A \circ B) \geq \prod_{i=k}^{n} \lambda_i(AB), \quad k = 1, 2, \ldots, n. \qquad (3.7.5)$$

Note that if $k = n$, then (3.7.5) is true by Theorem 3.7.3. If $k = 1$, then (3.7.5) reduces to $|A \circ B| \geq |AB|$, which is Corollary 3.7.6. Theorem 3.7.4 appears in Bapat (1987b), whereas Theorem 3.7.5 dates back to Oppenheim (1930).

3.8. Schur power matrix

If A and B are matrices of order $m \times n$ and $p \times q$, respectively, then the *Kronecker product* of A and B is denoted by $A \otimes B$. Thus $A \otimes B$ is the $mp \times nq$ matrix given by

$$A \otimes B = \begin{bmatrix} a_{11}B & a_{12}B & \cdots & a_{1n}B \\ \vdots & \vdots & \ddots & \vdots \\ a_{m1}B & a_{m2}B & \cdots & a_{mn}B \end{bmatrix}.$$

Lemma 3.8.1. *(i) If A, B, C, and D are matrices such that AC and BD are defined, then*

$$(A \otimes B)(C \otimes D) = AC \otimes BD.$$

(ii) If A and B are positive semidefinite, then $A \otimes B$ is positive semidefinite.

Proof. (i). This assertion is proved by carefully putting together the definitions of matrix multiplication and the Kronecker product.
(ii). Since A and B are positive semidefinite, we can write $A = UU^T$, $B = VV^T$ for some matrices U and V. Then by (i),

$$\begin{aligned} A \otimes B &= UU^T \otimes VV^T \\ &= (U \otimes V)(U^T \otimes V^T) \\ &= (U \otimes V)(U \otimes V)^T, \end{aligned}$$

which is positive semidefinite, and the proof is complete. ∎

Let us assume that the elements of S_n, the permutation group of degree n, have been ordered in some way. This ordering will be assumed fixed in the subsequent discussion. Let A be an $n \times n$ matrix. The *Schur power* of A, denoted by $\pi(A)$, is the $n! \times n!$ matrix whose rows as well as columns are indexed by S_n and whose (σ, τ)-entry is $\prod_{i=1}^{n} a_{\sigma(i)\tau(i)}$ if $\sigma, \tau \in S_n$.

As an illustration, suppose the elements of S_3 are ordered as 123, 132, 213, 231, 312, 321, and let

$$A = \begin{bmatrix} a & b & c \\ d & e & f \\ g & h & k \end{bmatrix}.$$

Then

$$
\pi(A) =
\begin{bmatrix}
aek & afh & bdk & bfg & cdh & ceg \\
afh & aek & bfg & bdk & ceg & cdh \\
bdk & cdh & aek & ceg & afh & bfg \\
cdh & bdk & ceg & aek & bfg & afh \\
bfg & ceg & afh & cdh & aek & bdk \\
ceg & bfg & cdh & afh & bdk & aek
\end{bmatrix}.
$$

We now make some elementary observations about $\pi(A)$. The diagonal entries of $\pi(A)$ are all equal to $a_{11} \cdots a_{nn}$, where A is $n \times n$. The sum of the entries in any row or column of $\pi(A)$ equals per A. In particular, per A is an eigenvalue of $\pi(A)$. If A is $n \times n$, then after a permutation of the rows and an identical permutation of the columns, $\pi(A)$ can be viewed as a principal submatrix of $\otimes^n A = A \otimes A \otimes \cdots \otimes A$, taken n times. If A is positive semidefinite, then $\otimes^n A$ is positive semidefinite and hence so is $\pi(A)$. Since per A is an eigenvalue of $\pi(A)$, we immediately have a proof of the fact that if A is positive semidefinite, then per $A \geq 0$.

It turns out that $|A|$ is also an eigenvalue of $\pi(A)$. To see this, define a vector ϵ of order $n!$ as follows. Index the elements of ϵ by S_n. If $\tau \in S_n$, then set $\epsilon(\tau) = 1$ if τ is even and -1 if τ is odd. Then for any $\sigma \in S_n$,

$$
\sum_{\tau \in S_n} \epsilon(\tau) \prod_{i=1}^{n} a_{\sigma(i)\tau(i)} = \sum_{\tau \in S_n} \epsilon(\tau) \prod_{i=1}^{n} a_{i\tau\sigma^{-1}(i)}
$$

$$
= \sum_{\rho \in S_n} \epsilon(\rho\sigma) \prod_{i=1}^{n} a_{i\rho(i)}
$$

$$
= \epsilon(\sigma) \sum_{\rho \in S_n} \epsilon(\rho) \prod_{i=1}^{n} a_{i\rho(i)}
$$

$$
= \epsilon(\sigma)|A|.
$$

Thus $\pi(A)\epsilon = |A|\epsilon$ and hence $|A|$ is an eigenvalue of $\pi(A)$ with ϵ as the corresponding eigenvector.

We now wish to prove a remarkable result of Schur that asserts that if A is positive semidefinite, then $|A|$ is in fact the smallest eigenvalue of $\pi(A)$. We first prove the following theorem:

Theorem 3.8.2. *Let A be an $n \times n$ positive semidefinite matrix. Then for any x,*

$$
x^T \pi(A) x \geq |A| x^T x.
$$

Proof. Let y be the vector of order n^n obtained by augmenting x with zero coordinates at appropriate places so that

$$x^T \pi(A)x = y^T(\otimes^n A)y. \qquad (3.8.1)$$

Since A is positive semidefinite, there exists a lower triangular matrix B such that $A = B^T B$. We have

$$y^T(\otimes^n A)y = y^T(\otimes^n(B^T B))y$$
$$= y^T(\otimes^n B^T)(\otimes^n B)y \qquad (3.8.2)$$

by Lemma 3.8.1. By the Cauchy-Schwarz Inequality, it follows from (3.8.2) that

$$y^T(\otimes^n A)y \le \{y^T(\otimes^n B^T)^2 y\}^{1/2}\{y^T(\otimes^n B)^2 y\}^{1/2}$$
$$= \{y^T(\otimes^n(B^2)^T)y\}^{1/2}\{y^T(\otimes^n B^2)y\}^{1/2}.$$

The relation (3.8.1) and the same relation with B^T and then B in place of A now give

$$x^T \pi(A)x \le \{x^T \pi(B^2)^T x\}^{1/2}\{x^T \pi(B^2)x\}^{1/2}. \qquad (3.8.3)$$

Since B^2 is lower triangular, the diagonal entries of $\pi(B^2)$ are equal to $|B|^2$. Also, $\pi(B^2)$ is lower triangular as well and therefore

$$x^T \pi(B^2)x = |B|^2 x^T x. \qquad (3.8.4)$$

Similarly,

$$x^T \pi(B^2)^T x = |B|^2 x^T x. \qquad (3.8.5)$$

Substituting (3.8.4) and (3.8.5) in (3.8.3) we get

$$x^T \pi(A)x \le |B|^2 x^T x = |A| x^T x,$$

and the proof is complete. ∎

Corollary 3.8.3. *If A is positive semidefinite, then $|A|$ is the smallest eigenvalue of $\pi(A)$.*

Proof. By Theorem 3.8.2,

$$|A| \le \min_{x \ne 0} \frac{x^T \pi(A)x}{x^T x},$$

and hence $|A|$ cannot exceed the smallest eigenvalue of $\pi(A)$. However, as observed earlier, $|A|$ is an eigenvalue of $\pi(A)$ and hence it must be the smallest eigenvalue. ∎

Theorem 3.8.2 provides a rich source of inequalities for the determinant of a positive definite matrix since we can make a judicious choice of x and get an inequality. For example, if x has all coordinates zero except one, then Theorem 3.8.2 reduces to the Hadamard Inequality. Another example is illustrated next.

Corollary 3.8.4. *Let A be an $n \times n$ positive semidefinite matrix, and let G be a subgroup of S_n. Then*

$$|A| \le \sum_{\sigma \in G} \prod_{i=1}^{n} a_{i\sigma(i)}.$$

Proof. Define the vector x of order $n!$ as follows: The elements of x are indexed by S_n. If $\sigma \in G$, then set $x(\sigma) = 1$, otherwise set $x(\sigma) = 0$. Then

$$x^T \pi(A)x = \sum_{\sigma, \tau \in S_n} x(\sigma)x(\tau) \prod_{i=1}^{n} a_{\sigma(i)\tau(i)}$$

$$= \sum_{\sigma, \tau \in G} \prod_{i=1}^{n} a_{\sigma(i)\tau(i)}$$

$$= \sum_{\sigma, \tau \in G} \prod_{i=1}^{n} a_{i\tau\sigma^{-1}(i)}. \tag{3.8.6}$$

Make the change of variable $\tau\sigma^{-1} = \rho$ in (3.8.6) to obtain

$$x^T \pi(A)x = O(G) \sum_{\rho \in G} \prod_{i=1}^{n} a_{i\rho(i)}, \tag{3.8.7}$$

where $O(G)$ is the order of G, i.e., the number of elements in G. Since $x^T x = O(G)$, the result follows from Theorem 3.8.2 and Equation (3.8.7). ∎

Note that in Corollary 3.8.4, if G is the subgroup consisting of the identity permutation only, then we get the Hadamard Inequality.

Theorem 3.8.2 is due to Schur (1918); a different proof and a generalization appear in Bapat and Sunder (1986). A very important unsolved problem is to settle the conjecture, due to Soules (1979), that if A is positive semidefinite, then per A is the largest eigenvalue of $\pi(A)$. Although the conjecture has been proved for matrices of order at most 3 [Bapat and Sunder (1986)], very little progress has been made in general; see also Soules (1994). A consequence of the conjecture is the *permanent-on-top conjecture* or the *permanental dominance conjecture*, which asserts that any *immanant* of a positive semidefinite matrix is dominated by the permanent [see Johnson (1987) and Merris (1987)].

3.9. Majorization inequalities for eigenvalues

We first develop some basic facts concerning the concept of majorization. Suppose x, y are the vectors given by

$$x = (2, 3, 3, 2, 2)^T, \qquad y = (1, 3, 7, 0, 1)^T.$$

The components of both x as well as y add up to 12. However, the components of x appear more clustered together than those of y. One way of making this concept more precise is given now. We use the following notation: If $x \in R^n$, then

$$x_{[1]} \geq \cdots \geq x_{[n]}$$

denote the components of x arranged in nonincreasing order.

Let $x, y \in R^n$. We say that x is *majorized* by y, if $\sum_{i=1}^n x_i = \sum_{i=1}^n y_i$ and

$$\sum_{i=1}^n x_{[i]} \leq \sum_{i=1}^n y_{[i]}, \quad k = 1, 2, \ldots, n-1.$$

The majorization ordering defined above appears in a variety of situations in areas such as combinatorics, group theory, statistics, and economics.

Lemma 3.9.1. *Let $x, y, z \in R^n$ and let x be majorized by y. Then*

$$\sum_{i=1}^n z_{[i]} x_{[i]} \leq \sum_{i=1}^n z_{[i]} y_{[i]}.$$

Proof. Define $z_{[n+1]} = 0$. Then we have

$$\sum_{i=1}^n z_{[i]} x_{[i]} = \sum_{k=1}^n (z_{[k]} - z_{[k+1]}) \sum_{i=1}^k x_{[i]}$$

$$\leq \sum_{k=1}^n (z_{[k]} - z_{[k+1]}) \sum_{i=1}^k y_{[i]}$$

$$= \sum_{i=1}^n z_{[i]} y_{[i]},$$

and the proof is complete. ∎

We now prove an important characterization of majorization usually referred to as the Hardy-Littlewood-Polya Theorem.

Theorem 3.9.2. *Let $x, y \in R^n$. Then the following conditions are equivalent:*

(i) x is majorized by y.

(ii) There exists a doubly stochastic matrix D such that $x = Dy$.

Proof. We first give a reformulation of (*ii*). If $\sigma \in S_n$, let P^σ denote the corresponding permutation matrix and let $y^\sigma = P^\sigma y$. Then by the Birkhoff–von Neumann Theorem (Theorem 2.1.6), condition (*ii*) is equivalent to the assertion that x is in the convex hull of $\{y^\sigma : \sigma \in S_n\}$. It is this reformulation of (*ii*) that we work with in this proof.

(*i*) $\Rightarrow$ (*ii*). Suppose (*ii*) does not hold. Then x is not in the convex hull of $\{y^\sigma : \sigma \in S_n\}$ and hence by the Separating Hyperplane Theorem [see, for example, Parthasarathy and Raghavan (1971)] there exists $z \in R^n$ such that

$$z^T x > z^T y^\sigma \quad \text{for all} \quad \sigma \in S_n. \tag{3.9.1}$$

Since $\sum_{i=1}^n z_{[i]} x_{[i]} \geq z^T x$ (see Exercise 12) we have, from (3.9.1),

$$\sum_{i=1}^n z_{[i]} x_{[i]} > \sum_{i=1}^n z_{[i]} y_{[i]},$$

which contradicts Lemma 3.9.1. Thus (*ii*) must hold.

(*ii*) $\Rightarrow$ (*i*). If x is in the convex hull of $\{y^\sigma : \sigma \in S_n\}$, then there exist $c(\sigma)$, $\sigma \in S_n$ such that

$$x = \sum_{\sigma \in S_n} c(\sigma) y^\sigma, \qquad c(\sigma) \geq 0, \qquad \sum_{\sigma \in S_n} c(\sigma) = 1.$$

Then, clearly, $\sum_{i=1}^n x_i = \sum_{i=1}^n y_i$. Also, for $k = 1, 2, \ldots, n-1$, we have

$$\sum_{i=1}^k x_{[i]} \leq \sum_{\sigma \in S_n} c(\sigma) \sum_{i=1}^k (y^\sigma)_{[i]}$$

$$= \sum_{\sigma \in S_n} c(\sigma) \sum_{i=1}^k y_{[i]}$$

$$= \sum_{i=1}^k y_{[i]}.$$

Therefore, x is majorized by y, and the proof is complete. ■

The next result indicates how we may get a family of inequalities if a majorization relation exists between two vectors.

Lemma 3.9.3. *Let $x, y \in R^n$, where x is majorized by y and let $\phi : R \to R$ be a convex function. Then*

$$\sum_{i=1}^n \phi(x_i) \leq \sum_{i=1}^n \phi(y_i).$$

Proof. By Theorem 3.9.2 there exists a doubly stochastic matrix D such that $x = Dy$. Therefore

$$\sum_{i=1}^{n} \phi(x_i) = \sum_{i=1}^{n} \phi\left(\sum_{j=1}^{n} d_{ij} y_j\right)$$

$$\leq \sum_{i=1}^{n} \sum_{j=1}^{n} d_{ij} \phi(y_j)$$

$$= \sum_{j=1}^{n} \phi(y_j) \sum_{i=1}^{n} d_{ij}$$

$$= \sum_{j=1}^{n} \phi(y_j),$$

and the proof is complete. ∎

Corollary 3.9.4. *Let $x \geq 0$, $y \geq 0$ be vectors in R^n such that x is majorized by y. Then $\prod_{i=1}^{n} x_i \geq \prod_{i=1}^{n} y_i$.*

Proof. We may assume $x > 0$, $y > 0$, and the general case will follow by a continuity argument. Let $\phi(x) = -\log x$. Then ϕ is a convex function on $(0, \infty)$ and, by Lemma 3.9.3,

$$-\sum_{i=1}^{n} \log x_i \leq -\sum_{i=1}^{n} \log y_i,$$

which is equivalent to $\prod_{i=1}^{n} x_i \geq \prod_{i=1}^{n} y_i$. ∎

If A is a symmetric $n \times n$ matrix, then an elegant result due to Schur asserts that the vector $(a_{11}, \ldots, a_{nn})^T$ is majorized by the vector $(\lambda_1(A), \ldots, \lambda_n(A))^T$. We now prove a result more general than that of Schur.

Theorem 3.9.5. *Let A and B be $n \times n$ matrices, where A is symmetric and B is positive semidefinite with $b_{ii} = 1$, $i = 1, 2, \ldots, n$. Then $(\lambda_1(A \circ B), \ldots, \lambda_n(A \circ B))^T$ is majorized by $(\lambda_1(A), \ldots, \lambda_n(A)^T$.*

Proof. Let P and Q be orthogonal matrices such that

$$A \circ B = P^T \begin{bmatrix} \lambda_1(A \circ B) & 0 & \cdots & 0 \\ 0 & \ddots & & 0 \\ \vdots & & \ddots & \vdots \\ 0 & & \cdots & \lambda_n(A \circ B) \end{bmatrix} P$$

and

$$
A = Q^T \begin{bmatrix} \lambda_1(A) & 0 & \cdots & 0 \\ 0 & \ddots & & 0 \\ \vdots & & \ddots & \vdots \\ 0 & & \cdots & \lambda_n(A) \end{bmatrix} Q.
$$

Then for $i = 1, 2, \ldots, n$,

$$
\lambda_i(A \circ B) = \sum_{k=1}^{n} \sum_{\ell=1}^{n} a_{k\ell} b_{k\ell} p_{ik} p_{i\ell}
$$

$$
= \sum_{k=1}^{n} \sum_{\ell=1}^{n} b_{k\ell} p_{ik} p_{i\ell} \sum_{j=1}^{n} \lambda_j(A) q_{jk} q_{j\ell}. \qquad (3.9.2)
$$

Define the matrix $D = (d_{ij})$ as

$$
d_{ij} = \sum_{k=1}^{n} \sum_{\ell=1}^{n} b_{k\ell} p_{ik} p_{i\ell} q_{jk} q_{j\ell}, \quad i, j = 1, 2, \ldots, n. \qquad (3.9.3)
$$

Then from (3.9.2) we have

$$
\begin{bmatrix} \lambda_1(A \circ B) \\ \vdots \\ \lambda_n(A \circ B) \end{bmatrix} = D \begin{bmatrix} \lambda_1(A) \\ \vdots \\ \lambda_n(A) \end{bmatrix}.
$$

Thus, by Theorem 3.9.2, the result will be proved if we show that D is doubly stochastic. First note that if we set $u_k = p_{ik} q_{jk}, k = 1, 2, \ldots, n$, then from (3.9.3),

$$
d_{ij} = \sum_{k=1}^{n} \sum_{\ell=1}^{n} b_{k\ell} u_k u_\ell,
$$

and since B is positive semidefinite, $d_{ij} \geq 0$ for all i, j. Now for $i = 1, 2, \ldots, n$,

$$
\sum_{j=1}^{n} d_{ij} = \sum_{k=1}^{n} \sum_{\ell=1}^{n} b_{k\ell} p_{ik} p_{i\ell} \sum_{j=1}^{n} q_{jk} q_{j\ell}. \qquad (3.9.4)
$$

Since Q is orthogonal, only terms corresponding to $k = \ell$ survive in (3.9.4) and thus

$$
\sum_{j=1}^{n} d_{ij} = \sum_{k=1}^{n} b_{kk} p_{ik}^2 = 1,
$$

since $b_{kk} = 1, k = 1, 2, \ldots, n$ and since P is orthogonal. Similarly, it can be shown that every column sum of D is 1 and therefore D is doubly stochastic. That completes the proof. ∎

Observe that if we set $B = I$ in Theorem 3.9.5, we recover the result of Schur mentioned earlier. Also note that in Theorem 3.9.5 if A were positive semidefinite as well, then the theorem together with Corollary 3.9.4 imply that

$$\prod_{i=1}^{n} \lambda_i(A \circ B) \geq \prod_{i=1}^{n} \lambda_i(A),$$

and this is equivalent to Oppenheim's Inequality.

A comprehensive account of majorization is given in Marshall and Olkin (1979). Ando (1989) provides an excellent survey. Theorem 3.9.5 is from Bapat and Sunder (1985); for an application of the theorem to multivariate statistical analysis, see Koyak (1987).

3.10. The parallel sum

If A and B are positive definite $n \times n$ matrices, then their *parallel sum*, denoted by $A : B$, is defined as

$$A : B = (A^{-1} + B^{-1})^{-1}.$$

This definition is motivated by parallel connections of resistors in electrical networks [see Anderson and Trapp (1975)]. Consider two resistors, having resistances a and b, which are positive numbers. If the resistors are connected in parallel, then their combined resistance is given by

$$\frac{ab}{a+b} = (a^{-1} + b^{-1})^{-1},$$

which is precisely the parallel sum of a and b. In general, one can consider an n-port network, which is essentially a black box with $2n$ terminals. These terminals are divided into n pairs called ports. The voltage and the current are measured at each port, giving rise to the voltage vector v and the current vector i. These vectors are related by the equation $v = Zi$, where the matrix Z is positive definite if the network contains only resistors. The parallel sum of two positive definite matrices then represents the combined resistance of two n-port networks connected in parallel. Several basic properties of the parallel sum thus have a network interpretation.

The main purpose of this section is to prove the Series-Parallel Inequality. We first develop some preliminaries.

Let A be an $n \times n$ positive definite matrix partitioned as

$$A = \begin{bmatrix} A_{11} & A_{12} \\ A_{21} & A_{22} \end{bmatrix}. \tag{3.10.1}$$

Recall that the Schur complement of A_{22} in A is defined as $\tilde{A}_{22} = A_{11} - A_{12}A_{22}^{-1}A_{21}$.

Lemma 3.10.1. *Let A be an $n \times n$ matrix partitioned as in (3.10.1), where A_{11} is $r \times r$. Suppose P is an $r \times r$ matrix such that*

$$\begin{bmatrix} A_{11} - P & A_{12} \\ A_{21} & A_{22} \end{bmatrix}$$

is positive semidefinite. Then $\tilde{A}_{22} \succeq P$.

Proof. Let x be an $r \times 1$ vector and let $y = -A_{22}^{-1}A_{21}x$. Then, using the hypothesis on P and the definition of $\tilde{A}_{22}$, we get

$$\begin{aligned}
x^T P x &= \begin{bmatrix} x^T, y^T \end{bmatrix} \begin{bmatrix} P & 0 \\ 0 & 0 \end{bmatrix} \begin{bmatrix} x \\ y \end{bmatrix} \\
&\leq \begin{bmatrix} x^T, y^T \end{bmatrix} \begin{bmatrix} A_{11} & A_{12} \\ A_{21} & A_{22} \end{bmatrix} \begin{bmatrix} x \\ y \end{bmatrix} \\
&= \begin{bmatrix} x^T, y^T \end{bmatrix} \begin{bmatrix} \tilde{A}_{22} & 0 \\ 0 & 0 \end{bmatrix} \begin{bmatrix} x \\ y \end{bmatrix} \\
&= x^T \tilde{A}_{22} x.
\end{aligned}$$

Since x is arbitrary, it follows that $\tilde{A}_{22} \succeq P$, and the proof is complete. ∎

Lemma 3.10.2. *Let A and B be $n \times n$ positive definite matrices, let $C = A + B$, and suppose A, B, and C are partitioned conformally as in (3.10.1). Then*

$$\tilde{C}_{22} \succeq \tilde{A}_{22} + \tilde{B}_{22}.$$

Proof. As in (*iii*) of Lemma 3.6.5, we see that

$$\begin{bmatrix} A_{11} - \tilde{A}_{22} & A_{12} \\ A_{21} & A_{22} \end{bmatrix}, \qquad \begin{bmatrix} B_{11} - \tilde{B}_{22} & B_{12} \\ B_{21} & B_{22} \end{bmatrix}$$

are positive semidefinite. Therefore their sum is positive semidefinite. The result now follows by Lemma 3.10.1. ∎

Lemma 3.10.3. *Let A and B be $n \times n$ positive definite matrices. Then $A : B$ is the Schur complement of $A + B$ in the $2n \times 2n$ matrix,*

$$\begin{bmatrix} A & A \\ A & A + B \end{bmatrix}. \tag{3.10.2}$$

Proof. It is easy to see, after taking the inverse of both sides, that

$$(A^{-1} + B^{-1})^{-1} = A(A + B)^{-1}B. \tag{3.10.3}$$

Thus, using (3.10.3),

$$\begin{aligned} A : B &= (A^{-1} + B^{-1})^{-1} \\ &= A(A + B)^{-1}(A + B - A) \\ &= A - A(A + B)^{-1}A, \end{aligned}$$

which is precisely the Schur complement of $A + B$ in (3.10.2). ∎

Lemma 3.10.4. *Let A be a positive definite $n \times n$ matrix. Then*

$$\lim_{\epsilon \downarrow 0} A : \epsilon I = 0.$$

Proof. As seen in the proof of Lemma 3.10.3,

$$A : \epsilon I = \epsilon A(A + \epsilon I)^{-1},$$

and the result is immediate. ∎

It follows from the definition of parallel sum that if A, B are positive definite, then so is $A : B$. Furthermore, the parallel sum is commutative ($A : B = B : A$) and associative ($A : (B : C) = (A : B) : C$). The next result gives some additional properties.

Theorem 3.10.5. *Let A, B, C, and D be $n \times n$ positive definite matrices. Then the following assertions hold:*

(i) $(A + B) : (C + D) \succeq A : C + B : D$.
(ii) If $A \succeq B$, then $A : C \succeq B : C$.

Proof. (*i*). This follows from Lemmas 3.10.2 and 3.10.3.

(*ii*). Since $A \succeq B$, $A - B$ is positive semidefinite. For a sufficiently small $\epsilon > 0$, we have, by (*i*),

$$\begin{aligned} A : C &= (B + A - B) : (C - \epsilon I + \epsilon I) \\ &\succeq B : (C - \epsilon I) + (A - B) : \epsilon I. \end{aligned} \tag{3.10.4}$$

The result follows after letting $\epsilon \downarrow 0$ in (3.10.4), in view of Lemma 3.10.4. ∎

A repeated application of Theorem 3.10.5 leads to the following important result:

Theorem 3.10.6 (Series-Parallel Inequality). *Let*
$$A_{ij}, \quad i = 1, 2, \ldots, p; \quad j = 1, 2, \ldots, q$$
be $n \times n$ positive definite matrices. Then
$$\left(\sum_{j=1}^{q} A_{1j} \right) : \cdots : \left(\sum_{j=1}^{q} A_{pj} \right) \succeq \sum_{j=1}^{q} (A_{1j} : \cdots : A_{pj}).$$

The parallel sum was introduced in Anderson and Duffin (1963). The concept is closely related to several other notions in many diverse areas such as shorted operators in electrical network theory [Anderson (1971), Anderson and Trapp (1975)], generalized inverse and the minus partial order, linear model in statistics, and rank additivity. Some references detailing the last three topics are Rao and Mitra (1971), Mitra and Odell (1986), and Mitra and Puri (1982); also see Chapter VI in Bapat (1993b). A proof of the Series-Parallel Inequality, based on statistical ideas, is given by Dey, Hande and Tiku (1994). For a unified treatment of parallel sum and some other concepts using linear programming techniques, see Morley (1989).

3.11. Symmetric function means

Let $A_1, \ldots, A_m$ be $n \times n$ positive definite matrices. The arithmetic mean of $A_1, \ldots, A_m$ is $\frac{1}{m}(A_1 + \cdots + A_m)$, whereas the harmonic mean is $m(A_1^{-1} + \cdots + A_m^{-1})^{-1}$, or, equivalently, $m(A_1 : \cdots : A_m)$. The arithmetic mean–harmonic mean inequality, which is well known for positive numbers, extends to positive definite matrices as follows.

Theorem 3.11.1. *Let $A_1, \ldots, A_m$ be $n \times n$ positive definite matrices. Then*
$$\frac{1}{m}(A_1 + \cdots + A_m) \succeq m(A_1^{-1} + \cdots + A_m^{-1})^{-1}.$$

Proof. Set $A_{ij} = A_{(i+j-1) \bmod m}$, $i = 1, 2, \ldots, p$, $j = 1, 2, \ldots, m$, and invoke the Series-Parallel Inequality. The result follows after a trivial simplification. ∎

Suppose $x = (x_1, \ldots, x_m)$ is a vector of positive numbers. Let $e_{r,m}(x)$ denote the r-th *elementary symmetric function* in $x_1, \ldots, x_m$, i.e.,
$$e_{r,m}(x) = \sum_{i_1 < \cdots < i_r} x_{i_1} \cdots x_{i_r}, \quad r = 1, 2, \ldots, m.$$
We set $e_{0,m}(x) = 1$. Define
$$T_{r,m}(x) = \frac{\binom{m}{r-1}}{\binom{m}{r}} \frac{e_{r,m}(x)}{e_{r-1,m}(x)}, \quad r = 1, 2, \ldots, m.$$

Observe that $T_{1,m}(x)$ is the arithmetic mean, whereas $T_{m,m}(x)$ is the harmonic mean of $x_1, \ldots, x_m$. Thus $T_{r,m}(x), r = 1, 2, \ldots, m$ can be thought of as a family of means, parametrized by r. These means have been called *symmetric function means*. Marcus and Lopes (1957) prove the following superadditivity property: If x and y are positive vectors of order m, then

$$T_{r,m}(x + y) \geq T_{r,m}(x) + T_{r,m}(y). \tag{3.11.1}$$

It is possible to represent $T_{r,m}(x)$ in terms of the arithmetic and the harmonic means of certain subvectors of x. To describe this representation we introduce the following notation. For $j = 1, 2, \ldots, m$, let $\widehat{x}_j$ denote the vector x with its j-th component removed. Then

$$T_{r,m}(x) = \sum_{j=1}^{m} \frac{1}{m - r + 1} x_j : \frac{1}{r - 1} T_{r-1,m}(\widehat{x}_j), \quad r = 1, 2, \ldots, m. \tag{3.11.2}$$

For example, with $m = 3, r = 2$, we have

$$T_{2,3}(x) = \frac{x_1 x_2 + x_1 x_3 + x_2 x_3}{x_1 + x_2 + x_3}$$

$$= \frac{1}{2} \{x_1 : (x_2 + x_3) + x_2 : (x_1 + x_3) + x_3 : (x_1 + x_2)\}.$$

There is a dual representation as well. Thus, for $r = 1, 2, \ldots, m$,

$$T_{r,m}(x) = \{rx_1 + (m - r)T_{r,m-1}(\widehat{x}_j)\} : \cdots : \{rx_m + (m - r)T_{r,m-1}(\widehat{x}_m)\}. \tag{3.11.3}$$

Again, as an example, with $m = 3, r = 2$,

$$\frac{1}{2} T_{2,3}(x) = \frac{1}{2} \frac{x_1 x_2 + x_1 x_3 + x_2 x_3}{x_1 + x_2 + x_3}$$

$$= \left\{x_1 + \frac{x_2 x_3}{x_2 + x_3}\right\} : \left\{x_2 + \frac{x_1 x_3}{x_1 + x_3}\right\} : \left\{x_3 + \frac{x_1 x_2}{x_1 + x_2}\right\}.$$

We now consider m-tuples of positive definite matrices. Let $\mathbf{A} = (A_1, \ldots, A_m)$ be an m-tuple of $n \times n$ positive definite matrices. We may use (3.11.2) and (3.11.3) to define two families of symmetric function means $\alpha_{r,m}(\mathbf{A})$, $\beta_{r,m}(\mathbf{A})$, $r = 1, 2, \ldots, m$. Set $\alpha_{1,m}(\mathbf{A}) = \frac{1}{m} \sum_{i=1}^{m} A_i$ and then define inductively

$$\alpha_{r,m}(\mathbf{A}) = \sum_{j=1}^{m} \frac{1}{m - r + 1} A_j : \frac{1}{r - 1} \alpha_{r-1,m}(\widehat{\mathbf{A}}_j), \quad r = 1, 2, \ldots, m,$$

where $\widehat{\mathbf{A}}_j$ denotes the $(m-1)$-tuple obtained by removing A_j from $\mathbf{A}$. Similarly, set $\beta_{m,m} = m(A_1 : \cdots : A_m)$ and then define inductively, for $r = 1, 2, \ldots, m$,

$$\beta_{r,m}(\mathbf{A}) = \{rA_1 + (m - r)\beta_{r,m-1}(\widehat{\mathbf{A}}_1)\} : \cdots : \{rA_m + (m - r)\beta_{r,m-1}(\widehat{\mathbf{A}}_m)\}.$$

Inequality (3.11.1) can be generalized to both $\alpha_{r,m}(\mathbf{A})$ and $\beta_{r,m}(\mathbf{A})$ (see Exercise 12). Besides superadditivity, not many nontrivial properties of these means are known, although there are certain unsolved problems in the area. In particular, we mention the following conjecture, posed in Anderson, Morley, and Trapp (1984):

Conjecture. Let $\mathbf{A} = (A_1, \ldots, A_m)$ be an m-tuple of $n \times n$ positive definite matrices. Then

(i) $\alpha_{r,m}(\mathbf{A}) \succeq \alpha_{r+1,m}(\mathbf{A})$, $\beta_{r,m}(\mathbf{A}) \succeq \beta_{r+1,m}(\mathbf{A})$, $r = 1, 2, \ldots, m$.
(ii) $\alpha_{r,m}(\mathbf{A}) \succeq \beta_{r,m}(\mathbf{A})$, $r = 1, 2, \ldots, m$.

For verification of some special cases of the conjecture, see Ando (1983) and Ando and Kubo (1989). A weighted version of symmetric function means of scalars has been considered in Bapat (1993a).

Exercises

1. Let A be a positive $n \times n$ matrix. Refer to the notation introduced in Section 3.2 and show that if $x \in P^m$, $y \in P^n$, then

$$\sum_i \sum_j a_{ij} x_i y_j \le \sum_i \sum_j a_{ij} \tilde{x}_i \tilde{y}_j.$$

2. Let A be a positive $m \times n$ matrix and suppose $(x, y) \in P^m \times P^n$ is a fixed point of the map ϕ_A introduced in Section 3.2. Show that the positive coordinates of (x, y) constitute optimal strategies for the corresponding submatrix of A.

3. Suppose A is a positive, symmetric $n \times n$ matrix. Show that there exist positive numbers $z_1, \ldots, z_n$ such that $(a_{ij} z_i z_j)$ is doubly stochastic. Furthermore, show that the doubly stochastic matrix obtained this way is unique.
[Hint: For any probability vector $x = (x_1, \ldots, x_n)$, let $\psi(x)$ denote the vector of row sums of $(a_{ij} x_i x_j)$, normalized to a probability vector. Then use the Kronecker Index Theorem. The uniqueness follows by Theorem 2.7.6.]

4. If $A > 0$ is an irreducible $n \times n$ matrix, then show that

$$r(A) = \sup_{x>0} \inf_{y>0} \frac{y^T A x}{y^T x} = \inf_{y>0} \sup_{x>0} \frac{y^T A x}{y^T x}.$$

If one replaces sup and inf by max and inf, respectively, is the resulting equality valid? [See Birkhoff and Varga (1958).]

5. Let $A \geq 0$, $B \geq 0$ be $n \times n$ matrices with right Perron vectors $y^{(1)}$, $y^{(2)}$ and left Perron vectors $x^{(1)}$, $x^{(2)}$, respectively, such that $x^{(1)} \circ y^{(1)} = x^{(2)} \circ y^{(2)}$. Show that $r(A + B) \geq r(A) + r(B)$. [See Elsner and Johnson (1989).]

6. Let $A \geq 0$ be an $n \times n$ matrix, let y be a right Perron vector of A, and suppose that $y_1 \geq \cdots \geq y_n$. Let B be obtained from A as follows: The i-th row of B is a rearrangement of the i-th row of A in nonincreasing order, $i = 1, 2, \ldots, n$. Then show that $By \geq r(A)y$ and conclude that $r(B) \geq r(A)$.

7. Let n^2 nonnegative numbers (not necessarily distinct) be given, and suppose that we want to find an $n \times n$ matrix A with these numbers as entries and with maximum Perron root. Show that it is sufficient to consider only those A for which every row and column is a nonincreasing sequence.
[Hint: Use Exercise 6; see Schwarz (1964).]

8. Let A be a nonnegative matrix with row sums at most one. Define $\mathcal{C}(A)$ as the set of all vectors $y \in R^n$ such that y is a left Perron vector of some stochastic matrix $B \geq A$. Show that $y \in \mathcal{C}(A)$ if and only if $Ay \leq y$. Hence conclude that $\mathcal{C}(A)$ has the following logarithmic convexity property: If $y \in \mathcal{C}(A)$, $z \in \mathcal{C}(A)$, then for $0 \leq \alpha \leq 1$, $y^{(\alpha)} \circ z^{(1-\alpha)} \in \mathcal{C}(A)$. [See Sahi (1993).]

9. Let A and B be nonnegative, sum-symmetric $n \times n$ matrices such that $A + B$ is doubly stochastic. Show that $r(A) + r(B) \geq 1$.

10. Show that a nonnegative matrix A is completely reducible if and only if every positive entry of A is contained in a positive circuit.

11. Let A be an $m \times n$ matrix. Show that

$$|AA^T| \leq \prod_{i=1}^{m} \sum_{j=1}^{n} a_{ij}^2.$$

12. Determine conditions under which equality occurs in the Hadamard Inequality.

13. Show that $A \circ B$ is a principal submatrix of the Kronecker product $A \otimes B$. Deduce the fact that the Hadamard product of positive semidefinite matrices is positive semidefinite.

14. Let A and B be nonnegative $n \times n$ matrices with right Perron vectors u and v, respectively. Show that $u \otimes v$ is a right Perron vector of $A \otimes B$. Deduce that $r(A \otimes B) = r(A)r(B)$.

15. Let A and B be $n \times n$ matrices, where A is positive semidefinite and B is symmetric. Let D be the diagonal matrix with $a_{11}, \ldots, a_{nn}$ along the diagonal. Show that the eigenvalues of $A \circ B$ are majorized by those of DB.

16. Let $x, z \in R^n$. Show that $\sum_{i=1}^{n} z_{[i]} x_{[i]} \geq z^T x$.

17. Let $\mathbf{A} = (A_1, \ldots, A_m)$, $\mathbf{B} = (B_1, \ldots, B_m)$ be m-tuples of $n \times n$ positive definite matrices. Show that $\alpha_{r,m}(\mathbf{A} + \mathbf{B}) \geq \alpha_{r,m}(\mathbf{A}) + \alpha_{r,m}(\mathbf{B})$ and $\beta_{r,m}(\mathbf{A} + \mathbf{B}) \geq \beta_{r,m}(\mathbf{A}) + \beta_{r,m}(\mathbf{B})$.
[Hint: Use the Series-Parallel Inequality and induction; see Anderson, Morley, and Trapp (1984).]

18. Let $M \geq 0$ be an $n \times n$ irreducible matrix with spectral radius $\rho(M)$. Let $Au = \rho(M)u$, $M^T v = \rho(M)v$, where $(u, v) = 1$. For any diagonal matrix $D = \text{diag}\{d_1, \ldots, d_n\}$ with positive diagonal

$$\rho(DM) = \rho(MD) \geq \prod_{i=1}^{n} d_i^{u_i v_i} \rho(M)$$

For any positive vector $x > 0$ we have

$$\prod_{i=1}^{n} \left[\frac{(Mx)_i}{x_i} \right]^{u_i v_i} \geq 1.$$

Equality holds only for positive scalar multiples of u.
(If M is doubly stochastic the first inequality follows easily by the concavity of the log function. The second one follows by the arithmetic mean geometric mean inequality. However the general case where M is just a nonnegative irreducible matrix is quite involved and the implications are far reaching. See: Friedland and Karlin (1975)).

4

Conditionally positive
definite matrices

The class of positive semidefinite matrices is important, both in theory and in applications, and is also well studied. A related class exhibiting interesting properties as well is that of symmetric matrices with exactly one eigenvalue of one sign and the remaining eigenvalues of the other sign. In this chapter we will be mainly concerned with symmetric matrices that are nonnegative and have exactly one positive eigenvalue. Certain closely related classes, such as those of distance matrices, conditionally positive definite matrices, and positive subdefinite matrices, are also studied. We illustrate how these matrices arise naturally in areas such as mathematical programming, numerical analysis, and statistics. In the last section we consider a generalization of the determinant and the permanent. Here a conditionally positive definite matrix based on the number of inversions of a permutation plays an important role.

4.1. Distance matrices

Let

$$H^n = \left\{ x \in R^n : \sum_{i=1}^{n} x_i = 0 \right\}.$$

A real, symmetric $n \times n$ matrix A is said to be *conditionally positive definite (conditionally negative definite)* if $x^T A x \geq (\leq 0)$ for any $x \in H^n$.

Strictly speaking, we should use the term "semidefinite" instead of "definite" in the definition above, but we follow standard terminology and continue to use "definite" for convenience. We abbreviate conditionally positive definite (conditionally negative definite) as c.p.d. (c.n.d.).

161

It is obvious that any positive semidefinite matrix is c.p.d.. The converse, however, is not true. The matrix

$$\begin{bmatrix} 0 & -1 \\ -1 & 0 \end{bmatrix}$$

is c.p.d. but is not positive semidefinite.

There are certain characterizations of c.p.d. matrices in terms of positive semidefinite matrices that we now proceed to develop.

Lemma 4.1.1. *Suppose $A = (a_{ij})$ is an $n \times n$ c.p.d. matrix. Then there exist real numbers $\alpha_1, \ldots, \alpha_n$ such that the matrix $(a_{ij} - \alpha_i - \alpha_j)$ is positive semidefinite.*

Proof. Let $x \in R^n$ and suppose $\sum_{i=1}^n x_i = n$. As usual, $\mathbf{1}$ denotes the column vector of order n with all entries equal to 1. Then $(x - \mathbf{1}) \in H^n$, and hence

$$(x - \mathbf{1})^T A (x - \mathbf{1}) \geq 0.$$

This leads to

$$x^T A x - 2x^T A \mathbf{1} + \mathbf{1}^T A \mathbf{1} \geq 0. \tag{4.1.1}$$

Let

$$\beta_i = (Ae)_i - \frac{\mathbf{1}^T A \mathbf{1}}{2n}.$$

Then, from (4.1.1),

$$x^T A x - \sum_{i=1}^n \beta_i x_i - \sum_{j=1}^n \beta_j x_j \geq 0. \tag{4.1.2}$$

Let $\alpha_i = \frac{\beta_i}{n}$, $i = 1, 2, \ldots, n$ and let $B = (b_{ij})$, where $b_{ij} = a_{ij} - \alpha_i - \alpha_j$. Then (4.1.2) shows that $x^T B x \geq 0$ for all $x \notin H^n$, since, for such x, after a suitable normalization we may assume that $\sum_{i=1}^n x_i = n$. Also, since A is c.p.d., we have $x^T B x \geq 0$ for $x \in H^n$ and thus B is positive semidefinite. This completes the proof. ∎

Lemma 4.1.2. *If A is positive semidefinite then the matrix $(e^{a_{ij}})$ is positive semidefinite.*

Proof. By a repeated application of Lemma 3.7.1, we see that for any positive integer k, (a_{ij}^k) is positive semidefinite, and since

$$e^{a_{ij}} = 1 + a_{ij} + \frac{a_{ij}^2}{2!} + \cdots,$$

the proof is complete. ∎

Theorem 4.1.3. *Let A be a symmetric n × n matrix. Then the following conditions are equivalent:*

(1) A is c.p.d..
(2) There exist numbers $\alpha_1, \ldots, \alpha_n$ such that the matrix $(a_{ij} - \alpha_i - \alpha_j)$ is positive semidefinite.
(3) For any $\beta > 0$, the matrix $(e^{\beta a_{ij}})$ is positive semidefinite.

Proof. (1) $\Rightarrow$ (2): This is contained in Lemma 4.1.1.
(2) $\Rightarrow$ (3): If (2) holds, then by Lemma 4.1.2 it follows that the matrix

$$(e^{\beta(a_{ij}-\alpha_i-\alpha_j)})$$

is positive semidefinite for any $\beta > 0$. Let D be the $n \times n$ diagonal matrix with its i-th diagonal entry equal to $e^{-\beta\alpha_i}$, $i = 1, 2, \ldots, n$. Then

$$e^{\beta a_{ij}} = D(e^{\beta(a_{ij}-\alpha_i-\alpha_j)})D$$

and hence $(e^{\beta a_{ij}})$ must be positive semidefinite.
(3) $\Rightarrow$ (1): Let $x \in H^n$. Since

$$e^{\beta a_{ij}} = 1 + \beta a_{ij} + \frac{\beta^2}{2!}a_{ij}^2 + \cdots,$$

we have

$$\sum_{i=1}^n \sum_{j=1}^n e^{\beta a_{ij}} x_i x_j = \beta \sum_{i=1}^n \sum_{j=1}^n a_{ij} x_i x_j + \frac{\beta^2}{2!} \sum_{i=1}^n \sum_{j=1}^n a_{ij}^2 x_i x_j + \cdots. \quad (4.1.3)$$

If $x^T A x < 0$, then for sufficiently small $\beta > 0$, the right-hand side of (4.1.3) is negative while the left-hand side must be nonnegative, which is a contradiction. Hence $x^T A x \geq 0$ and the proof is complete. ■

Lemma 4.1.4. *If A is a c.n.d. matrix, then A has at most one positive eigenvalue.*

Proof. Suppose A has two positive eigenvalues, λ and μ, with x and y as the corresponding eigenvectors, respectively. Here, λ and μ may be distinct, or $\lambda = \mu$, in which case it is an eigenvalue of multiplicity greater than one. Thus, we may assume that $x^T A y = 0$.
Since $x^T A x = \lambda x^T x > 0$, then $x \notin H^n$. Similarly, $y \notin H^n$, and by a suitable normalization we assume that $\sum_{i=1}^n x_i = \sum_{i=1}^n y_i$. Then $x - y \in H^n$, but

$$(x - y)^T A(x - y) = x^T A x + y^T A y$$
$$= \lambda x^T x + \mu y^T y > 0,$$

and this contradicts the fact that A is c.n.d.. Therefore, A has at most one positive eigenvalue, counting multiplicities. ■

We now turn to matrices that are c.n.d. and nonnegative. First, we note a simple consequence of Lemma 4.1.4.

Corollary 4.1.5. *Let A be a nonnegative, nonzero matrix that is c.n.d.. Then A has exactly one positive eigenvalue.*

Proof. Because A is a symmetric matrix, its eigenvalues are all real. Since A is nonzero, the eigenvalues cannot all be zero, and further, since A has nonnegative trace, at least one of its eigenvalues must be positive. However, since A is c.n.d., by Lemma 4.1.4, it has at most one positive eigenvalue, and the result is proved. ∎

If $x \in R^s$, recall that the Euclidean norm of x is defined as

$$||x|| = \left(\sum_{i=1}^{s} x_i^2 \right)^{\frac{1}{2}}.$$

If $x^1, \ldots, x^n$ are points in an Euclidean space, their "squared distance matrix" may be defined as the matrix of inter-point squared distances $(||x^i - x^j||^2)$. Note that such a matrix is symmetric and has zeros on the main diagonal. An additional property of such matrices is proved in the next result.

Lemma 4.1.6. *Let $x^1, \ldots, x^n \in R^s$ for some s, and let $A = (a_{ij})$ be the $n \times n$ matrix defined as $a_{ij} = ||x^i - x^j||^2$, $i, j = 1, 2, \ldots, n$. Then A is c.n.d..*

Proof. We have

$$a_{ij} = ||x^i - x^j||^2$$
$$= ||x^i||^2 + ||x^j||^2 - 2\langle x^i, x^j \rangle,$$

where, as usual, $\langle x^i, x^j \rangle$ is the inner product of x^i, x^j. The matrix $((x^i, x^j))$ is positive semidefinite and now the result follows from Theorem 4.1.3. ∎

Lemma 4.1.6 has an interesting converse, which is attributed to Schoenberg [see, for example, Blumenthal (1970)]. This is proved next.

Theorem 4.1.7. *Let A be a symmetric $n \times n$ matrix with zero diagonal entries. Then there exist vectors $x^1, \ldots, x^n \in R^s$ for some s such that $a_{ij} = ||x^i - x^j||^2$ for all i, j if and only if A is c.n.d..*

Proof. The "if" part was proved in Lemma 4.1.6. So now suppose that A is c.n.d. with zero diagonal entries. By Theorem 4.1.3, there exist real numbers

$\alpha_1, \ldots, \alpha_n$ and a positive semidefinite matrix B such that

$$a_{ij} = \alpha_i + \alpha_j - b_{ij}, \quad i, j = 1, 2, \ldots, n.$$

Since B is positive semidefinite, there exist vectors $y^1, \ldots, y^n \in R^s$ for some s such that

$$b_{ij} = \langle y^i, y^j \rangle, \quad i, j = 1, 2, \ldots, n.$$

Now, since $a_{ii} = 0$ for all i, we have

$$\alpha_i = \frac{b_{ii}}{2} = \frac{\|y^i\|^2}{2}, \quad i = 1, 2, \ldots, n,$$

and thus

$$a_{ij} = \frac{\|y^i\|^2}{2} + \frac{\|y^j\|^2}{2} - \langle y^i, y^j \rangle$$

$$= \|x^i - x^j\|^2, \quad i, j = 1, 2, \ldots, n,$$

where $x^i = \frac{y^i}{\sqrt{2}}, i = 1, 2, \ldots, n$. That completes the proof. ∎

Blumenthal (1970) gives a detailed account of distance matrices. Parthasarathy and Schmidt (1972) deal with c.p.d. kernels and their applications in probability theory.

4.2. Quasi-convex quadratic forms

An important concept in optimization theory is that of convexity. Recall that a real-valued function ϕ defined on the convex set S is said to be *convex* if for any $x, y \in S$ and for any $\lambda \in [0, 1]$,

$$\phi(\lambda x + (1 - \lambda)y) \le \lambda \phi(x) + (1 - \lambda)\phi(y).$$

A weaker concept, which is also relevant in mathematical programming and in economics, is that of quasi-convexity.

A real-valued function ϕ defined on the convex set S is said to be *quasi-convex* if for any $x, y \in S$ and for any $\lambda \in [0, 1]$,

$$\phi(\lambda x + (1 - \lambda)y) \le \max\{\phi(x), \phi(y)\}.$$

Alternatively, ϕ is quasi-convex on S if the level sets $\{x \in S : \phi(x) \le \alpha\}$ are convex for all α. If $-\phi$ is quasi-convex, ϕ is said to be *quasi-concave*. Clearly, if ϕ is convex, then it is quasi-convex. The following function is quasi-convex but not convex:

$$\phi(x) = \begin{cases} 0, & x < 0 \\ 1, & x \ge 0. \end{cases}$$

For a differentiable function ϕ of $x_1, \ldots, x_n$, we denote the gradient vector of ϕ at x, by $\nabla \phi(x)$, defined by

$$(\nabla \phi(x))_i = \frac{\partial \phi(x)}{\partial x_i}, \quad i = 1, 2, \ldots, n.$$

For differentiable functions, a related concept is that of pseudo-convexity. A real-valued differentiable function ϕ defined on the convex set S is said to be *pseudo-convex* if

$$\nabla \phi(x)^T (y - x) \geq 0 \implies \phi(y) \geq \phi(x)$$

for all $x, y \in S$.

We will be interested in functions of the form $\phi(x) = x^T A x$, where A is a symmetric matrix. For such functions the definitions of quasi-convexity and pseudo-convexity can be made more transparent as is indicated in the next two results.

Lemma 4.2.1. *Let A be a symmetric $n \times n$ matrix and let $\phi(x) = x^T A x, x \in R^n$. A necessary and sufficient condition for ϕ to be quasi-convex on the convex set S is that $y^T A y \leq x^T A x$ implies $x^T A y \leq x^T A x$ for all $x, y \in S$.*

Proof. We first prove the sufficiency. Let $x, y \in S, \lambda \in [0, 1]$, and suppose $y^T A y \leq x^T A x$. Then $x^T A y \leq x^T A x$ and

$$(\lambda x + (1 - \lambda)y)^T A(\lambda x + (1 - \lambda)y) = \lambda^2 x^T A x + 2\lambda(1 - \lambda)x^T A y$$
$$+ (1 - \lambda)^2 y^T A y \leq x^T A x.$$

Similarly, if $x^T A x \leq y^T A y$, then

$$(\lambda x + (1 - \lambda)y)^T A(\lambda x + (1 - \lambda)y) \leq y^T A y,$$

and so ϕ is quasi-convex. Conversely, suppose ϕ is quasi-convex on S. Let $x, y \in S$ with $y^T A y \leq x^T A x$. For any $\lambda \in (0, 1)$, we have

$$(\lambda x + (1 - \lambda)y)^T A(\lambda x + (1 - \lambda)y) \leq x^T A x.$$

Hence

$$\lambda^2 x^T A x + 2\lambda(1 - \lambda)x^T A y + (1 - \lambda)^2 y^T A y \leq x^T A x,$$

and thus

$$2\lambda(1 - \lambda)x^T A y + (1 - \lambda)^2 y^T A y \leq (1 - \lambda^2)x^T A x. \tag{4.2.1}$$

Dividing (4.2.1) throughout by $\lambda(1 - \lambda)$ and letting $\lambda \to 1$, it follows that $x^T A y \leq x^T A x$. ∎

Lemma 4.2.2. *Let A be a symmetric $n \times n$ matrix and let $\phi(x) = x^T Ax, x \in R^n$. Then ϕ is pseudo-convex on the convex set S if and only if for all $x, y \in S$, $x^T Ax \leq y^T Ax$ implies $x^T Ax \leq y^T Ay$.*

Proof. The gradient vector of $\phi(x) = x^T Ax$ is seen to be $\nabla\phi(x) = 2Ax$. The result now follows from the definition of pseudo-convexity. ∎

We summarize the previous two results in the next lemma in a form that will prove to be convenient later.

Lemma 4.2.3. *Let A be a symmetric $n \times n$ matrix and let $S \subset R^n$ be convex. Then $Q(x) = x^T Ax$ is*
(a) *convex on S if and only if for all $x, y \in S$,*

$$2(x - y)^T Ax \geq x^T Ax - y^T Ay;$$

(b) *quasi-convex on S if and only if for all $x, y, \in S$,*

$$x^T Ax \geq y^T Ay \implies x^T Ax \geq y^T Ax;$$

(c) *pseudo-convex on S if and only if for all $x, y \in S$,*

$$x^T Ax > y^T Ay \implies x^T Ax > y^T Ax.$$

We now introduce some additional definitions. A symmetric $n \times n$ matrix A is said to be *positive subdefinite* if for any v, $v^T Av < 0$ implies that either $Av \geq 0$ or $Av \leq 0$. A symmetric $n \times n$ matrix A is said to be *strictly positive subdefinite* if for any v, $v^T Av < 0$ implies either $Av > 0$ or $Av < 0$.

The definitions of negative subdefinite and strictly negative subdefinite matrices are similar. The symmetric $n \times n$ matrix A is said to be *merely positive subdefinite (strictly merely positive subdefinite)* if it is positive subdefinite (strictly positive subdefinite) but not positive semidefinite. The proof of the next result is immediate from these definitions.

Lemma 4.2.4. *(a) If A is positive subdefinite, then*

$$x \geq 0, \quad x^T Ax < 0 \implies Ax \leq 0.$$

(b) If A is strictly positive subdefinite, then

$$x \geq 0, \quad x^T Ax < 0 \implies Ax < 0.$$

Lemma 4.2.5. *A merely positive subdefinite matrix has at least one negative eigenvalue. An eigenvector belonging to a negative eigenvalue is either ≥ 0 or ≤ 0.*

Proof. Suppose A is merely positive subdefinite. Then, since A is not positive semidefinite, it has at least one negative eigenvalue. Let λ be a negative eigenvalue of A and y the corresponding eigenvector. Then $Ay = \lambda y$ implies $y^T A y = \lambda y^T y < 0$, and therefore Ay is either ≥ 0 or ≤ 0. It follows that $y \geq 0$ or $y \leq 0$, and the proof is complete. ∎

Lemma 4.2.6. *Let A be an $n \times n$ merely positive subdefinite matrix. Let λ be a negative eigenvalue of A with y as the corresponding eigenvector such that $y \geq 0$ and $y^T y = 1$. Then the matrix*

$$D = \begin{bmatrix} A & \lambda y \\ \lambda y^T & \lambda \end{bmatrix}$$

is merely positive subdefinite.

Proof. Note that by Lemma 4.2.5 there is no loss of generality in assuming that $y \geq 0$, and by a suitable normalization we can make $y^T y = 1$. Let $w = (v^T, \alpha)^T$ be a vector in R^{n+1} and let $z = v + \alpha y$. Then $w^T D w = z^T A z$. Also,

$$Dw = \begin{bmatrix} Az \\ y^T Az \end{bmatrix}.$$

Thus,

$$w^T Dw < 0 \Rightarrow z^T A z < 0$$
$$\Rightarrow Az \geq 0 \text{ or } Az \leq 0$$
$$\Rightarrow Dw \geq \text{ or } Dw \leq 0.$$

Hence, D is positive subdefinite. Since D has a negative diagonal entry λ, it cannot be positive semidefinite and hence it is merely positive subdefinite. ∎

Lemma 4.2.7. *If A is merely positive subdefinite and has a negative diagonal entry, then $A \leq 0$.*

Proof. Let $a_{11} < 0$ and denote the i-th unit vector by e_i. Since $e_1^T A e_1 = a_{11} < 0$, it follows by Lemma 4.2.4 that $Ae_1 \leq 0$. Thus $a_{1j} \leq 0$ for all j. Now suppose $a_{22} > 0$. If $a_{12} = 0$, let

$$x = (a_{22} - 2a_{11})e_1 - 2a_{11}e_2 \geq 0.$$

Then

$$x^T A x = a_{11}\left(a_{22}^2 + 4a_{11}^2\right) < 0,$$

whereas

$$(Ax)_2 = -2a_{11}a_{22} > 0,$$

in contradiction with Lemma 4.2.4. If $a_{12} < 0$, we get a similar contradiction by letting $x = a_{22}e_1 - 2a_{12}e_2$. Thus, we conclude that $a_{22} \leq 0$. Similarly, $a_{jj} \leq 0$ for all j. Now fix j, k. If a_{jj} or a_{kk} is negative, it follows, as in the beginning of the proof, that $a_{jk} \leq 0$. Suppose $a_{jj} = a_{kk} = 0$. If $a_{jk} > 0$, let $x = e_j - e_k$. Then $x^T A x = 2a_{jk} < 0$ and $(Ax)_j = -a_{jk} < 0$, $(Ax)_k = a_{jk} > 0$, which contradicts the fact that A is merely positive subdefinite. That completes the proof. ∎

The assumption that A has a negative diagonal entry can be dropped in Lemma 4.2.7 as we prove next.

Theorem 4.2.8. *If A is merely positive subdefinite, then $A \leq 0$.*

Proof. If A is merely positive subdefinite then, by Lemma 4.2.6, so is the matrix D defined there. But D has a negative diagonal entry λ. Hence by Lemma 4.2.7, $D \leq 0$, which implies that $A \leq 0$, and the proof is complete. ∎

The next result gives a necessary and sufficient condition for a symmetric matrix to be merely positive subdefinite.

Theorem 4.2.9. *The symmetric $n \times n$ matrix A is merely positive subdefinite if and only if $A \leq 0$ and A has exactly one (simple) negative eigenvalue.*

Proof. We remark that the word "simple" in parentheses is always applicable when we talk of exactly one negative (or positive) eigenvalue in this chapter, although it may not be explicitly mentioned on each occasion. First, suppose that A is merely positive subdefinite. By Theorem 4.2.8, $A \leq 0$. Suppose λ and μ are two negative eigenvalues of A. Let x and y be the corresponding eigenvectors, so that $x^T y = 0$. We assume that $x^T x = y^T y = 1$. By Lemma 4.2.5 we may assume that $x \geq 0$, $y \geq 0$. Now,

$$(x - y)^T A(x - y) = x^T A x + y^T A y$$
$$= \lambda + \mu < 0,$$

and hence, $z = A(x - y)$ is either ≥ 0 or ≤ 0. However, $x^T z = \lambda < 0$ and $y^T z = \mu > 0$. This is a contradiction since $x \geq 0$, $y \geq 0$ and hence A has exactly one negative eigenvalue.

Conversely, suppose $A \leq 0$ and has exactly one negative eigenvalue. Then A is not positive semidefinite and hence there is a vector x such that $x^T A x < 0$. We want to show that for any such x, either $Ax \geq 0$ or $Ax \leq 0$. Suppose Ax has both positive and negative components. Then there exists a vector $y > 0$

such that $y^T A x = 0$. By the Spectral Theorem, there exists an orthogonal matrix P and a diagonal matrix D such that $A = PDP^T$, where the diagonal entries of D are the eigenvalues of A. We assume, without loss of generality, that $d_{11} < 0, d_{ii} \geq 0, i = 2, \ldots, n$. Let $u = P^T x, v = P^T y$. Then

$$u^T D u = x^T A x < 0,$$
$$v^T D u = y^T A x = 0,$$

and

$$v^T D v = y^T A y < 0.$$

Also,

$$u^T D u = \sum_{i=1}^{n} d_{ii} u_i^2 < 0 \Rightarrow u_1 \neq 0$$

$$v^T D v = \sum_{i=1}^{n} d_{ii} v_i^2 < 0 \Rightarrow v_1 \neq 0.$$

Let $\alpha = \frac{u_1}{v_1}$. Then

$$(u - \alpha v)^T D(u - \alpha v) = u^T D u - 2\alpha v^T D u + \alpha^2 v^T D v < 0,$$

whereas

$$(u - \alpha v)^T D(u - \alpha v) = \sum_{i=1}^{n} d_{ii} (u_i - \alpha v_i)^2 \geq 0.$$

This contradiction shows that either $Ax \geq 0$ or $Ax \leq 0$, and the proof is complete. ∎

The next result gives a simple criterion to distinguish the strictly merely positive subdefinite matrices.

Theorem 4.2.10. *Let A be a merely positive subdefinite matrix. Then A is strictly merely positive subdefinite if and only if it does not contain a zero row.*

Proof. If A is merely positive subdefinite, then there is a vector v with $v^T A v < 0$. If A contains a zero row, then Av contains a zero component and thus A cannot be strictly positive subdefinite.

Now suppose that A is merely positive subdefinite and has no zero row. By Theorem 4.2.9, A has a unique negative eigenvalue. Let it be λ, and let y be the corresponding eigenvector with $y \geq 0$ and $y^T y = 1$. By the Spectral Theorem we can write

$$A = \lambda y y^T + B,$$

where B is positive semidefinite. Suppose y has a zero component and, without loss of generality, let it be y_1. Then $a_{11} = \lambda y_1^2 + b_{11} = b_{11}$. Since $A \leq 0$, we have $b_{11} \leq 0$. Since B is positive semidefinite, the first row of B, and hence that of A, must have all zeros. This is a contradiction and so $y > 0$.

If A is not strictly merely positive subdefinite, then there exists a vector z such that $z^T A z < 0$ and either $Az \geq 0$ or $Az \leq 0$, with some component of Az, say, the first, being equal to zero. We assume, without loss of generality, that $Az \geq 0$ and that the second component of Az is positive.

Let a_1^T, a_2^T denote the first two rows of A and let $b_1 = z + \alpha y, b_2 = z - \alpha y$, where $\alpha > 0$ is so small that the following inequalities hold:

$$b_1^T A b_1 = z^T A z + 2\alpha \lambda z^T y + \alpha^2 \lambda < 0$$
$$b_2^T A b_2 = z^T A z + 2\alpha \lambda z^T y + \alpha^2 \lambda < 0$$
$$a_2^T b_1 = a_2^T z + \alpha a_2^T y > 0$$
$$a_2^T b_2 = a_2^T z - \alpha a_2^T y > 0 .$$

Since A is positive subdefinite,

$$a_1^T b_1 = a_1^T z + \alpha a_1^T y$$
$$= a_1^T z + \alpha \lambda y_1$$
$$= \alpha \lambda y_1 \geq 0$$

and

$$a_1^T b_2 = a_1^T z - \alpha a_1^T y$$
$$= a_1^T z - \alpha \lambda y_1$$
$$= -\alpha \lambda y_1 \geq 0.$$

Thus $y_1 = 0$, which is a contradiction, and the proof is complete. ∎

Theorem 4.2.11. *The quadratic form $Q(x) = x^T A x$ is convex in R^n if and only if A is positive semidefinite.*

Proof. If $Q(x)$ is convex, then by setting $y = 0$ in Lemma 4.2.3 (a), we see that $x^T A x \geq 0$ for all $x \in R^n$ and hence A is positive semidefinite. Conversely, suppose A is positive semidefinite and let $x, y \in R^n$. Then $(x - y)^T A(x - y) \geq 0$; this is equivalent to

$$2(x - y)^T A x \geq x^T A x - y^T A y.$$

It follows that $Q(x)$ is convex by Lemma 4.2.3. ∎

We now give a characterization of quadratic forms that are quasi-convex over the nonnegative orthant. The result is analogous to Theorem 4.2.11.

Theorem 4.2.12. *The quadratic form $Q(x) = x^T A x$ is quasi-convex for every nonnegative x if and only if A is positive subdefinite.*

Proof. First, suppose A is positive subdefinite, let $x, y > 0$, and let

$$x^T A x \geq y^T A y. \tag{4.2.2}$$

If

$$(x - y)^T A(x - y) \geq 0, \tag{4.2.3}$$

then by adding (4.2.2) to (4.2.3) we see that $x^T A x \geq y^T A x$.

If $(x - y)^T A(x - y) < 0$, then because A is positive subdefinite, $A(x - y)$ is either ≥ 0 or ≤ 0. However, from (4.2.2),

$$(x + y)^T A(x - y) \geq 0,$$

and hence $A(x - y) \geq 0$. It follows that $x^T A x \geq y^T A x$. Thus the quasi-convexity of $Q(x)$ is proved over the positive orthant. The quasi-convexity over the nonnegative orthant follows by continuity.

Conversely, suppose that A is not positive subdefinite. Then there exists a vector $z \geq 0$ such that $z^T A z < 0$ and that Az has components of both signs. Thus there exists a vector $w > 0$ such that $w^T A z = 0$.

Choose $\alpha > 0$ so small that $u = \frac{1}{2}z + \frac{\alpha}{2}w \geq 0$, and let $v = \frac{1}{2}z - \frac{\alpha}{2} \geq 0$. Then $u + v = z, u - v = \alpha w$ and therefore

$$\alpha^2 w^T A w = (u - v)^T A(u - v) < 0, \tag{4.2.4}$$
$$\alpha z^T A w = (u + v)^T A(u - v) = 0. \tag{4.2.5}$$

Adding (4.2.4) and (4.2.5) we see that $u^T A u < u^T A v$. However, from (4.2.5), $u^T A u = v^T A v$. Thus, $Q(x)$ is not quasi-convex in view of Lemma 4.2.3 (b), and the proof is complete. ∎

A characterization of quadratic forms that are pseudo-convex over the non-negative orthant (with the origin removed) is given in Exercise 4 and can be proved using similar techniques.

The development in this section is largely based on Martos (1969) and Cottle and Ferland (1972). The theory can be extended to quasi-convex quadratic functions and also to differentiable functions. We refer to Martos (1971) and to Ferland (1981) for details.

4.3. An interpolation problem

Suppose $x^1, \ldots, x^n$ are distinct points in the plane and we want to interpolate the data $y^1, \ldots, y^n$ at $x^1, \ldots, x^n$ by the function

$$f(x) = \sum_{i=1}^{n} c_i \sqrt{1 + ||x - x^i||^2},$$

i.e., we want to determine constants $c_1, \ldots, c_n$ such that the function f satisfies $f(x^i) = y^i, i = 1, 2, \ldots, n$. The main purpose of this section is to prove a result of Micchelli (1986) that a unique choice of $c_1, \ldots, c_n$ always exists. In the process we will prove some results of independent interest.

The equations $f(x^i) = y^i, i = 1, 2, \ldots, n$ can be expressed as linear equations in $c_1, \ldots, c_n$ with the coefficient matrix

$$(\sqrt{1 + ||x^i - x^j||^2});$$

thus, it will be sufficient to show that this matrix is nonsingular. We will, in fact, prove a more general result.

The *Vandermonde matrix* of order n is a matrix of the form

$$V(x_1, \ldots, x_n) = \begin{bmatrix} 1 & x_1 & x_1^2 & \cdots & x_1^{n-1} \\ \vdots & \vdots & \vdots & \cdots & \vdots \\ 1 & x_n & x_n^2 & \cdots & x_n^{n-1} \end{bmatrix},$$

where $x_1, \ldots, x_n$ are real numbers.

Lemma 4.3.1. *The determinant of a Vandermonde matrix is given by*

$$|V(x_1, \ldots, x_n)| = \prod_{i>j}(x_i - x_j).$$

Proof. A proof of the lemma based on the divisibility of polynomials can be found in several sources [see, for example, Bellman (1970)]. We sketch a different but well-known proof.

In $V(x_1, \ldots, x_n)$, subtract the first row from each of the remaining rows and then evaluate the determinant in terms of the first column, to get

$$|V(x_1, \ldots, x_n)| = (x_2 - x_1) \cdots (x_n - x_1) \begin{vmatrix} 1 & x_2 + x_1 & x_2^2 + x_1 x_2 + x_1^2 & \cdots \\ \vdots & \vdots & \vdots & \\ 1 & x_n + x_1 & x_n^2 + x_1 x_n + x_1^2 & \cdots \end{vmatrix}.$$

Since the determinant is a multilinear function of its columns, the determinant on the right-hand side of the equation above can be expressed as a sum of several determinants. All these determinants can be seen to be zero except a Vandermonde determinant $|V(x_2, \ldots, x_n)|$. Thus

$$|V(x_1, \ldots, x_n)| = (x_2 - x_1) \cdots (x_n - x_1)|V(x_2, \ldots, x_n)|.$$

Now the proof is completed by an inductive argument.   ■

Corollary 4.3.2. $V(x_1, \ldots, x_n)$ *is nonsingular if and only if* $x_1, \ldots, x_n$ *are distinct. Furthermore, if* $x_1 < \cdots < x_n$, *then* $|V(x_1, \ldots, x_n)| > 0$.

Theorem 4.3.3. *Let* A *be an* $n \times n$ *positive semidefinite matrix with distinct rows. Then* $(e^{a_{ij}})$ *is positive definite.*

Proof. Since A is positive semidefinite, $A = BB^T$ for some $n \times n$ matrix B. Because the rows of A are distinct, those of B must also be distinct. For any positive integer k, let $A^{(k)}$ denote the matrix (a_{ij}^k). Now,

$$(e^{a_{ij}}) = J + A + \frac{1}{2!}A^{(2)} + \cdots + \frac{1}{k!}A^{(k)} + \cdots,$$

where J is the matrix of all ones.

Note that

$$A^{(2)} = A \circ A$$
$$= BB^T \circ BB^T$$
$$= (B * B)(B * B)^T,$$

where, if X and Y are matrices of order $n \times p$ and $n \times q$, respectively, then $X * Y$ is the $n \times pq$ matrix obtained by taking the Schur product of every column of X with every column of Y.

Define $B^{*(k)} = B * B * \cdots * B$, taken k times, for any positive integer k. Then it can be seen that

$$A^{(k)} = B^{*(k)} B^{*(k)^T}, \quad k = 1, 2, \ldots.$$

Hence $(e^{a_{ij}}) = DD^T$, where D is the infinite matrix given by

$$D = \left(\mathbf{1}, B, \frac{1}{\sqrt{2!}} B * B, \ldots, \frac{1}{\sqrt{k!}} B^{*(k)}, \cdots\right)$$

with $\mathbf{1}$ as the vector of all ones. Thus, the result will be proved if we show that the following infinite matrix has n linearly independent columns:

$$C = (\mathbf{1}, B, B * B, \ldots, B^{*(k)}, \ldots).$$

Since B has distinct rows, there exists a vector π such that $z^T = \pi^T B$ has distinct components. Therefore, the Vandermonde matrix

$$V = \begin{bmatrix} 1 & z_1 & z_1^2 & \cdots & z_1^{n-1} \\ \vdots & \vdots & \vdots & \cdots & \vdots \\ 1 & z_n & z_n^2 & \cdots & z_n^{n-1} \end{bmatrix}$$

is nonsingular. Again, since the determinant is a multilinear function of the columns, $|V|$ can be expanded as a linear combination of determinants of certain $n \times n$ submatrices of C. Since $|V|$ is nonzero, it follows that C has a nonsingular $n \times n$ submatrix, and the proof is complete. ∎

Lemma 4.3.4. Let $x^1, \ldots, x^n$ be distinct vectors in R^s and let A be the $n \times n$ matrix defined by

$$a_{ij} = e^{-\|x^i - x^j\|^2}, \quad i, j = 1, 2, \ldots, n.$$

Then A is positive definite.

Proof. We have

$$\|x^i - x^j\|^2 = \|x^i\|^2 + \|x^j\|^2 - 2\langle x^i, x^j \rangle.$$

Since $x^1, \ldots, x^n$ are distinct, by Theorem 4.3.3,

$$\left(e^{\langle x^i, x^i \rangle} \right)$$

is positive definite. Let D be the $n \times n$ diagonal matrix with the i-th diagonal entry equal to

$$e^{-\|x^i\|^2}, \quad i = 1, 2, \ldots, n.$$

Then

$$\left(e^{-\|x^i - x^j\|^2} \right) = D\left(e^{2\langle x^i, x^j \rangle} \right) D,$$

and hence the result is proved. ∎

Lemma 4.3.5. Let A be an $n \times n$ c.n.d. matrix that is not negative semidefinite and suppose $c^T A c = 0$ for $c \in H^n$ if and only if $c = 0$. Then A is nonsingular with exactly one positive eigenvalue.

Proof. Suppose $Ax = 0$ for some x. If $x \in H^n$, then because $x^T A x = 0$ we have $x = 0$ by the hypothesis on A. Let $\sum_{i=1}^n x_i \neq 0$. Without loss of generality, let $\sum_{i=1}^n x_i > 0$. Since A is not negative semidefinite, there exists y such that

$y^T A y > 0$. Again, $\sum_{i=1}^n y_i$ must be nonzero; without loss of generality, let it be positive. Let

$$\alpha = \frac{\sum_{i=1}^n x_i}{\sum_{i=1}^n y_i}.$$

Then $(x - \alpha y) \in H^n$ and $(x - \alpha y)^T A (x - \alpha y) = \alpha^2 y^T A y > 0$, which is a contradiction. Thus $x = 0$ and hence A must be nonsingular. By Lemma 4.1.4, A has at most one positive eigenvalue. Since A is not negative semidefinite, it has at least one positive eigenvalue. Thus A has exactly one positive eigenvalue, and the proof is complete. ∎

Let $F: (0, \infty) \to R$ be a function that is infinitely differentiable, and suppose

$$(-1)^k F^{(k)}(x) \geq 0, k = 0, 1, \ldots,$$

for all x, where $F^{(k)}$ denotes the k-th derivative of F and $F^{(0)} = F$. Then F is said to be *completely monotonic*.

A simple example of a completely monotonic function is $F(x) = e^{-x}, x > 0$. An important theorem of Bernstein characterizes completely monotonic functions as follows [see, for example, Phelps (1966), p. 11]:

Theorem 4.3.6. *Suppose $F : (0, \infty) \to R$ is infinitely differentiable. Then F is completely monotonic if and only if*

$$F(t) = \int_0^\infty e^{-t\sigma} d\mu(\sigma), \quad t > 0,$$

where $d\mu(\sigma)$ is a Borel measure on $(0, \infty)$.

We now prove one of the main results of this section.

Theorem 4.3.7. *Let F be positive on $(0, \infty)$ and continuous on $[0, \infty]$. Let F' be completely monotonic but not constant on $(0, \infty)$. Then for any distinct vectors $x^1, \ldots, x^n$ in R^s, the matrix $(F \|x^i - x^j\|^2)$ is nonsingular.*

Proof. By Theorem 4.3.6,

$$F'(t) = \int_0^\infty e^{-t\sigma} d\mu(\sigma), \quad t > 0,$$

where $d\mu(\sigma)$ is a Borel measure on $(0, \infty)$. Then

$$-F(t) = \int_0^\infty \frac{e^{-t\sigma}}{\sigma} d\mu(\sigma), \quad t > 0.$$

Therefore

$$F(t + \epsilon) - F(\epsilon) = \int_0^\infty \frac{e^{-\epsilon\sigma}}{\sigma}(1 - e^{-t\sigma})d\mu(\sigma).$$

Let $c \in H^n$. Setting $t = ||x^i - x^j||^2$ in the above equation, multiplying both sides by $c_i c_j$, and summing over i, j gives

$$\sum_{i=1}^n \sum_{j=1}^n c_i c_j F(||x^i - x^j||^2 + \epsilon) = -\int_0^\infty \frac{e^{-\epsilon\sigma}}{\sigma} \sum_{i=1}^n \sum_{j=1}^n c_i c_j e^{-||x^i-x^j||^2\sigma} d\mu(\sigma).$$

By Lemma 4.3.4, $(e^{-||x^i-x^j||^2\sigma})$ is positive definite; so letting $\epsilon \downarrow 0$, we get

$$\sum_{i=1}^n \sum_{j=1}^n c_i c_j F(||x^i - x^j||^2) \leq 0,$$

with equality if and only if $c = 0$. The matrix $(F||x^i - x^j||^2)$ is clearly not negative semidefinite and so, by Lemma 4.3.5, it is nonsingular. ■

Corollary 4.3.8. *If $x^1, \ldots, x^n$ are distinct vectors in R^s, then the matrix*

$$(\sqrt{1 + ||x^i - x^j||^2})$$

is nonsingular.

Proof. The function $F(t) = \sqrt{1+t}$ has the property that $F'(y) = \frac{1}{2}(1+t)^{1/2}$ is completely monotonic. The result then follows from Theorem 4.3.7. ■

For results concerning distance matrices that use the p-norm distance, see Baxter (1991).

4.4. A characterization theorem

We now study the class of positive, symmetric matrices with exactly one positive eigenvalue. This class will be denoted by $\mathcal{A}$. Of course, quite a few of the properties that we discuss hold, with little or no modification, for the class of nonnegative, symmetric matrices with exactly one positive eigenvalue. This extension can be achieved either by a continuity argument or by following the same proof technique as that employed for $\mathcal{A}$.

In the next result we record a simple relationship between $\mathcal{A}$ and c.n.d. matrices. It follows from Corollary 4.1.5 that if A is a positive, c.n.d. matrix, then $A \in \mathcal{A}$. Now we show that if $A \in \mathcal{A}$, then a normalized version of A is c.n.d..

Lemma 4.4.1. *Let $A \in \mathcal{A}$ and let v be a Perron eigenvector of A. Then the matrix $(\frac{a_{ij}}{v_i v_j})$ is c.n.d..*

Proof. Suppose A is $n \times n$. Since A is positive, so is v. By the Spectral Theorem,

$$A = \lambda v v^T + B,$$

where λ is the Perron root of A. Since A has exactly one positive eigenvalue, for any x with $x^T v = 0$, $x^T B x \leq 0$. Note that

$$\left(\frac{a_{ij}}{v_i v_j} \right) = \lambda J + \left(\frac{b_{ij}}{v_i v_j} \right),$$

where J is the matrix of all ones. For any $y \in H^n$, we have

$$\sum_{i=1}^{n} \sum_{j=1}^{n} \frac{a_{ij}}{v_i v_j} y_i y_j = \sum_{i=1}^{n} \sum_{j=1}^{n} b_{ij} \frac{y_i}{v_i} \frac{y_j}{v_j} \leq 0,$$

and hence A is c.n.d.. This completes the proof. ∎

In the next few results we illustrate some applications of Theorem 4.3.6.

Theorem 4.4.2. *Let A be a positive, c.n.d. matrix and let $F : (0, \infty) \to R$ be completely monotonic. Then the matrix $(F(a_{ij}))$ is positive semidefinite.*

Proof. By Theorem 4.3.6,

$$F(a_{ij}) = \int_0^\infty e^{-a_{ij}\sigma} d\mu(\sigma), \quad i, j = 1, 2, \ldots, n,$$

where $d\mu(\sigma)$ is a Borel measure on $(0, \infty)$. Since A is c.n.d., by Theorem 4.1.3, for each $\sigma > 0$, the matrix $(e^{-a_{ij}\sigma})$ is positive semidefinite. Because a nonnegative linear combination of positive semidefinite matrices is positive semidefinite, it follows that $(F(a_{ij}))$ is positive semidefinite. ∎

Corollary 4.4.3. *Let A be a nonnegative, symmetric matrix. Then A is c.n.d. if and only if for all $\lambda > 0$, the matrix $((1 + \lambda a_{ij})^{-1})$ is positive semidefinite.*

Proof. Note that for any $\lambda > 0$, the function

$$F(t) = \frac{1}{1 + \lambda t}, \quad t > 0$$

is completely monotonic. If A is c.n.d., it follows by a simple continuity argument from Theorem 4.4.2 that $(F(a_{ij}))$ is positive semidefinite.

Conversely, if for all $\lambda > 0$, the matrix $((1 + \lambda a_{ij})^{-1})$ is positive semidefinite, then so is the matrix $((1 + \lambda a_{ij})^{-k})$ for all positive integers k, by Lemma 3.7.1.

Letting $k \to \infty$ we see that $(e^{-\lambda a_{ij}})$ is also positive semidefinite, and hence, by Theorem 4.1.3, A must be c.n.d.. ∎

Theorem 4.4.4. *If $A \in \mathcal{A}$, then $(\log a_{ij})$ is c.n.d..*

Proof. By Theorem 4.1.3, it is sufficient to show that for any $\lambda > 0$, the matrix

$$(e^{-\lambda \log a_{ij}}) = \left(a_{ij}^{-\lambda}\right)$$

is positive semidefinite. First suppose that A is c.n.d.. Since for any $\lambda > 0$, the function $F(t) = t^{-\lambda}$ is completely monotonic, by Theorem 4.4.2 the matrix $(a_{ij}^{-\lambda})$ is positive semidefinite. In general, if v is a Perron eigenvector of A, then by Lemma 4.4.1, $(a_{ij}/v_i v_j)$ is c.n.d. and again for any $\lambda > 0$, the matrix

$$\left(\left(\frac{a_{ij}}{v_i v_j}\right)^{-\lambda}\right)$$

is positive semidefinite. It follows that for any $A \in \mathcal{A}$ and for any $\lambda > 0$, the matrix $(a_{ij}^{-\lambda})$ is positive semidefinite. This completes the proof. ∎

In the process of proving Theorem 4.4.4 we have also proved the following:

Corollary 4.4.5. *If $A \in \mathcal{A}$, then (a_{ij}^{-1}) is positive semidefinite.*

Suppose A is a positive, symmetric matrix. What additional condition on A would be necessary and sufficient in order that $A \in \mathcal{A}$? In the next result we collect several such conditions. It must be emphasized that some of these conditions are equivalent under weaker assumptions. For example, the assumption of positivity of A may be relaxed in some cases. Some such assertions will be given in the Exercises. Although the formulation of Theorem 4.4.6 is new, many of the equivalent assertions have been inspired by similar results in Martos (1969), Cottle and Ferland (1972), Ferland (1972), and Micchelli (1986).

Theorem 4.4.6. *Let A be a positive, symmetric $n \times n$ matrix. Let*

$$\Omega = \{x \in R^n : x^T A x > 0\},$$

and let v be the Perron eigenvector of A with $v^T v = 1$. Let

$$K = \{x \in \Omega : v^T x \geq 0\}.$$

Then the following conditions are equivalent:

(1) $A \in \mathcal{A}$, i.e., A has exactly one positive eigenvalue.
(2) K is convex.

(3) *For any $x \in R^n$ and any $y \in \Omega$, $(y^T Ax)^2 \geq (y^T Ay)(x^T Ax)$.*

(4) *For any $x \in K$, $y \in K$; $y^T Ax > 0$.*

(5) *K is convex and $-x^T Ax$ is quasi-convex on K.*

(6) *K is convex and $-x^T Ax$ is pseudo-convex on K.*

(7) *$x \in R^n$, $y \in \Omega$, $y^T Ax = 0 \Rightarrow x^T Ax \leq 0$.*

(8) *For any $x \in K$, $v^T x > 0$.*

(9) *A and all its principal minors have exactly one positive eigenvalue.*

(10) *For any $r \times r$ principal minor B of A, $(-1)^{r-1}|B| \geq 0, r = 1, 2, \ldots, n$.*

(11) *$\left(\frac{a_{ij}}{v_i v_j}\right)$ is c.n.d..*

(12) *$\left(e^{\frac{-\lambda a_{ij}}{v_i v_j}}\right)$ is positive semidefinite for any $\lambda > 0$.*

(13) *$(\alpha_i \beta_j + \beta_i \alpha_j - a_{ij})$ is positive semidefinite for some $\alpha, \beta \in R^n$.*

(14) *$\begin{pmatrix} A & v \\ v^T & 0 \end{pmatrix}$ has exactly one positive eigenvalue.*

(15) *A is negative subdefinite.*

(16) *$x^T Ax$ is concave on $\{x \in R^n : v^T x = 1\}$.*

(17) *$\left(\frac{1}{v_i v_j + \lambda a_{ij}}\right)$ is positive semidefinite for any $\lambda > 0$.*

(18) *The (unique) doubly stochastic matrix of the form $(a_{ij} z_i z_j)$, where $z_1, \ldots, z_n$ are positive, is c.n.d..*

Proof. We first make some preliminary remarks. Since $v \in \Omega$, Ω is nonempty. The condition $v^T v = 1$ is only a normalizing condition and the theorem could be proved even without it. By the Spectral Theorem, there exists an orthogonal matrix P, with first row equal to v^T, such that $PAP^T = D$, where D is the diagonal matrix with the eigenvalues of A, say $d_1, \ldots, d_n$, along its diagonal. We assume that $d_1 > 0$ and that $d_1 \geq d_2 \geq \cdots \geq d_n$.

(1) $\Rightarrow$ (2): If $x \in K$ and $\alpha > 0$, then clearly $\alpha x \in K$. So it is sufficient to show that K is closed under addition. Suppose $x \in K$, $y \in K$ and let $u = Px$, $v = Py$. Then $u_1 \geq 0$, $v_1 \geq 0$, $u^T Du > 0$, and $v^T Dv > 0$. Since $d_2 \leq 0$, we have

$$\sum_{i=2}^{n} d_i u_i v_i \leq -\sum_{i=2}^{n} d_i |u_i||v_i|$$

$$\leq \left(\sum_{i=2}^{n}(-d_i)u_i^2\right)^{1/2} \left(\sum_{i=2}^{n}(-d_i)v_i^2\right)^{1/2}$$

$$< \left(d_1 u_1^2\right)^{1/2} \left(d_1 v_1^2\right)^{1/2}$$

$$= d_1 u_1 v_1,$$

where the second inequality follows by the Cauchy-Schwarz Inequality. Hence

$u^T D v > 0$. Now,

$$(x + y)^T A(x + y) = (u + v)^T D(u + v)$$
$$= u^T Du + v^T Dv + 2u^T Dv > 0.$$

Thus $x + y \in \Omega$. Also, $v^T(x + y) = v^T x + v^T y > 0$ and so $x + y \in K$.

(2) $\Rightarrow$ (1): If $d_2 > 0$, let $u^T = (0, 1, 0, \ldots, 0)$ and $v^T = (0, -1, 0, \ldots, 0)$. Let $x = P^T u$, $y = P^T v$. Then $x^T Ax = y^T Ay = d_2 > 0$. Hence $x, y \in K$, and since K is convex, $x + y \in K$. However,

$$(x + y)^T A(x + y) = (u + v)^T D(u + v)$$
$$= 0,$$

since $u + v = 0$, which is a contradiction.

(1) $\Rightarrow$ (3): Let $x \in R^n$, $y \in \Omega$, and without loss of generality, suppose $y \in K$ (otherwise consider $-y$). If $x^T Ax \leq 0$, then (3) clearly holds, so suppose $x^T Ax > 0$ and, again, without loss of generality, suppose $x \in K$. Let $u = Px$, $v = Py$. Then $u_1 \geq 0$, $v_1 \geq 0$. If $u_1 = 0$, then, since $d_2 \leq 0$,

$$x^T Ax = u^T Du = \sum_{i=2}^{n} d_i u_i^2 \leq 0,$$

which contradicts that $x \in \Omega$. Thus $u_1 > 0$. Similarly, $v_1 > 0$. Let

$$g(\alpha) = (y + \alpha x)^T A(y + \alpha x).$$

Then $g(0) = y^T Ay > 0$. Also, if $\bar{\alpha} = -u_1/v_1$, then

$$g(\bar{\alpha}) = (y + \bar{\alpha} x)^T A(y + \bar{\alpha} x)$$
$$= (v + \bar{\alpha} u)^T D(v + \bar{\alpha} u)$$
$$= \sum_{i=2}^{n} d_i (v_i + \bar{\alpha} u_i) \leq 0.$$

Hence $g(\alpha) = 0$ has a real solution. The discriminant of $g(\alpha)$ is therefore nonnegative, and that gives (3).

(3) $\Rightarrow$ (1): If (1) is not true, then there exist $\alpha > 0$, $\beta > 0$ and vectors x, y such that $Ax = \alpha x$, $Ay = \beta y$, and $y^T Ax = 0$. (It is possible that $\alpha = \beta$.) Then $x^T Ax > 0$, $y^T Ay > 0$, and this contradicts (3).

(1) $\Rightarrow$ (4): The proof of this is contained in that of (1) $\Rightarrow$ (2).

(4) $\Rightarrow$ (1): As noted earlier, if (1) is not true, then there exist $\alpha > 0$, $\beta > 0$ and vectors x, y such that $Ax = \alpha x$, $Ay = \beta y$, and $y^T Ax = 0$. Then $x^T Ax > 0$, $y^T Ay > 0$ and, without loss of generality, $x \in K$. This contradicts (4).

(1) $\Rightarrow$ (5): If (1) holds, then it was shown in the proof of (1) $\Rightarrow$ (2) that K is convex. Suppose $x \in K$, $y \in K$. We must show that

$$-y^T Ax \leq \max(-x^T Ax, -y^T Ay).$$

Suppose $y^T A x < x^T A x$ and $y^T A x < y^T A y$. Since (4) must be true, $y^T A x > 0$. Hence

$$(y^T A x)^2 < (x^T A x)(y^T A x) < (x^T A x)(y^T A y),$$

which contradicts (3), and we have shown that (1) $\Rightarrow$ (3).

(1) $\Rightarrow$ (6): Let $x \in K$, $y \in K$ and suppose $y^T A x \le x^T A x$. Then we must show that $y^T A y \le x^T A x$. Suppose $y^T A x \le x^T A x < y^T A y$. Since (4) must hold, $y^T A x > 0$. Therefore,

$$(y^T A x)^2 \le (y^T A x)(x^T A x) < (y^T A y)(x^T A x),$$

which contradicts (3).

(5) $\Rightarrow$ (1) and (6) $\Rightarrow$ (1) are clear since it was shown that (2) $\Rightarrow$ (1).

(3) $\Rightarrow$ (7): Suppose $x \in R^n$, $y \in \Omega$ and $y^T A x = 0$. If $x^T A x > 0$, then, since $y^T A y > 0$, it would contradict (3). So $x^T A x \le 0$.

(7) $\Rightarrow$ (1): If (1) is not true, then there exist $\alpha > 0$, $\beta > 0$ and vectors x, y such that $A x = \alpha x$, $A y = \beta y$, and $y^T A x = 0$. (As before, it is possible that $\alpha = \beta$.) Then $x^T A x > 0$, $y^T A y > 0$; this contradicts (7).

(1) $\Rightarrow$ (8): This was proved during the course of proving (1) $\Rightarrow$ (3).

(8) $\Rightarrow$ (1): Suppose $d_2 > 0$. Let $u^T = (0, 1, 0, \ldots, 0)$ and let $x = P^T u$. Then $x^T A x = d_2 > 0$. Hence $x \in K$. But $u_1 = 0$, and this contradicts (8).

(1) $\Rightarrow$ (9): Let B be a principal submatrix of A of order $n - 1$ and let $\mu_1 \ge \mu_2 \ge \cdots \ge \mu_{n-1}$ be the eigenvalues of B. By the well-known interlacing property of the eigenvalues of symmetric matrices [see, for example, Strang (1980), p. 350],

$$d_1 \ge \mu_1 \ge d_2 \ge \mu_2.$$

Since $d_2 \le 0$, B has at most one positive eigenvalue. Also, since B is positive it follows that it has exactly one positive eigenvalue. The assertion can be proved for principal submatrices of order less than $n - 1$ by backward induction.

(9) $\Rightarrow$ (1) is obvious.

(9) $\Rightarrow$ (10): This follows from the fact that the determinant is equal to the product of the eigenvalues.

(10) $\Rightarrow$ (1): We use Descarte's rule of signs [see, for example, Polya and Szegö (1976), p. 41], which asserts that the number of positive roots of the equation

$$a_n z^n + a_{n-1} z^{n-1} + \cdots + a_1 z + a_0 = 0$$

is not greater than the number of variations of sign in the sequence $a_0, a_1, \ldots, a_n$. Consider the characteristic polynomial of A,

$$|A - \lambda I| = c_n (-\lambda)^n + c_{n-1} (-\lambda)^{n-1} + \cdots + c_1 (-\lambda) + c_0,$$

where c_i is the sum of the principal minors of A of order $n-i$, $i = 0, 1, \ldots, n-1$, and $c_n = 1$. The hypotheses on A implies that

$$c_0, c_1(-1), c_2(-1)^2, \ldots, c_{n-1}(-1)^{n-1}, c_n(-1)^n$$

has precisely one variation of sign, and so by Descarte's rule, A has at most one positive eigenvalue. Since A is positive, it has exactly one positive eigenvalue.

(1) $\Rightarrow$ (11): This follows from Lemma 4.4.1.

(11) $\Rightarrow$ (1): Since $(\frac{a_{ij}}{v_i v_j})$ is c.n.d., by Corollary 4.1.5 it has exactly one positive eigenvalue. It follows that A has exactly one positive eigenvalue.

The equivalence of (11), (12), and (13) follows from Theorem 4.1.3.

(1) $\Rightarrow$ (14): By the interlacing property mentioned earlier,

$$\begin{bmatrix} A & v \\ v^T & 0 \end{bmatrix}$$

has at most two positive eigenvalues. First, suppose that A is nonsingular. We have

$$\begin{vmatrix} A & v \\ v^T & 0 \end{vmatrix} = |A|(-v^T A^{-1} v).$$

Since $v^T A^{-1} v = v^T v / d_1 > 0$, it follows that $|A|$ and

$$\begin{vmatrix} A & v \\ v^T & 0 \end{vmatrix}$$

have opposite signs and hence (14) holds. The case of singular A is settled by a continuity argument.

(14) $\Rightarrow$ (1): By the interlacing property, A has at most one positive eigenvalue, and because A is positive, it has at least one.

The equivalence of (1) and (15) follows from Theorem 4.2.9.

(1) $\Rightarrow$ (16): Let $x, y \in R^n$ with $v^T x = v^T y = 1$. Then we must show that

$$\left(\frac{x+y}{2}\right)^T A \left(\frac{x+y}{2}\right) \geq \frac{x^T A x + y^T A y}{2}.$$

This simplifies to $(x - y)^T A(x - y) \leq 0$. By (11), $(\frac{a_{ij}}{v_i v_j})$ is c.n.d. and

$$\sum_{i=1}^{n} \sum_{j=1}^{n} \frac{a_{ij}}{v_i v_j} v_i(x_i - y_i) v_j(x_j - y_j) \leq 0.$$

Hence $(x - y)^T A(x - y) \leq 0$.

(16) $\Rightarrow$ (1): Suppose $Ax = \alpha x, Ay = \beta y$, where $\alpha > 0, \beta > 0$, and $y^T Ax = 0$. We assume, without loss of generality, that $v^T x = v^T y = 1$. Then

$$(x - y)^T A(x - y) = x^T Ax + y^T Ay$$
$$= \alpha x^T x + \beta y^T y > 0,$$

which contradicts (16).

The equivalence of (1) and (17) follows from Corollary 4.4.3.

(1) $\Rightarrow$ (18): Since A is symmetric and positive, there exist positive numbers $z_1, \ldots, z_n$ such that $(a_{ij} z_i z_j)$ is doubly stochastic. This well-known fact [see Brualdi (1974) and Bapat (1982)] may be proved using Brouwer's Fixed Point Theorem (see Exercise 2 in Chapter 6). By Exercise 5, $(a_{ij} z_i z_j)$ also has exactly one positive eigenvalue, and since its Perron eigenvector has all components equal, the result follows by (11).

(18) $\Rightarrow$ (1): If $(a_{ij} z_i z_j)$ is c.n.d., by Corollary 4.1.5 it has exactly one positive eigenvalue. By Exercise 5, A has exactly one positive eigenvalue. ∎

4.5. Log-concavity and discrete distributions

A sequence of positive numbers $\alpha_0, \alpha_1, \alpha_2, \ldots$ is said to be *log-concave* if $\alpha_k^2 \geq \alpha_{k-1} \alpha_{k+1}, k = 1, 2, \ldots$. Such sequences arise frequently in combinatorics and in statistics. If a sequence $\alpha_0, \alpha_1, \alpha_2, \ldots$ is log-concave, then it must be *unimodal,* i.e., either it is nondecreasing or it is nonincreasing, or for some m, $\alpha_1 \leq \alpha_2 \leq \cdots \leq \alpha_m \geq \alpha_{m+1} \geq \cdots$. This can be seen as follows. From the definition of log-concavity we see that $\frac{\alpha_k}{\alpha_{k-1}}$ is decreasing in k. Thus if $\frac{\alpha_2}{\alpha_1} \leq 1$, then the sequence is nonincreasing. Similarly, if for some finite m, $\frac{\alpha_m}{\alpha_{m-1}} \geq 1$ and $\frac{\alpha_{m+1}}{\alpha_m} \leq 1$, then $\alpha_1 \leq \alpha_2 \leq \cdots \leq \alpha_m \geq \alpha_{m+1} \geq \cdots$. It turns out that in many applications one is interested in demonstrating unimodality, which is not a very tractable property. Log-concavity implies unimodality and is mathematically more well behaved. A function f defined on $0, 1, 2, \ldots$ is said to be log-concave if $f(0), f(1), f(2), \ldots$ is log-concave.

If $\alpha_0, \alpha_1, \alpha_2, \ldots$ is a log-concave sequence, then the symmetric matrix

$$\begin{bmatrix} \alpha_{k-1} & \alpha_k \\ \alpha_k & \alpha_{k+1} \end{bmatrix} \tag{4.5.1}$$

has nonpositive determinant and hence it has exactly one positive eigenvalue. This observation serves two purposes. It justifies discussing log-concavity in this chapter and it serves as a basis for generalization to higher dimensions as will be seen later.

We now introduce some elementary concepts related to discrete distributions. A random variable X is said to be *discrete* if it takes values only in a countable set $\{x_0, x_1, x_2, \ldots\}$ of real numbers. The probability that X is equal to x is denoted by $Pr(X = x)$. The function $f(x) = Pr(X = x)$ is known as the *probability density function (p.d.f.)* of X.

The *binomial distribution* is an important discrete distribution; it arises as follows. Suppose a coin turns up heads with probability p on any single toss. Let the coin be tossed n times and let X denote the number of heads obtained in the n tosses. Then it can be seen that X has the p.d.f.

$$Pr(X = x) = \binom{n}{x} p^x (1 - p)^{n-x}, \ x = 0, 1, \ldots, n. \qquad (4.5.2)$$

In this situation, X is said to have the binomial distribution with parameters n, p. It is a simple exercise to check that the p.d.f. of X in (4.5.2) is log-concave.

Suppose X is a discrete random variable that takes values $\alpha_1, \ldots, \alpha_k$ with probabilities $p_1, \ldots, p_k$, respectively. The *entropy* of this density is defined by

$$I(p) = -\sum_{i=1}^{k} p_i \log p_i,$$

where we define $0 \log 0 = 0$. Note that the entropy $I(p)$ as a function of $p = (p_1, \ldots, p_k)$ is maximized when $p_1 = p_2 = \cdots = p_k = \frac{1}{k}$, and thus, the entropy of X measures the closeness of X to the uniform distribution. In other words, the higher the entropy, the more random is the distribution of X. In the extreme case, when only one of the p_is is equal to one, the remaining being zero, X has no randomness and its entropy is the least possible.

If $X_1, \ldots, X_r$ are discrete random variables, then $X = (X_1, \ldots, X_r)$ is a discrete random vector and we define its p.d.f. as

$$Pr(X = x) = Pr(X_1 = x_1, \ldots, X_r = x_r), \ \text{where } x = (x_1, \ldots, x_r).$$

In the remainder of this section, we study a certain generalization of the binomial distribution to random vectors and we identify the parameter values at which its entropy is maximized. In the process we make use of results from the theory of permanents and c.n.d. matrices developed earlier.

To begin with, note that the binomial distribution of (4.5.2) can be expressed in terms of permanents as follows:

$$Pr(X = x) = \binom{n}{x} p^x (1 - p)^{n-x}$$

$$= \frac{1}{x!(n - x)!} \operatorname{per} \begin{bmatrix} p & \cdots & p & 1-p & \cdots & 1-p \\ \vdots & \ddots & \vdots & \vdots & \ddots & \vdots \\ p & \cdots & p & 1-p & \cdots & 1-p \end{bmatrix}, \qquad (4.5.3)$$

where the column of ps is repeated x times and the column of $(1 - p)$s is repeated $n - x$ times. The expression (4.5.3) admits generalizations. For example, suppose n coins, not necessarily identical, are tossed once, and let X be the number of heads obtained. If p_i is the probability of heads on a single toss of the i-th coin, $i = 1, 2, \ldots, n$, then it can be verified that $Pr(X = x)$ is given by

$$\frac{1}{x!(n-x)!} \text{ per} \begin{bmatrix} p_1 & \cdots & p_1 & 1-p_1 & \cdots & 1-p_1 \\ \vdots & \ddots & \vdots & \vdots & \ddots & \vdots \\ p_n & \cdots & p_n & 1-p_n & \cdots & 1-p_n \end{bmatrix}, x = 0, 1, \ldots, n,$$

(4.5.4)

where, again, the columns are repeated x and $n - x$ times, respectively. We now consider a further generalization. Thus, suppose an experiment can result in any one of r possible outcomes and suppose n trials of the experiment are performed. Let p_{ij} be the probability that the experiment results in the j-th outcome at the i-th trial, $i = 1, 2, \ldots, n; j = 1, 2, \ldots, r$. Let P denote the $n \times n$ matrix (p_{ij}), which, of course, is stochastic, i.e., has all row sums equal to one. We assume throughout that P is positive and that $n \geq 2, r \geq 2$. Let X_j denote the number of times the j-th outcome is obtained in the n trials, $j = 1, 2, \ldots, r$, and let $X = (X_1, \ldots, X_r)$. In this setup X is said to have the *multiparameter multinomial distribution* with the parameter matrix P. If the rows of P are all identical, then X is said to have the multinomial distribution. Let

$$\mathcal{K}_{n,r} = \left\{ k = (k_1, \ldots, k_r) : k_i \geq 0, \text{ integers}, \sum_{i=1}^{r} k_i = n \right\}.$$

If A is an $m \times r$ matrix and if $k \in \mathcal{K}_{n,r}$, then $A(k)$ denotes the $m \times n$ matrix obtained by taking k_j copies of the j-th column of A, $j = 1, 2, \ldots, r$. If $k \in \mathcal{K}_{n,r}$, we define $k! = k_1! \cdots k_r!$.

If X has the multiparameter multinomial distribution with the parameter matrix P, then as a simple generalization of (4.5.3), we have

$$Pr(X = k) = \frac{1}{k!} \text{per } P(k), \quad k \in \mathcal{K}_{n,r}. \tag{4.5.5}$$

If $k \in \mathcal{K}_{n-2,r}$, we define $k_{ij} = k + e_i + e_j, i, j = 1, 2, \ldots, r$, where e_i is the i-th row of the $r \times r$ identity matrix. The first of the next two definitions is motivated by the corresponding property of the matrix in (4.5.1).

A function $\phi : \mathcal{K}_{n,r} \to (0, \infty)$ is said to be *generalized log-concave* if for any $k \in \mathcal{K}_{n-2,r}$, the $r \times r$ positive, symmetric matrix $(\phi(k_{ij}))$ has exactly one positive eigenvalue. A function $\phi : \mathcal{K}_{n,r} \to R$ is conditionally positive definite if for any $k \in \mathcal{K}_{n-2,r}$, the matrix $(\phi(k_{ij}))$ is c.p.d..

We now consider a multivariate analogue of the majorization ordering considered earlier in Section 3.9. If P and Q are $n \times r$ matrices, we say that Q is *majorized* by P if Q is obtained from P by a repeated averaging of rows. More formally, let $\mathcal{D}$ denote the class of $n \times n$ doubly stochastic matrices that can be written as a finite product of matrices of the form $tI + (1 - t)T$, where $0 \leq t \leq 1$, I is the $n \times n$ identity matrix, and T is an $n \times n$ permutation matrix that interchanges only two coordinates (i.e., corresponds to a transposition). Then if P and Q are $n \times r$ matrices, Q is majorized by P if and only if $Q = DP$ for some $D \in \mathcal{D}$.

A function g defined on the set of $n \times r$ matrices is said to be (multivariate) *Schur-concave* if $g(Q) \geq g(P)$ whenever Q is majorized by P.

We now need a technical result.

Lemma 4.5.1. *Let A be an $n \times r$ matrix with its i-th row equal to a_i, $i = 1, 2, \ldots, n$, and let $\phi : \mathcal{K}_{n,r} \to R$. Then for any $z \in R^r$,*

$$\sum_{k \in \mathcal{K}_{n,r}} \frac{\phi(k)}{k!} \, \mathrm{per} \begin{bmatrix} z \\ z \\ a_3 \\ \vdots \\ a_n \end{bmatrix}(k) = \sum_{\ell \in \mathcal{K}_{n-2,r}} \frac{z'(\phi(\ell_{ij}))z}{\ell!} \, \mathrm{per} \begin{bmatrix} a_3 \\ \vdots \\ a_n \end{bmatrix}(\ell). \qquad (4.5.6)$$

Proof. Recall that $B(k)$ denotes the matrix obtained by taking k_j copies of the j-th column of B for all j. The lemma is proved by verifying that the coefficients of $z_i z_j$ for any $i \leq j$ are the same on both sides of (4.5.6). We leave it as an exercise. ∎

Theorem 4.5.2. *Let $\phi : \mathcal{K}_{n-r} \to R$ be c.p.d. and, for any $n \times r$ positive stochastic matrix P, let*

$$g(P) = \sum_{k \in \mathcal{K}_{n,r}} \phi(k) Pr(X = k),$$

where X has the multiparameter multinomial distribution with the parameter matrix P. Then g is Schur-concave.

Proof. Let $0 \leq t \leq 1$ and let Q be the $n \times r$ matrix defined as follows:

$$q_{1j} = tp_{1j} + (1 - t)p_{2j}, \quad j = 1, 2, \ldots, r$$
$$q_{2j} = (1 - t)p_{1j} + tp_{2j}, \quad j = 1, 2, \ldots, r$$
$$q_{ij} = p_{ij}, \quad i = 3, \ldots, n.$$

The result is proved if we show that $g(Q) \geq g(P)$. Let $X(Y)$ have the multiparameter multinomial distribution with the parameter matrix $P(Q)$. Let z denote the vector defined by $z_j = p_{1j} - p_{2j}$, $j = 1, 2, \ldots, r$. Also, let p_i denote the i-th row of P, $i = 1, 2, \ldots, n$. Then

$$k!\{Pr(Y = k) - Pr(X = k)\} = \text{per } Q(k) - \text{per } P(k)$$

$$= t(1-t) \text{ per} \begin{bmatrix} z \\ z \\ p_3 \\ \vdots \\ p_n \end{bmatrix}, \qquad (4.5.7)$$

where the second equality follows after some simple algebra. Multiply both sides of (4.5.7) by $\frac{\phi(k)}{k!}$ and sum over $k \in \mathcal{K}_{n,r}$ to get

$$g(Q) - g(P) = t(1-t) \sum_{k \in \mathcal{K}_{n,r}} \frac{\phi(k)}{k!} \text{ per} \begin{bmatrix} z \\ z \\ p_3 \\ \vdots \\ p_n \end{bmatrix}. \qquad (4.5.8)$$

Now, by Lemma 4.5.1, we can rewrite the right-hand side of (4.5.8) as

$$\sum_{\ell \in \mathcal{K}_{n-2,r}} \frac{z^T(\phi(\ell_{ij}))z}{\ell!} \text{ per} \begin{bmatrix} p_3 \\ \vdots \\ p_n \end{bmatrix} (\ell).$$

The result follows since $z \in H^r$ and for each $\ell \in \mathcal{K}_{n-2,r}$, $(\phi(\ell_{ij}))$ is c.p.d.. ∎

The next result can be derived using the Alexandroff Inequality proved in Chapter 2. The proof is outlined in Exercise 14.

Theorem 4.5.3. *Let A be a positive $n \times (n-2)$ matrix, $n \geq 2$, and let $x^1, \ldots, x^r$ be positive vectors in R^n. Define the $r \times r$ matrix $B = (b_{ij})$ as*

$$b_{ij} = \text{per}(A, x^i, x^j), \quad i, j = 1, 2, \ldots, r.$$

Then B has exactly one positive eigenvalue.

The representation (4.5.5) and Theorem 4.5.3 lead to part (a) of the next result. Part (b) then follows from Exercise 13.

Theorem 4.5.4. *Let X have the multiparameter multinomial distribution with the $n \times r$ parameter matrix P and let $k \in \mathcal{K}_{n-2,r}$. Then*

(a) $(k_{ij}! Pr(X = k_{ij}))$ *has exactly one positive eigenvalue.*
(b) $(Pr(X = k_{ij}))$ *has exactly one positive eigenvalue.*

Now we can state the main result of this section.

Theorem 4.5.5. *Let P be a positive, stochastic $n \times r$ matrix. Let $g(P)$ denote the entropy function*

$$g(P) = - \sum_{k \in \mathcal{K}_{n,r}} Pr(X = k) \log Pr(X = k),$$

where X has the multiparameter multinomial distribution with parameter matrix P. Then

(a) *g is Schur-concave.*
(b) *If Q is the $n \times r$ matrix with $nq_{ij} = \sum_{\ell=1}^{n} p_{\ell j}$ for all i, j, then $g(Q) \geq g(P)$.*

Proof. (a). For $k \in \mathcal{K}_{n,r}$, define

$$\phi(k) = - \log Pr(X = k).$$

Then ϕ is c.p.d. in view of Theorem 4.5.4 (b) and Theorem 4.4.4. The result then follows by Theorem 4.5.2. Part (b) is a simple consequence of (a). ∎

According to Theorem 4.5.5, the entropy of a multiparameter multinomial distribution is maximized when the probability distribution is the same for all trials. Theorem 4.5.5 is based on Bapat (1986b), and it resolves a conjecture in Karlin and Rinott (1981). Lemma 4.5.1 and Theorem 4.5.2 are also from Karlin and Rinott (1981). Certain other standard distributions on $\mathcal{K}_{n,r}$ also have densities that are generalized log-concave. Some of these are covered in Exercise 15 [also see Bapat (1987c)]. For more applications of the Alexandroff Inequality in probability and statistics, see Bapat (1990).

4.6. The q-permanent

As usual, let S_n denote the set of permutations on n symbols. If $\sigma \in S_n$, then $\ell(\sigma)$ denotes the number of inversions of σ. Recall that an *inversion* of σ is a pair (i, j) such that $1 \leq i < j \leq n$ and $\sigma(i) > \sigma(j)$. As an example, the permutation

$$\begin{pmatrix} 1 & 2 & 3 & 4 & 5 & 6 \\ 3 & 6 & 4 & 2 & 1 & 5 \end{pmatrix}$$

in S_6 has 9 inversions. The identity permutation has zero inversions. The maximum number of inversions in S_n is $\frac{n(n-1)}{2}$, attained at the permutation

$n, n-1, \ldots, 2, 1$. We refer to Comtet (1974, pp. 236–40) for elementary properties of $\ell(\sigma)$. In particular, we note that $\ell(\sigma)$ is even (odd) if σ is an even (odd) permutation.

If A is an $n \times n$ matrix and q a real number, then we define the *q-permanent* of A, denoted by $\text{per}_q(A)$, as

$$\text{per}_q(A) = \sum_{\sigma \in S_n} q^{\ell(\sigma)} \prod_{i=1}^{n} a_{i\sigma(i)}.$$

Observe that $\text{per}_{-1}(A) = |A|$, $\text{per}_0(A) = \prod_{i=1}^{n} a_{ii}$, and $\text{per}_1(A) = \text{per } A$. Here we have made the usual convention that $0^0 = 1$. The q-permanent thus provides a parametric generalization of both the determinant and the permanent. The q-permanent appears to be a function with a very rich structure but at the same time it does not lend itself to manipulations very easily.

The main purpose of this section is to show that if A is a positive semidefinite matrix and $-1 \le q \le 1$, then $\text{per}_q(A) \ge 0$. This result has been proved by Bożejko and Speicher (1991) in connection with a problem in mathematical physics dealing with parametric generalizations of Brownian motion. The proof that we give is based on c.n.d. matrices.

If q is a real number, define the $n! \times n!$ matrix Γ_q as follows. (Although Γ_q depends on n we do not use n in the notation for convenience.) The rows and columns of Γ_q are indexed by S_n. If $\sigma, \tau \in S_n$, then the (σ, τ)-entry of Γ_q is $q^{\ell(\tau \sigma^{-1})}$. We set

$$\pi_q(A) = \pi(A) \circ \Gamma_q,$$

where $\pi(A)$ is the Schur power matrix introduced in Section 3.8 and $\circ$ denotes the Schur product. Observe that

$$\text{per}_q(A) = \frac{1}{n!} \langle \Pi_q(A)\mathbf{1}, \mathbf{1} \rangle, \tag{4.6.1}$$

where $\mathbf{1}$ is the column vector of all ones.

A function $f : S_n \to R$ is said to be positive definite on S_n if the $n! \times n!$ matrix

$$(f(\tau \sigma^{-1}))_{\sigma, \tau \in S_n}$$

is positive semidefinite. If $f(\sigma) = 1$ for all $\sigma \in S_n$, then clearly f is positive definite on S_n. Another simple example is given in the following.

Lemma 4.6.1. *Let $f(\sigma) = 1$ if σ is even and $f(\sigma) = -1$ otherwise. Then f is positive definite on S_n.*

Proof. Recall that $f(\alpha\beta) = f(\alpha)f(\beta)$ for any $\alpha, \beta \in S_n$. For any vector x indexed by S_n,

$$\sum_\sigma \sum_\tau f(\sigma\tau^{-1})x_\sigma x_\tau = \sum_\sigma \sum_\tau f(\sigma)f(\tau^{-1})x_\sigma x_\tau$$

$$= \sum_\sigma \sum_\tau f(\sigma)f(\tau)x_\sigma x_\tau$$

$$= \left\{ \sum_\sigma f(\sigma)x_\sigma \right\}^2$$

$$\geq 0,$$

and the proof is complete. ∎

Lemma 4.6.2. *If f and g are positive definite on S_n, then so is $h = f \circ g$, defined by $h(\sigma) = f(\sigma)g(\sigma)$.*

Proof. Observe that

$$(h(\sigma\tau^{-1})) = (f(\sigma\tau^{-1})) \circ (g(\sigma\tau^{-1})).$$

The result then follows from Lemma 3.7.1.

We proved in Section 3.8 that if A is positive semidefinite, then so is $\pi(A)$. This is just another way of saying that if A is positive semidefinite, then the function $f(\sigma) = \prod_{i=1}^n a_{i\sigma(i)}$ is positive definite on S_n.

We now introduce additional notation. Let

$$T_n = \{(i, j), 1 \leq i, j \leq n : i < j\}$$

and, for $\sigma \in S_n$, let

$$\sigma(T_n) = \{(\sigma(i), \sigma(j)) : (i, j) \in T_n\}.$$

Then $\ell(\sigma) = \text{card}\,(\sigma(T_n) \setminus T_n)$, where card denotes cardinality. ∎

Lemma 4.6.3. *It is true that*

$$2\ell(\sigma\tau^{-1}) = card\,(\sigma^{-1}(T_n)\Delta\tau^{-1}(T_n)),$$

where Δ denotes symmetric difference.

Proof. We have

$$\ell(\sigma) = \ell(\sigma^{-1}) = card\,(\sigma^{-1}(T_n) \setminus T_n) = card\,(T_n \setminus \sigma(T_n)).$$

Therefore,

$$
\begin{aligned}
2\ell(\sigma\tau^{-1}) &= \ell(\sigma\tau^{-1}) + \ell(\sigma\tau^{-1}) \\
&= \mathrm{card}\ (\sigma\tau^{-1}(T_n) \setminus T_n) + \mathrm{card}\ (T_n \setminus \sigma\tau^{-1}(T_n)) \\
&= \mathrm{card}\ (\tau^{-1}(T_n) \setminus \sigma^{-1}(T_n)) + \mathrm{card}\ (\sigma^{-1}(T_n) \setminus \tau^{-1}(T_n)) \\
&= \mathrm{card}\ (\sigma^{-1}(T_n) \triangle \tau^{-1}(T_n)),
\end{aligned}
$$

and the proof is complete.   ■

Lemma 4.6.4. *Let $W_1, W_2, \dots, W_r$ be subsets of a finite set W and let $A = (a_{ij})$ be the $r \times r$ matrix defined by $a_{ij} = \mathrm{card}\ (W_i \cap W_j), i, j = 1, 2, \dots, r$. Then A is positive semidefinite.*

Proof. Suppose card $(W) = m$. Each W_i can be represented by a $(0, 1)$ vector, say w_i, of order m, where the entries are indexed by W and an entry is 1 or 0 depending upon whether the corresponding element is in W_i or otherwise. Now note that a_{ij} is simply the inner product $\langle w_i, w_j \rangle$; therefore, A is positive semidefinite.   ■

The next result establishes the crucial link between the number of inversions and c.n.d. matrices.

Theorem 4.6.5. *If $q \in [-1, 1]$, then the $n! \times n!$ matrix $(\ell(\sigma\tau^{-1}))$ is c.n.d..*

Proof. By Lemma 4.6.3,

$$
\begin{aligned}
2\ell(\sigma\tau^{-1}) &= \mathrm{card}\ (\sigma^{-1}(T_n) \triangle \tau^{-1}(T_n)) \\
&= \mathrm{card}\ (\sigma^{-1}(T_n)) + \mathrm{card}\ (\tau^{-1}(T_n)) \\
&\quad - 2\mathrm{card}\ (\sigma^{-1}(T_n) \cap \tau^{-1}(T_n)).
\end{aligned}
\tag{4.6.2}
$$

By Lemma 4.6.4 the matrix

$$
(\mathrm{card}\ (\sigma^{-1}(T_n) \cap \tau^{-1}(T_n)))
$$

is positive semidefinite, and the result follows from Theorem 4.1.3 and Equation (4.6.2).   ■

Theorem 4.6.6. *If $q \in [-1, 1]$, then $q^{\ell(\sigma)}$ is positive definite on S_n.*

Proof. First suppose that $0 \le q \le 1$. Then the function $F(x) = q^x$ is completely monotonic. We must show that the matrix Γ_q is positive semidefinite. This is immediate from Theorems 4.4.2 and 4.6.5. If $-1 \le q < 0$, then write

$$
q^{\ell(\sigma)} = (-1)^{\ell(\sigma)}(-q)^{\ell(\sigma)}.
\tag{4.6.3}
$$

As remarked earlier, $\ell(\sigma)$ has the same parity as σ and therefore, by Lemma 4.6.1, $(-1)^{\ell(\sigma)}$ is positive semidefinite on S_n. By the first part of the proof, $(-q)^{\ell(\sigma)}$ is positive semidefinite on S_n, and the result follows from Lemma 4.6.2 and Equation (4.6.3). ∎

We now prove the main result of this section.

Theorem 4.6.7. *If A is a positive semidefinite $n \times n$ matrix and if $-1 \le q \le 1$, then $per_q(A) \ge 0$.*

Proof. By Theorem 4.6.6, the matrix Γ_q is positive semidefinite. Therefore, $\pi_q(A) = \pi(A) \circ \Gamma_q$ is positive semidefinite. The result now follows by (4.6.1). ∎

For inequalities and unsolved problems concerning the q-permanent, see Bapat and Lal (1993b), Bapat (1992a), and Lal (1993). In particular, it has been conjectured that if A is positive semidefinite, then $per_q(A)$ as a function of q is monotonically increasing in $[-1, 1]$. A formula for the determinant of Γ_q and certain conjectures regarding the inverse of Γ_q are contained in Zagier (1992) and Stanciu (1992).

Exercises

1. If A is an $n \times n$ c.p.d. matrix, show that $B = (b_{ij})$ is positive semidefinite, where $b_{ij} = a_{ij} - a_{in} - a_{nj} + a_{nn}$ for all i, j.

2. A real $n \times n$ matrix is said to be *quasi-positive definite* if $x^T A x > 0$ for any nonzero $x \in R^n$. Show that the Schur product of a quasi-positive definite matrix and a positive definite matrix is quasi-positive definite.

3. If ϕ is a real-valued differentiable function defined on a convex set S in R^n, show that ϕ is quasi-convex on S if and only if for all x, y in S, $\phi(y) \le \phi(x) \Rightarrow \nabla\phi(x)^T(y - x) \le 0$.

4. Let A be a symmetric $n \times n$ matrix. Show that the quadratic form $Q(x) = x^T A x$ is pseudo-convex for every nonnegative, nonzero x if and only if A is strictly positive subdefinite [see Martos (1969)].

5. For an $n \times n$ matrix B, let $\delta(B)$ denote the number of positive eigenvalues of B.

(a) If A is symmetric and C nonsingular, show that $\delta(A) = \delta(C^T A C)$

(b) If A is symmetric, then for any C (perhaps rectangular) show that $\delta(C^T A C) \leq \delta(A)$.
[Remark: The result (a) is sometimes referred to as Sylvester's law of inertia. See Rathore and Chetty (1981) and the references contained therein.]

6. Let A be a symmetric $n \times n$ matrix and let $\Omega = \{x \in R^n : x^T A x > 0\}$ be nonempty. Let $d_1 > 0$ be the largest eigenvalue of A with u as a corresponding eigenvector and let $K = \{x \in \Omega : u^T x \geq 0\}$. Show that conditions (1) through (8) of Theorem 4.4.6 remain equivalent in this general setup.

7. Let $f(x_1, x_2) = x_1^2 + 2x_2^2 - 6x_1x_2$. Determine subsets of R^2 on which f is quasi-convex.
(Hint: Use Exercise 6.)

8. Let A be a symmetric, nonnegative, nonzero matrix. Show that conditions (9) and (10) of Theorem 4.4.6 are equivalent under these hypotheses.

9. Let $0 < x_1 < \cdots < x_r$ and let $n_1, \ldots, n_r$ be distinct positive integers. Show that the matrix $(x_i^{n_j})$ is nonsingular.
(Hint: Use Descarte's rule of signs).

10. Let $\alpha_1, \alpha_2, \cdots$ be log-concave.

(a) Show that $\alpha_1, \alpha_2, \cdots$ must be unimodal, i.e., either it is nondecreasing, or it is nonincreasing, or for some k, $\alpha_1 \leq \alpha_2 \leq \cdots \leq \alpha_k \geq \alpha_{k+1} \geq \cdots$.

(b) Show that for any $i < j$, $\alpha_i \alpha_j \leq \alpha_{i+1} \alpha_{j-1}$.

(c) If $\beta_k = \sum_{i=1}^{k} \alpha_i$ for all k, show that $\beta_1, \beta_2, \ldots$ is log-concave.

11. If $\alpha_1, \alpha_2 \cdots$ and $\beta_1, \beta_2, \ldots$ are log-concave, show that the convolution sequence $\gamma_1, \gamma_2, \ldots$ is log-concave, where $\gamma_k = \sum_{i=1}^{k} \alpha_i \beta_{k-i+1}$ for all k.
[See, for example, Dharmadhikari and Joag-dev (1988).]

12. There are two coins with probabilities of heads p and q, respectively. The first coin is tossed until a head is obtained, then the second coin is tossed until a head is obtained. Let X denote the total number of tosses. Show that the p.d.f. of X is log-concave.

13. Let A be a positive, symmetric matrix of order r and let $0 < \theta_i \le 1, i = 1, 2, \ldots, r$. Let B be the matrix defined by $b_{ii} = \theta_i a_{ii}, i = 1, 2, \ldots, r$ and $b_{ij} = a_{ij}, i \ne j$. If $A \in \mathcal{A}$, show that $B \in \mathcal{A}$. Use this and Exercise 5 (a) to prove part (b) of Theorem 4.5.4.

14. Prove Theorem 4.5.3 using the Alexandroff Inequality for permanents given in Chapter 2 and the equivalence of (1) and (3) in Theorem 4.4.6.

15. Show that the following densities defined on $\mathcal{K}_{n,r}$ are generalized log-concave:

(a) hypergeometric: $N_i > n$ are positive integers, $i = 1, 2, \ldots, R$, and

$$f(k) = \frac{\binom{N_1}{k_1} \cdots \binom{N_r}{k_r}}{\binom{N_1 + \cdots + N_r}{n}},$$

(b) negative hypergeometric: $\alpha_1, \ldots, \alpha_r$ are positive integers and

$$f(k) = \frac{\prod_{i=1}^{r} \binom{k_i + \alpha_i + 1}{k_i}}{\binom{n + \sum \alpha_i + r}{n}},$$

(c) multinomial : $\theta_1, \ldots, \theta_r$ are positive numbers adding to 1 and

$$f(k) = n! \prod_{i=1}^{n} \frac{\theta_i^{k_i}}{k_i!}.$$

16. Consider a die that falls on value i with probability $p_i, i = 1, 2, \ldots, 6$. Suppose the die is rolled n times and let X_i be the number of times i occurs. Show that the entropy of X is maximized when $p_i = \frac{1}{6}$ for all i.
[Hint: Use Exercise 15(c) and Theorem 4.5.5.]

5

Topics in combinatorial theory

In this chapter we discuss certain combinatorial topics where positive semidefinite matrices and nonnegative matrices appear. In the first section we give a quick introduction to matroids and prove a result due to Rado, which includes Hall's Theorem on systems of distinct representatives as a special case. Then we discuss basic properties of the mixed discriminant, a function that allows a unified treatment of the theory of permanent of a nonnegative matrix and the determinant of a positive semidefinite matrix. The Alexandroff Inequality for mixed discriminants is proved and it is used to settle a special case of a conjecture of Mason. The next section deals with a topic in the area of spectra of graphs. Graphs whose adjacency matrices have Perron root less than 2 are characterized. These graphs correspond to the well-known Coxeter graphs (or Dynkin Diagrams). It is also shown that these graphs are precisely the ones giving rise to a finite Weyl group. The next section focuses on matrices over an algebraic structure called max algebra. As far as the eigenproblem is concerned, such matrices behave somewhat like nonnegative matrices. The last section deals with Boolean matrices. The main emphasis is on characterization of Boolean matrices that admit Moore-Penrose inverse.

5.1. Matroids

The theory of matroids has been formulated and studied to understand the notions of linear dependence and independence in an abstract setting. We give only a brief introduction to this vast and rapidly expanding area.

A *matroid* $M = (S, \mathcal{I})$ consists of a finite set S and a collection $\mathcal{I}$ of subsets of S satisfying the following conditions:

  (i) $\phi \in \mathcal{I}$.
 (ii) If $X \in \mathcal{I}$ and $Y \subset X$, then $Y \in \mathcal{I}$.
(iii) If $X, Y \in \mathcal{I}$ and card $(X) <$ card (Y), then there exists $y \in Y$ such that $y \notin X$ and $X \cup \{y\} \in \mathcal{I}$.

196

The sets in $\mathcal{I}$ are called *independent sets*. A set that is not independent is said to be *dependent*.

Example 5.1. Let A be an $m \times n$ matrix over a field F. Let $S = \{1, 2, \ldots, n\}$, and let $\mathcal{I}$ be the collection of subsets T of S satisfying the property that the columns of A indexed by T are linearly independent. Then $(S, \mathcal{I})$ is a matroid. This can be seen as follows. By convention, $\phi \in \mathcal{I}$. If a set is linearly independent, then so is any subset of the set. Suppose $T, V \in \mathcal{I}$ with card $(T) <$ card (V). If every column of A indexed by an element in V can be expressed as a linear combination of columns of A indexed by T, then the subspace spanned by the columns indexed by V must be contained in the subspace spanned by the columns indexed by T. This is a contradiction since the dimension of the subspace spanned by the columns indexed by V (respectively, T) is card (V) (respectively, card (T)) whereas we have card $(T) <$ card (V).

Example 5.2. Let A be a positive semidefinite $n \times n$ matrix. Let $S = \{1, 2, \ldots, n\}$, and let $\mathcal{I}$ be the collection of subsets T of S for which the principal submatrix of A formed by the rows and columns indexed by T is positive definite. Then $(S, \mathcal{I})$ is a matroid. This can be seen as follows. We have $A = X^T X$ for some matrix X. Then a subset T of S is independent if and only if the corresponding columns of X are linearly independent. Thus this example is essentially the same as Example 5.1.

Example 5.3. Let S be a set of cardinality n, and let $\mathcal{I}$ be the collection of all subsets of S of cardinality at most k, where k is a fixed positive integer. Then $(S, \mathcal{I})$ is a matroid, called the *uniform matroid*, with parameters k, n, denoted by $U_{k,n}$. In particular, $U_{n,n}$ is the matroid in which every subset is independent.

Example 5.4. Let $G = (V, E)$ be a graph. Thus, V is a finite set called the set of vertices and E is a finite set whose elements, called edges, are (unordered) pairs of elements of V. For basic graph theory concepts, not defined here, see Biggs (1974), Bondy and Murty (1976), and Lovász (1979). Declare $F \subset E$ to be independent if the graph (V, F) does not contain a cycle. Then we obtain an important example of a matroid, known as a *graphic matroid*. We only sketch a proof of the fact that we do indeed have a matroid. We restrict ourselves to *simple graphs*, i.e., graphs with no loops or multiple edges. Suppose each edge of G is given an arbitrary, but fixed, orientation. The *vertex-edge incidence matrix B* of G is defined in an obvious way. The order of B is card $(V) \times$ card (E). The rows of B are indexed by the vertices, columns of B by the edges. The (i, j) entry of B is 0 if the i-th vertex and the j-th edge are not incident. It is 1 if the

j-th edge emanates from the i-th vertex and -1 if the j-th edge terminates at the i-th vertex. Clearly, the column sums of B are zero. The matrix C obtained by deleting a row of B, say the last row, is called the *reduced incidence matrix*. It can be shown that a set of columns of C are linearly independent if and only if the graph formed by the corresponding edges does not contain a cycle. Thus we have an instance of Example 5.1 again.

Let $M = (S, \mathcal{I})$ be a matroid and let $X \subset S$. The *rank* of X, denoted by $\rho(X)$, is defined as

$$\rho(X) = \max\{\operatorname{card}(V) : V \subset X, V \in \mathcal{I}\}.$$

The rank of the matroid M is defined to be the rank of S.

Note that in Example 5.2, if $T \subset S$, then $\rho(T)$ turns out to be the usual rank of the submatrix of A formed by the columns indexed by T.

It follows from the definition that the rank of a set is at most equal to the cardinality of the set whereas the rank of an independent set equals the cardinality of the set. Also, if $A \subset B$, then $\rho(A) \leq \rho(B)$. The next property, which shows that the rank function is "submodular," is less obvious.

Theorem 5.1.1. *Let $M = (S, \mathcal{I})$ be a matroid. Then for any $A, B \subset S$,*

$$\rho(A \cup B) + \rho(A \cap B) \leq \rho(A) + \rho(B).$$

Proof. Let $\rho(A \cup B) = p, \rho(A \cap B) = q$. If $p = q$, the result is obvious, so let $p > q$. There exists an independent subset X of $A \cap B$ of cardinality q. By repeated application of condition (iii) in the definition of a matroid, we can construct an independent set Y of cardinality p such that $X \subset Y \subset A \cup B$. Let $Y = X \cup V \cup W$, where $V \subset A \backslash B, W \subset B \backslash A$. Then $X \cup V, X \cup W$ are independent subsets of A, B, respectively. Therefore

$$\begin{aligned}
\rho(A) + \rho(B) &\geq \operatorname{card}(X \cup V) + \operatorname{card}(X \cup W) \\
&= 2\operatorname{card}(X) + \operatorname{card}(V) + \operatorname{card}(W) \\
&= \operatorname{card}(Y) + \operatorname{card}(X) \\
&= \rho(A \cup B) + \rho(A \cap B),
\end{aligned}$$

and the proof is complete.   ∎

If $A_1, \ldots, A_k$ is a family of sets, then for any $T \subset \{1, \ldots, k\}$, we define $A(T) = \cup_{i \in T} A_i$. We now prove an important result in the area of transversal theory due to Rado [see Welsh (1976), p. 98].

Theorem 5.1.2. *Let $M = (S, \mathcal{I})$ be a matroid, and let $A_1, \ldots, A_k$ be a family of subsets of S. Then the following conditions are equivalent:*

(i) *There exist distinct $x_i \in A_i, i = 1, 2, \ldots, k$ such that $\{x_1, \ldots, x_k\}$ is an independent set.*

(ii) *For any $T \subset \{1, 2, \ldots, k\}, \rho(A(T)) \geq card\ (T)$.*

Proof. It is easy to see that (i) $\Rightarrow$ (ii). We now prove (ii) $\Rightarrow$ (i). We denote the singleton set $\{x\}$ simply by x. Suppose (ii) holds. Then $\rho(A_i) \geq 1, i = 1, 2, \ldots, k$, and hence each A_i is nonempty. If each A_i is a singleton, then the result is trivial. So suppose, without loss of generality, that card $(A_1) \geq 2$. We claim that there exists $x \in A_1$ such that the family of subsets $A_1 \setminus \{x\}, A_2, \ldots, A_k$ also satisfies (ii) of the theorem. For, otherwise, if x, y are distinct elements of A_1, there must exist $T_1, T_2 \subset \{2, \ldots, k\}$ such that

$$\rho((A_1 \setminus x) \cup A(T_1)) < card\ (T_1) + 1, \quad \rho((A_1 \setminus y) \cup A(T_2)) < card\ (T_2) + 1.$$

Let $X = (A_1 \setminus x) \cup A(T_1), Y = (A_1 \setminus y) \cup A(T_2)$. Then by Theorem 5.1.1,

$$\rho(X \cup Y) + \rho(X \cap Y) \leq \rho(X) + \rho(Y) \leq card\ (T_1) + card\ (T_2).$$

Since $X \cup Y = A_1 \cup A(T_1 \cup T_2)$ and $X \cap Y \supset A(T_1 \cap T_2)$ we have

$$\rho(A_1 \cup A(T_1 \cup T_2)) + \rho(A(T_1 \cap T_2)) \leq card\ (T_1) + card\ (T_2).$$

This implies, in view of (ii), that

$$1 + card\ (T_1 \cup T_2) + card\ (T_1 \cap T_2) \leq card\ (T_1) + card\ (T_2),$$

which is a contradiction. Thus we may delete elements from A_1, until we arrive at a singleton in such a way that the new family still satisfies (ii). We then delete elements from A_2 until we get a singleton. Continuing this way we arrive at a family of singletons that satisfies (ii) and hence (i). That completes the proof. ∎

Let S be a set of cardinality n and consider the uniform matroid $U_{n,n}$ defined on S. If we apply Theorem 5.1.2 to this matroid then we recover the well-known theorem of Hall on systems of distinct representatives, which is equivalent to the Frobenius-König Theorem (Theorem 2.1.4) and is stated next.

Theorem 5.1.3 (Hall). *Let $A_1, \ldots, A_k$ be a family of subsets of the finite set S. Then the following conditions are equivalent:*

(i) *There exist distinct $x_i \in A_i, i = 1, 2, \ldots, k$.*

(ii) *For any $T \subset \{1, 2, \ldots, k\}, card\ (A(T)) \geq card\ (T)$.*

For an introduction to the theory of matroids the reader is referred to Welsh (1976), White (1986), and Recski (1989). A wealth of information concerning various extensions and refinements of Hall's Theorem is contained in Mirsky (1971).

5.2. Mixed discriminants

If $A^k = (a_{ij}^k)$ are $n \times n$ matrices, $k = 1, 2, \ldots, n$, then their *mixed discriminant*, denoted by $D(A^1, \ldots, A^n)$, is defined as

$$D(A^1, \cdots, A^n) = \frac{1}{n!} \sum_{\sigma \in S_n} \begin{vmatrix} a_{11}^{\sigma(1)} & \cdots & a_{1n}^{\sigma(n)} \\ \vdots & \ddots & \vdots \\ a_{n1}^{\sigma(1)} & \cdots & a_{nn}^{\sigma(n)} \end{vmatrix}, \qquad (5.2.1)$$

where S_n denotes, as usual, the set of permutations of $1, 2, \ldots, n$. (Throughout this section, A^k should not be confused with the k-th power of A.) Thus, if $A = (a_{ij})$, $B = (b_{ij})$ are 2×2 matrices, then

$$D(A, B) = \frac{1}{2}(a_{11}b_{22} - a_{21}b_{12} - a_{12}b_{21} + a_{22}b_{11}).$$

If $A^k = A$, $k = 1, 2, \ldots, n$, then clearly, $D(A^1, \ldots, A^n) = |A|$. Also, if each A_k is a diagonal matrix,

$$A^k = \begin{bmatrix} a_{11}^k & & \\ & \ddots & \\ & & a_{nn}^k \end{bmatrix},$$

then $D(A^1, \ldots, A^n)$ equals $\frac{1}{n!}$per B, where $B = (b_{ij}) = (a_{ii}^j)$. Thus the mixed discriminant provides an interesting generalization of both the determinant and the permanent.

We now consider some properties of mixed discriminants of positive semidefinite matrices. First note that if A^k, $k = 1, 2, \ldots, n$ are $n \times n$ matrices and if $A^1 = \beta B^1 + \gamma C^1$ for reals β, γ, then

$$D(A^1, A^2, \ldots, A^n) = \beta D(B^1, A^2, \ldots, A^n) + \gamma D(C^1, A^2, \ldots, A^n).$$
$$(5.2.2)$$

Thus, if A^k, $k = 1, 2, \ldots, n$ are positive semidefinite, then we can write each A^k as a sum of rank one positive semidefinite matrices, and then by a repeated application of (5.2.2), $D(A^1, \ldots, A^n)$ can be expressed as a sum of mixed discriminants of rank one positive semidefinite matrices. Also, it is easy to see from the definition that if $x_1, \ldots, x_n$ are vectors in R^n, then

$$D\left(x_1 x_1^T, \ldots, x_n x_n^T\right) = \frac{1}{n!}|(x_1, \ldots, x_n)|^2.$$

Thus we have proved the following lemma.

Lemma 5.2.1. *Let A^k, $k = 1, 2, \ldots, n$ be positive semidefinite $n \times n$ matrices, and suppose $A^k = X_k X_k^T$ for each k. Then*

$$D(A^1, \ldots, A^n) = \frac{1}{n!} \sum |(x_1, \ldots, x_n)|^2,$$

where the sum is over all choices $\{x_1, \ldots, x_n\}$ such that x_k is a column of $X_k, k = 1, 2 \ldots, n$.

As an immediate consequence of Lemma 5.2.1 we conclude that the mixed discriminant of positive semidefinite matrices is nonnegative. In fact, the mixed discriminant of a set of positive definite matrices must be positive (see Exercise 1). This last statement also follows from the next result.

Theorem 5.2.2. *Let $A^k, k = 1, 2, \ldots, n$ be $n \times n$ positive semidefinite matrices. Then the following conditions are equivalent:*

(i) $D(A^1, \ldots, A^n) > 0$.
(ii) For any $T \subset \{1, 2, \ldots, n\}$,

$$\text{card } (T) + \dim \left\{ \bigcap_{i \in T} \{\text{Null space of } A^i\} \right\} \leq n.$$

(iii) For any $T \subset \{1, 2, \ldots, n\}$,

$$\text{rank} \left(\sum_{i \in T} A^i \right) \geq |T|.$$

Proof. If A, B are positive semidefinite, then

$$(A + B)x = 0 \Rightarrow x^T(A + B)x = 0$$
$$\Rightarrow x^T Ax = 0, \qquad x^T Bx = 0$$
$$\Rightarrow Ax = 0, \qquad Bx = 0,$$

and therefore the null space of $A + B$ equals the intersection of the null space of A and the null space of B. Thus for any $T \subset \{1, 2, \ldots, n\}$,

$$\bigcap_{i \in T} \{\text{Null space of } A^i\} = \text{Null space of } \sum_{i \in T} A^i, \qquad (5.2.3)$$

and therefore (ii) and (iii) are equivalent. Let $A^k = X_k X_K^T, k = 1, 2, \ldots, n$. By Theorem 5.2.2, $D(A^1, \ldots, A^n) > 0$ if and only if there exist columns $x_1, x_2, \ldots, x_n$ of $X_1, X_2, \ldots, X_n$, respectively, that are linearly independent. By Theorem 5.1.2, a necessary and sufficient condition for this to happen is that for any $T \subset \{1, 2, \ldots, n\}$, the dimension of the space spanned by $\{X_i, i \in T\}$ is at least card (T). This is clearly equivalent to (iii), and the proof is complete. ∎

Theorem 5.2.2 generalizes the Frobenious-König Theorem. This can be seen as follows. Suppose $A = (a_{ij})$ is a nonnegative $n \times n$ matrix. Define A^k to be the diagonal matrix with $a_{1k}, a_{2k}, \ldots, a_{nk}$ along the diagonal, $k = 1, 2, \ldots, n$.

Then A^k is positive semidefinite and Theorem 5.2.2 is applicable. It is evident that $D(A^1, \ldots, A^n) > 0$ if and only if per $A > 0$, whereas (*ii*) of Theorem 5.2.2 is equivalent to saying that if A has a zero submatrix of order $r \times s$, then $r + s \le n$.

Let $\mathcal{D}_n$ denote the set of all n-tuples $\mathbf{A} = (A^1, A^2, \ldots, A^n)$ of $n \times n$ positive semidefinite matrices satisfying trace $A^i = 1, i = 1, 2, \ldots, n; \sum_{i=1}^{n} A^i = I$. Then by the process of identifying a nonnegative $n \times n$ matrix with an n-tuple of diagonal matrices described in the preceding paragraph, $\mathcal{D}_n$ can be viewed as a generalization of the class of $n \times n$ doubly stochastic matrices. The permanent function on Ω_n, the polytope of $n \times n$ doubly stochastic matrices, is generalized to the mixed discriminant over $\mathcal{D}_n$. In this context, the next result generalizes the fact that the permanent of a doubly stochastic matrix is positive. Compare the proof of the next result to that of Lemma 2.1.5.

Theorem 5.2.3. *If* $\mathbf{A} = (A^1, \ldots, A^n) \in \mathcal{D}_n$, *then* $D(A^1, A^2, \ldots, A^n) > 0$.

Proof. By Theorem 5.2.2, $D(A^1, A^2, \ldots, A^n) \ge 0$. If equality occurs, then by Theorem 5.2.2, there exists $T \subset \{1, 2, \ldots, n\}$ such that

$$t := \text{rank} \left(\sum_{i \in T} A^i \right) < \text{card } (T) := s. \tag{5.2.4}$$

We assume, without loss of generality, that $T = \{1, 2, \ldots, s\}$. The hypothesis as well as the conclusion of the theorem are unaffected if each A^k is replaced by $P A^k P^T$ for some orthogonal matrix P. Here we make use of the observation that if P is orthogonal, then

$$D(P A^1 P^T, \ldots, P A^n P^T) = D(A^1, \ldots, A^n).$$

Therefore, we can assume that the null space of $\sum_{i \in T} A^i$ is spanned by the first $n - t$ basis vectors of the standard basis. Hence, by (5.2.3), each $A^i, i = 1, 2, \ldots, s$ can be expressed in the form

$$A^i = \begin{bmatrix} 0 & 0 \\ 0 & D_i \end{bmatrix},$$

where D_i is $t \times t$. Let $A^i, i = s + 1, \ldots, n$ be conformally partitioned as

$$A^i = \begin{bmatrix} B_i & C_i \\ C_i^T & D_i \end{bmatrix}.$$

Now

$$\text{trace} \sum_{i=s+1}^{n} B_i = \text{trace } I_{(n-t) \times (n-t)} = n - t,$$

whereas

$$\text{trace} \sum_{i=s+1}^{n} B_i \leq \text{trace} \sum_{i=s+1}^{n} A^i \leq n - s.$$

Thus $n - t \leq n - s$, which contradicts (5.2.4), and the proof is complete. ∎

To make further progress along these lines one might consider the following questions, left as open problems:

(1) What are the extreme points of $\mathcal{D}_n$? The answer would provide a natural generalization of the Birkhoff–von Neumann Theorem.

(2) What is the minimum of $D(A^1, \ldots, A^n)$ over $\mathcal{D}_n$? In particular, is it attained when each A^i is the diagonal matrix

$$\begin{bmatrix} \frac{1}{n} & & \\ & \ddots & \\ & & \frac{1}{n} \end{bmatrix}?$$

An affirmative answer to this question would clearly generalize the Van der Waerden–Egorychev–Falikman Inequality (Theorem 2.8.7).

The material in this section is based on Bapat (1989b). Egorychev (1990) and Panov (1985a,b) provide additional details concerning the mixed discriminant.

5.3. The Alexandroff Inequality

We now proceed to prove the Alexandroff Inequality for mixed discriminants, which includes the Alexandroff Inequality for permanents proved in Chapter 2 as a special case. We first develop some preliminaries.

An instructive way to define the mixed discriminant is to say that $n!D$ $(A^1, \ldots, A^n)$ is the coefficient of $x_1 \cdots x_n$ in the expansion of the determinant $|\sum_{i=1}^{n} x_i A^i|$. The equivalence of this definition with the earlier one can be seen by using the multilinearity of the determinant as a function of each column. More generally, one can prove the following result. The proof is omitted.

Lemma 5.3.1. *Let $A^k, k = 1, 2, \ldots, n$ be $n \times n$ matrices, and let $r_1, \ldots, r_n$ be nonnegative integers adding to n. Then the coefficient of $x_1^{r_1} \cdots x_n^{r_n}$ in $|\sum_{i=1}^{n} x_i A^i|$ is equal to*

$$\frac{n!}{r_1! \cdots r_n!} D(\underbrace{A^1, \ldots, A^1}_{r_1}, \ldots, \underbrace{A^n, \ldots, A^n}_{r_n}).$$

We also need the following result; see Hardy, Littlewood, and Pólya (1952, p. 104) for a proof.

Lemma 5.3.2. *Suppose the polynomial*

$$f(x) = \sum_{i=0}^{n} \binom{n}{i} \alpha_i x^i$$

with real coefficients has only real roots. Then $\alpha_i^2 \geq \alpha_{i-1}\alpha_{i+1}$, $1 \leq i \leq n-1$ *(i.e., the sequence* $\alpha_1, \ldots, \alpha_n$ *is log-concave). Furthermore, if* $f(0) \neq 0$, *then equality occurs for any i if and only if all the roots of* $f(x)$ *are equal.*

Theorem 5.3.3 (Alexandroff Inequality). *Let* $A^1, \ldots, A^{n-r}, A$ *be* $n \times n$ *positive definite matrices* $(1 \leq r \leq n-1)$. *Then*

$$\{D(A^1, \ldots, A^{n-r}, \underbrace{A, \ldots, A}_{r})\}^2 \geq D(A^1, \ldots, A^{n-r}, A^{n-r}, \underbrace{A, \ldots, A}_{r-1})$$

$$\times D(A^1, \ldots, A^{n-r-1}, \underbrace{A, \ldots, A}_{r+1}),$$

where equality holds if and only if $A = \alpha A^{n-r}$ *for some real* α.

Proof. Let us introduce the notation

$$D_m(r_1, \ldots, r_m) = D(\underbrace{A^1, \ldots, A^1}_{r_1}, \underbrace{A^2, \ldots, A^2}_{r_2}, \ldots, \underbrace{A^m, \ldots, A^m}_{r_m}),$$

where $A^1, \ldots, A^m$ are $n \times n$ matrices. By Lemma 5.3.1,

$$\left| \sum_{i=1}^{m} x_i A^i \right| = \sum \frac{n!}{r_1! \cdots r_m!} D_m(r_1, \ldots, r_m) x_1^{r_1} \cdots x_m^{r_m},$$

where the summation is over nonnegative integers $r_1, \ldots, r_m$ adding to n. We must prove that if $A^1, \ldots, A^m$ are positive definite, then for $r_{m-1} \geq 1, r_m \geq 1$,

$$\{D_m(r_1, \ldots, r_m)\}^2 \geq D_m(r_1, \ldots, r_{m-2}, r_{m-1} + 1, r_m - 1)$$

$$\times D_m(r_1, \ldots, r_{m-2}, r_{m-1} - 1, r_m + 1).$$

We proceed by induction with respect to m. Suppose $m = 2$. The roots of the polynomial

$$f(x) = |xA^1 + A^2| = \sum_{i=0}^{n} \binom{n}{i} D_2(i, n-i) x^i$$

are the negatives of the eigenvalues of $(A^1)^{-\frac{1}{2}}A^2(A^1)^{-\frac{1}{2}}$ and hence they are real. Since A^2 is positive definite, $f(0) \neq 0$. It follows from Lemma 5.3.2 that the coefficients must satisfy

$$\{D_2(i, n-i)\}^2 \geq D_2(i+1, n-i-1)D_2(i-1, n-i+1), \ 1 \leq i \leq n-1.$$

Furthermore, equality occurs for any i only if all the roots of $f(x)$ are equal. Thus in case of equality for any i, $(A^1)^{-\frac{1}{2}}A^2(A^1)^{-\frac{1}{2}}$ must be a scalar matrix, or in other words, $A^2 = \alpha A^1$ for some real α.

Now suppose the theorem holds for $m \geq 2$ and let $A^1, \ldots, A^{m+1}$ be given positive definite matrices. Let $B_y = yA^m + A^{m+1}$, where y is a real parameter. Then

$$\left| \sum_{i=1}^{m-1} x_i A^i + x_m B_y \right| = \sum \frac{n!}{r_1! \cdots r_m!} D_m(r_1, \ldots, r_m)(y) x_1^{r_1} \cdots x_m^{r_m}, \quad (5.3.1)$$

where the coefficients, which depend on y, are defined as

$$D_m(r_1, \ldots, r_m)(y) = D(\underbrace{A^1, \ldots, A^1}_{r_1}, \ldots, \underbrace{A^{m-1}, \ldots, A^{m-1}}_{r_{m-1}}, \underbrace{B_y, \ldots, B_y}_{r_m}).$$

The left-hand side of (5.3.1) can also be written as

$$\left| \sum_{i=1}^{m-1} x_i A^i + x_m y A^m + x_m A^{m+1} \right| = \sum \frac{n!}{r_1! \cdots r_{m-1}! i! (r_m - i)!}$$
$$\times D_{m+1}(r_1, \ldots, r_{m-1}, i, r_m - i) y^i x_1^{r_1} \cdots x_m^{r_m}. \quad (5.3.2)$$

Comparing the coefficients of $x_1^{r_1} \cdots x_m^{r_m}$ in (5.3.1) and (5.3.2) and writing $r_m = k$ we get

$$D_m(r_1, \ldots, r_{m-1}, k)(y) = \sum_{i=0}^{k} \binom{k}{i} D_{m+1}(r_1, \ldots, r_{m-1}, i, k-i) y^i. \quad (5.3.3)$$

Equation (5.3.3) holds for any tuple $r_1, \ldots, r_{m-1}$ satisfying $r_1 + \cdots + r_{m-1} \leq n$, $k = n - r_1 - \cdots - r_{m-1}$ and for any real y. We show that if A^m and A^{m+1} are not proportional, then $D_m(r_1 \ldots, r_{m-1}, k)(y)$ has all its roots real and distinct. This is trivial if $r_1 + \cdots + r_{m-2} = n$ or $n - 1$. So we assume $r_1 + \cdots + r_{m-2} \leq n-2$. Define

$$Q_k(y) = D_m(r_1, \ldots, r_{m-2}, s - k, k)(y),$$

where $s = n - r_1 - \cdots - r_{m-2}$. Since the theorem is assumed true for m, we have

$$\{Q_k(y)\}^2 \geq Q_{k-1}(y)Q_{k+1}(y), \quad (5.3.4)$$

for $1 \leq k \leq s - 1$ and all real y. Suppose A^m and A^{m+1} are not proportional. Let y_0 be a root of Q_k. Let, if possible,

$$0 = Q_k(y_0) = Q_{k-1}(y_0) Q_{k+1}(y_0).$$

Then by the induction assumption, $B_{y_0} = \alpha A^{m-1}$ for some real α. Since A^m and A^{m+1} are not proportional, $\alpha \neq 0$. If $\alpha > 0$, then B_{y_0} is positive definite. But then $Q_k(y_0)$ would be a mixed discriminant of positive definite matrices, which must be positive by Theorem 5.2.2. We get a similar contradiction if $\alpha < 0$. We therefore conclude that strict inequality must occur in (5.3.4) at any root of Q_k.

We now give an argument that proves very useful in dealing with sequences of polynomials that form what is called a "Strum sequence" [see, for example, Polya and Szegö (1976)]. Let us formulate the auxiliary claim: The polynomial $Q_k(y)$ has k different real roots ($1 \leq k \leq s$). Furthermore, the roots of $Q_{k-1}(y)$ separate (interlace) the roots of $Q_k(y)$, $2 \leq k \leq s$.

We prove the claim by induction on k. Note that $Q_1(y)$ is linear and that the coefficient of y in $Q_1(y)$ is positive. Thus $Q_1(y)$ has a unique root, say $q^{(1)}$. By (5.3.4) and the subsequent discussion it follows that

$$0 = \{Q_1(q^{(1)})\}^2 > Q_0(q^{(1)}) Q_2(q^{(1)}).$$

Since Q_0 is positive, $Q_2(q^{(1)}) < 0$. Since $Q_2(y)$ is a quadratic in y that goes to ∞ as $y \to \infty$, it follows that it has two distinct real roots, say $q_1^{(2)} < q_2^{(2)}$, which are separated by $q^{(1)}$. Suppose the claim is true for k and $k - 1$, $2 \leq k \leq s - 1$. Let $q_i^{(k-1)}$, $q_i^{(k)}$ denote the roots of $Q_{k-1}(y)$, $Q_k(y)$, respectively, so that by the induction assumption

$$q_1^{(k)} < q_1^{(k-1)} < q_2^{(k)} < q_2^{(k-1)} < \cdots < q_{k-1}^{(k-1)} < q_k^{(k)}.$$

Set $q_0^{(k-1)} = -\infty$, $q_k^{(k-1)} = \infty$. Since the highest coefficient of $Q_{k-1}(y)$ is positive,

$$\operatorname{sgn} Q_{k-1}(y) = (-1)^{r+k+1}, \quad y \in \left(q_r^{(k-1)}, q_{r+1}^{(k-1)}\right),$$

for $0 \leq r \leq k - 1$. By applying (5.3.4) at $y = q_i^{(k)}$, $1 \leq i \leq k$, and invoking the subsequent remark regarding strict inequality, we can show that in $(-\infty, \infty)$, $Q_{k+1}(y)$ changes sign $k + 1$ times, in fact,

$$\operatorname{sgn} Q_{k+1}\left(q_i^{(k)}\right) = (-1)^{i+k+1}, \quad 1 \leq i \leq k.$$

Therefore, $Q_{k+1}(y)$ has $k + 1$ different real roots. It also follows that the roots are separated by those of $Q_k(y)$, and the claim is proved. The claim, together with Lemma 5.3.2 and Equation (5.3.3), show that

$$\{D_{m+1}(r_1, \ldots, r_{m+1})\}^2 > D_{m+1}(r_1, \ldots, r_{m-1}, r_m + 1, r_{m+1} - 1)$$
$$\times D_{m+1}(r_1, \ldots, r_{m-1}, r_m - 1, r_{m+1} + 1). \quad (5.3.5)$$

When $A^{m+1} = \alpha A^m$ for some real α, the inequality (5.3.5) becomes an equality and the proof is complete. ∎

We now illustrate an application of Theorem 5.3.3 in the theory of matroids. An $m \times n$ matrix whose entries are 0, 1, or -1 is said to be *totally unimodular* if every minor of the matrix is 0, 1, or -1. A matroid $M = (S, \mathcal{I})$ is said to be *unimodular* if there exists a totally unimodular matrix A with card (S) columns such that $T \subset S$ is independent if and only if the columns of A corresponding to T are linearly independent. We say that the matroid M has been represented by the matrix A.

Let $G = (V, E)$ be a graph. Recall the definition of the vertex-edge incidence matrix B given in Section 5.1.

Lemma 5.3.4. *The matrix B is totally unimodular.*

Proof. Let X be a submatrix of B of order r. If $r = 1$, then the determinant of X is clearly 0, 1, or -1. The proof will be given by induction on r. If every column of X consists of two nonzero entries, then these must be 1 and -1, in which case the column sums of X are zero and therefore the determinant of X is zero. If X has a column consisting of a single nonzero entry, then we may expand the determinant of X along that column and use the induction assumption. Finally, if X has no nonzero entry, then its determinant is zero. ∎

As a consequence of Lemma 5.3.4 we have the fact that every graphic matroid is unimodular.

Let $M = (S, \mathcal{I})$ be a matroid of rank n. Suppose the elements of S are colored so that each element gets one of s possible colors, denoted by $1, 2, \ldots, s$. Let $r_1, r_2, \ldots, r_s$ be a tuple of nonnegative integers adding to n. An independent set is called a *base* of M if it is not properly contained in another independent set. How many bases does M have comprised of precisely r_i elements of color $i, i = 1, 2, \ldots, s$? Suppose this number is denoted by $B(r_1, \ldots, r_s)$. We define

$$\tilde{B}(r_1, \ldots, r_s) = \frac{r_1! \cdots r_s!}{n!} B(r_1, \ldots, r_s).$$

We now show that for unimodular matroids, $B(r_1, \ldots, r_s)$ can be expressed in terms of a mixed discriminant. This can be seen as follows. Let M be unimodular and suppose it is represented by the totally unimodular matrix A. Without loss of generality, suppose A is partitioned as

$$A = [X_1, \ldots, X_s],$$

where the columns of X_i correspond to elements that are colored i, $i = 1, 2, \ldots, s$. Consider the sum

$$\sum |(x_1, \ldots, x_n)|^2 \tag{5.3.6}$$

extending over all choices $(x_1, \ldots, x_n)$ such that the first r_1 x_is are columns of X_1, the next r_2 x_is are columns of X_2, and so on. Since A is totally unimodular, every term in the summation is 0 or 1. Furthermore, it is 1 precisely when $(x_1, \ldots, x_n)$ is a base. Thus the sum (5.3.6) counts the number of bases of the required type. By Lemma 5.2.1, the expression (5.3.6) equals

$$\frac{n!}{r_1! \cdots r_s!} D(\underbrace{A^1, \ldots, A^1}_{r_1}, \ldots, \underbrace{A^s, \ldots, A^s}_{r_s}), \tag{5.3.7}$$

where $A^i = X_i X_i^T$, $i = 1, 2, \ldots, s$.

Lemma 5.3.5. *Let $M = (S, \mathcal{I})$ be a unimodular matroid of rank n, and let $r_1, \ldots, r_s$ be nonnegative integers adding to n. If $r_{s-1} \geq 1, r_s \geq 1$, then*

$$\{B(r_1, \ldots, r_s)\}^2 \geq B(r_1, \ldots, r_{s-2}, r_{s-1} + 1, r_s - 1)$$
$$\times B(r_1, \ldots, r_{s-2}, r_{s-1} - 1, r_s + 1).$$

Proof. We have seen that $B(r_1, \ldots, r_s)$ equals the expression in (5.3.6). Therefore

$$\tilde{B}(r_1, \ldots, r_s) = D(\underbrace{A^1, \ldots, A^1}_{r_1}, \ldots, \underbrace{A^s, \ldots, A^s}_{r_s}).$$

The result now follows by an application of Theorem 5.3.3.   ∎

It is illuminating to specialize the preceding discussion to graphic matroids. Let $G = (V, E)$ be a graph and let C be the reduced vertex-edge incidence matrix of an oriented version of G. Then the graphic matroid on E is represented by C. (See Example 5.4 in Section 5.1.) Suppose the edges of G are colored so that each edge gets one of s possible colors $1, 2 \ldots, s$. Let

$$C = [C_1, \ldots, C_s]$$

be the corresponding partitioning of C and let $A^i = C_i C_i^T$, $i = 1, 2, \ldots, s$. A subgraph of G is said to be a *spanning tree* if it is a connected subgraph with vertex set V and has no cycle. Note that a set of edges form a base if and only if they constitute a spanning tree of G. Therefore, we conclude that the number of spanning trees of G in which there are r_i edges of color i, $i = 1, 2, \ldots, s$, is given by (5.3.6). By Lemma 5.3.1 it follows that the total number of spanning trees of G is precisely $|CC^T|$; this is the well-known "Matrix-Tree Theorem" of Kirchhoff [see, for example, Biggs (1974)].

Let $M = (S, \mathcal{I})$ be a matroid of rank n. Let α_i denote the number of independent sets of M of cardinality i, $0 \le i \le n$, where we set $\alpha_0 = 1$. A conjecture due to Mason (1972) asserts that the sequence $\alpha_0, \alpha_1, \ldots, \alpha_n$ is log-concave, i.e., $\alpha_i^2 \ge \alpha_{i-1}\alpha_{i+1}$, $i = 1, 2, \ldots, n-1$. We refer to Chapter 4 for basic properties of log-concavity. We now show that Mason's conjecture can be proved if the matroid is unimodular. We first need some definitions. If $M_1 = (S_1, \mathcal{I}_1)$ and $M_2 = (S_2, \mathcal{I}_2)$ are two matroids, then their *direct sum,* denoted by $M_1 + M_2$, is defined to be the matroid $(S, \mathcal{I})$, where $S = S_1 \cup S_2$ and where a set is independent if it is the union of an independent set of M_1 and an independent set of M_2. It can be verified that a base of M is obtained by taking the union of a base of M_1 and a base of M_2.

Let $M = (S, \mathcal{I})$ be a matroid of rank n and let $0 \le k \le n$. The *rank k truncation* of M is a matroid defined on the ground set S where a set is declared to be independent if it is an independent set of M of cardinality at most k.

Theorem 5.3.6. *Let $M = (S, \mathcal{I})$ be a unimodular matroid of rank n. Then the sequence $\alpha_0, \alpha_1, \ldots, \alpha_n$ is log-concave.*

Proof. Let $U_{n,n}$ be the uniform matroid with parameters n, n and let N be the rank n truncation of the direct sum $M + U_{n,n}$. A base for N is obtained by taking the union of an independent set of M, say with i elements, with any $n - i$ elements of X. Color the elements of N with two colors 1 and 2 so that elements in S are colored 1 and elements in X are colored 2. Then

$$B(i, n - i) = \alpha_i \binom{n}{n - i} = \alpha_i \binom{n}{i},$$

and therefore $\tilde{B}(i, n - i) = \alpha_i$. The result now follows from Theorem 5.3.5. ∎

Our proof of the Alexandroff Inequality is based on Schneider (1966). The inequality is of fundamental importance in the theory of mixed volumes in geometry of convex sets [see Burago and Zalgaller (1988) and Schneider (1993)]. Theorem 5.3.6 is from Stanley (1981). Constantine (1987) provides another proof of the Matrix-Tree Theorem. Spanning trees in which there are a prescribed number of edges of each color have been treated in Bapat and Constantine (1992) and Bapat (1992b).

5.4. Coxeter graphs

As in the previous section, we consider only simple graphs. The *adjacency matrix* of the graph $G = (V, E)$, denoted by $A(G)$, is defined as follows. The

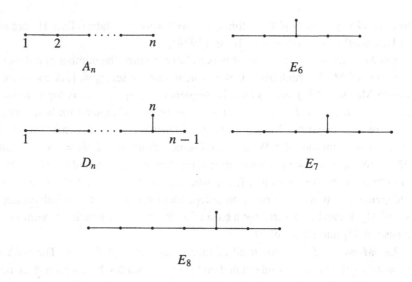

Figure 5.1. $\mathcal{C}_1$

order of $A(G)$ is card $(V) \times$ card (V). The rows as well as the columns of $A(G)$ are indexed by the vertices of G arranged in the same order. The (i, j) entry of $A(G)$ is 1 or 0 depending upon whether the i-th and the j-th vertices are adjacent or nonadjacent, respectively.

Clearly, $A(G)$ is a symmetric $(0, 1)$-matrix with diagonal entries equal to zero. If the ordering of the vertices is changed then we get a different matrix that is permutation similar to $A(G)$. Because we are interested in the spectral properties of $A(G)$, the difference in the ordering does not change the results that we obtain. The *index* of the graph G is defined to be the Perron root of $A(G)$. We denote the index of G by $\rho(G)$.

Let $\mathcal{C}_1$ and $\mathcal{C}_2$ denote the classes of graphs shown in Figures 5.1 and 5.2, respectively. In the first part of this section we show that a connected graph G satisfies $\rho(G) < 2$ (respectively, $\rho(G) = 2$) if and only if $G \in \mathcal{C}_1$ (respectively, $G \in \mathcal{C}_2$).

Lemma 5.4.1. *The adjacency matrix $A(G)$ is irreducible if and only if the graph G is connected.*

Proof. If $A(G)$ is irreducible, then for any (i, j), there exist $i = i_0, i_1, \ldots, i_{k-1}$, $i_k = j$ such that the (i_t, i_{t+1}) entry of $A(G)$ is positive, $t = 0, 1, \ldots, k - 1$. Thus there exists a path from vertex i to vertex j and G must be connected. The converse is proved similarly. ∎

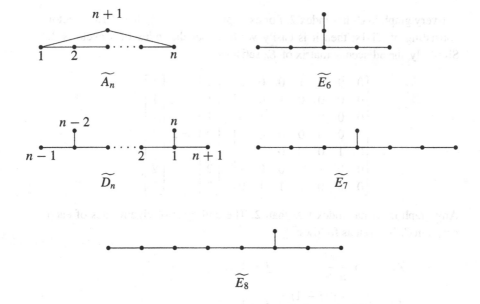

Figure 5.2. C_2

Let $G_1 = (V_1, E_1), G_2 = (V_2, E_2)$ be graphs such that $V_2 \subset V_1, E_2 \subset E_1$. Then G_2 is said to be a *subgraph* of G_1. We also say that G_1 is a *supergraph* of G_2.

Lemma 5.4.2. *If $G_2 = (V_2, E_2)$ is a subgraph of $G_1 = (V_1, E_1)$, then $\rho(G_2) \leq \rho(G_1)$. The inequality is strict if G_1 is connected and V_2 is a proper subset of V_1.*

Proof. Suppose card $(V_1) = n$, card $(V_2) = m \leq n$. If $m < n$, then we may add $n - m$ isolated vertices to G_2, corresponding to vertices in $V_1 \setminus V_2$, and thus, without loss of generality, we may assume that $m = n$. Suppose G_1 is connected, so that $A(G_1)$ is irreducible. Choose $\epsilon > 0$, sufficiently small, so that $A(G_2) + \epsilon A(G_1) \leq A(G_1)$, with strict inequality for some entry. Now, as in Exercise 13, Chapter 1, setting v to be a right Perron vector of $A(G_1)$ and u to be a left Perron vector of $A(G_2) + \epsilon A(G_1)$, we see that

$$u^T A(G_1) v > u^T (A(G_2) + \epsilon A(G_1)) v,$$

and, therefore, $\rho(A(G_1)) > \rho(A(G_2) + \epsilon A(G_1))$. By a similar argument, $\rho(A(G_2) + \epsilon A(G_1)) \geq \rho(A(G_2))$ and, therefore, $\rho(G_1) > \rho(G_2)$. If G_1 is not connected, then a similar argument shows that $\rho(G_1) \geq \rho(G_2)$, and the proof is complete. ∎

Every graph in C_2 has index 2. For example, the cycle $\widetilde{A}_n$ has Perron vector consisting of all 1s; then it is easily verified that the index of a cycle is 2. Similarly, the adjacency matrix of $\widetilde{E}_6$ satisfies

$$
\begin{bmatrix}
0 & 0 & 0 & 1 & 0 & 0 & 0 \\
0 & 0 & 0 & 0 & 1 & 0 & 0 \\
0 & 0 & 0 & 0 & 0 & 1 & 0 \\
1 & 0 & 0 & 0 & 0 & 0 & 1 \\
0 & 1 & 0 & 0 & 0 & 0 & 1 \\
0 & 0 & 1 & 0 & 0 & 0 & 1 \\
0 & 0 & 0 & 1 & 1 & 1 & 0
\end{bmatrix}
\begin{bmatrix} 1 \\ 1 \\ 1 \\ 2 \\ 2 \\ 2 \\ 3 \end{bmatrix}
= 2
\begin{bmatrix} 1 \\ 1 \\ 1 \\ 2 \\ 2 \\ 2 \\ 3 \end{bmatrix}.
$$

Any graph in C_1 has index less than 2. The entire set of eigenvalues of each graph in C_1 is given as follows:

$$A_n \quad 2\cos\frac{\pi j}{n+1}, \qquad j = 1, \ldots, n,$$

$$D_n \quad 2\cos\frac{\pi(2j-1)}{2n-2}, \quad j = 1, \ldots, n,$$

$$E_6 \quad 2\cos\frac{\pi m j}{12}, \qquad m = 1, 4, 5, 7, 8, 11,$$

$$E_7 \quad 2\cos\frac{\pi m j}{18}, \qquad m = 1, 5, 7, 9, 11, 13, 17,$$

$$E_6 \quad 2\cos\frac{\pi m j}{12}, \qquad m = 1, 7, 11, 13, 17, 19, 23, 29.$$

Theorem 5.4.3. *If a connected graph G satisfies $\rho(G) < 2$ $(\rho(G) = 2)$, then $G \in C_1$ (C_2).*

Proof. We have already observed that every graph in C_2 has index 2. Suppose a connected graph G has index less than 2. By Lemma 5.4.2, G cannot be a supergraph of any graph in C_2. Thus

(i) G cannot contain a cycle and hence G is a tree.
(ii) G cannot have a vertex of degree 4 or more, for then G will be a supergraph of $\tilde{D}_4$. So every vertex of G has degree at most 3.
(iii) If there are two vertices of degree 3, then G will be a supergraph of some $\tilde{D}_n, n \geq 5$ and hence G has at most one vertex of degree 3.
(iv) Suppose G has a vertex, say v_0, of degree 3. Then G consists of precisely three "branches" emanating from v_0. Let $p \leq q \leq r$ be the lengths (number of edges) of the three branches. Since G cannot be a supergraph of $\tilde{E}_6, \tilde{E}_7,$

or $\tilde{E}_8$, we conclude that either $p = q = 1$ or $p = 1, q = 2$, and $2 \le r \le 4$. That is, G is D_n for some $n \ge 4$ or G is E_6, E_7, or E_8.

(v) If G has no vertex of degree 3, then G is A_n for some n.

We conclude that G is in $\mathcal{C}_1$. Conversely, as already noted, if G is in $\mathcal{C}_1$, then $\rho(G) < 2$. Similarly, it can be shown that a connected graph G has index 2 if and only if G is in $\mathcal{C}_2$. ∎

Let $s_1, s_2, \ldots, s_n$ be indeterminates. A *word* in $s_1, s_2, \ldots, s_n$ is an expression of the type $s_{i_1} s_{i_2} \cdots s_{i_k}$, where $i_1, i_2, \ldots, i_k$ are indices in $1, 2, \ldots, n$; which are not necessarily distinct. Words can be multiplied by concatenation:

$$(s_{i_1} s_{i_2} \cdots s_{i_k})(s_{j_1} s_{j_2} \cdots s_{j_l}) = s_{i_1} s_{i_2} \cdots s_{i_k} s_{j_1} s_{j_2} \cdots s_{j_l}.$$

We also have an empty word, denoted by 1, which acts as multiplicative identity. Let G be a graph with n vertices. We now denote the adjacency matrix of G simply by A, rather than $A(G)$, for convenience. Impose the following relations on $s_1, s_2, \ldots, s_n$:

$$s_i^2 = 1; \quad (s_i s_j)^{a_{ij}+2} = 1, \quad i \ne j, \quad i, j = 1, \ldots, n. \tag{5.4.1}$$

The group of all words in $s_1, s_2, \ldots, s_n$, subject to relations (5.4.1), is called the *Coxeter group* associated with the graph G. The basic question that we consider is: What are the graphs that give rise to a finite Coxeter group? It turns out that this class of graphs is the same as the class $\mathcal{C}_1$. Such graphs are sometimes referred to as Coxeter graphs. The remainder of this section is mainly devoted to a proof of this fact.

It will be instructive to consider an example. Let G be A_4, the path on four vertices. We claim that the Coxeter group of G is (isomorphic to) S_5, the group of permutations of 5 symbols. This can be seen as follows. Any permutation can be written as a product of disjoint cycles. Recall that a *transposition* is a permutation that interchanges only two coordinates, keeping the rest fixed. If two adjacent coordinates are interchanged, the transposition is *elementary*. Let $s_1 = (12)$, $s_2 = (23)$, $s_3 = (34)$, and $s_4 = (45)$. Here (12), for example, denotes the elementary transposition that interchanges 1 and 2. It is well known that any permutation in S_5 can be written as a product of s_1, s_2, s_3, and s_4. Also, it is easily checked that $s_i^2 = id$, the identity permutation, $(s_1 s_3)^2 = (s_1 s_4)^2 = (s_2 s_4)^2 = id$, and $(s_1 s_2)^3 = (s_2 s_3)^3 = (s_3 s_4)^3 = id$. Thus (5.4.1) is satisfied and hence the Coxeter group of G is S_5. The argument can be extended to show that the Coxeter group of the path on n vertices is S_{n+1}.

The *Cartan matrix* associated with G is defined as $C = I - \frac{1}{2}A$. With s_i, we associate the $n \times n$ matrix B_i, $i = 1, 2, \ldots, n$, given by

$$
B_i = \begin{bmatrix} I_1 & 0 & 0 \\ a_i & -1 & b_i \\ 0 & 0 & I_2 \end{bmatrix},
$$

where I_1 and I_2 are identity matrices of order $i - 1$ and $n - i$, respectively and where a_i, b_i are row vectors such that for $j \neq i$, the j-th entry of $(a_i, -1, b_i)$ is $-2c_{ij}$.

It can be verified in a tedious but routine fashion that

$$
B_i^2 = I, \quad (B_i B_j)^{a_{ij}+2} = I, \quad i \neq j, \quad i, j = 1, 2, \ldots, n.
$$

Thus the Coxeter group of G is isomorphic to the group of matrices of the form $B_{i_1} B_{i_2} \cdots B_{i_k}$. We work with this formulation of the Coxeter group.

Lemma 5.4.4. *Let $U = (u_{ij})$ be the $n \times n$ matrix defined as*

$$
u_{ij} = \begin{cases} 1 & \text{if } i = j \\ 0 & \text{if } i > j \\ 2c_{ij} & \text{if } i < j \end{cases}.
$$

Then $-U^{-1}U^T = B_1 B_2 \cdots B_n$.

Proof. For each i, the i-th row of U is $(\mathbf{0}, 1, -b_i)$ and the i-th row of U^T is $(-a_i, 1, \mathbf{0})$, where $\mathbf{0}$ denotes the zero vector of order $i-1$ and $n-i$, respectively. For each i, define L_i and U_i by

$$
U = \begin{pmatrix} L_i^T & X_i \\ 0 & U_i \end{pmatrix}
$$

so that

$$
U^T = \begin{pmatrix} L_i & 0 \\ X_i^T & U_i^T \end{pmatrix},
$$

where L_i and U_i are of order i and $n - i$, respectively. We claim that for $i = 0, 1, \ldots, n$,

$$
U B_1 B_2 \cdots B_i = \begin{pmatrix} -L_i & 0 \\ 0 & U_i \end{pmatrix}.
$$

This is obvious for $i = 0$. Assume it to be true for $i - 1$ and proceed by induction. Thus

$$U B_1 B_2 \cdots B_i = \begin{pmatrix} -L_{i-1} & 0 \\ 0 & U_{i-1} \end{pmatrix} B_i$$

$$= \begin{pmatrix} -L_{i-1} & 0 & 0 \\ 0 & 1 & -b_i \\ 0 & 0 & U_i \end{pmatrix} \begin{pmatrix} I & 0 & 0 \\ a_i & -1 & -b_i \\ 0 & 0 & I \end{pmatrix}$$

$$= \begin{pmatrix} -L_{i-1} & 0 & 0 \\ a_i & -1 & 0 \\ 0 & 0 & U_i \end{pmatrix}$$

$$= \begin{pmatrix} -L_i & 0 \\ 0 & U_i \end{pmatrix}.$$

Therefore the claim is true. Thus for $i = n$ we have $U B_1 B_2 \cdots B_n = -U^T$, and the result is proved. ∎

An $n \times n$ matrix A is said to be *diagonalizable* if there exists a nonsingular matrix S such that $S A S^{-1}$ is a diagonal matrix. Let λ be an eigenvalue of A. Recall that the *algebraic multiplicity* of λ is the number of times λ occurs as a root of the characteristic polynomial of A. The *geometric multiplicity* of λ is the dimension of the space of eigenvectors of A corresponding to λ. We state the following result, the first part of which was noted in Chapter 1 as well.

Lemma 5.4.5. *(i) For any eigenvalue, the algebraic multiplicity is never less than the geometric multiplicity.*
(ii) A matrix is diagonalizable if and only if the algebraic multiplicity and the geometric multiplicity of each eigenvalue are equal.
(iii) Suppose A satisfies $A^N = I$ for some positive integer N. Then every eigenvalue of A has modulus 1, and A is diagonalizable.

The proof of Lemma 5.4.5 can be given using the Jordan canonical form. We refer to Gantmacher (1959) for a derivation of this form. We now obtain a result on M-matrices that will be useful. Throughout this section, by an M-matrix, we mean a nonsingular M-matrix.

Lemma 5.4.6. *Let M be an $n \times n$ M-matrix such that $M + M^T$ is irreducible. Then the following assertions are true:*

(i) If $M + M^T$ is not positive definite, then $-M^{-1}M^T$ has an eigenvalue $\lambda \geq 1$.

(ii) If $M + M^T$ is not positive semidefinite, then $-M^{-1}M^T$ has an eigenvalue $\lambda > 1$.

(iii) If $M + M^T$ is positive semidefinite but not positive definite, then $-M^{-1}M^T$ is not diagonalizable.

Proof. Let D be the diagonal matrix with $m_{11}, \ldots, m_{nn}$ along the diagonal and let $E = D^{-\frac{1}{2}}$. For $0 \leq \mu \leq 1$, let

$$P(\mu) = (1 + \mu)I - EM^T E - \mu EME.$$

Since M is an M-matrix, so is EME, and for $\mu > 0, i \neq j$, the (i, j) entry of $P(\mu)$ is zero if and only if the (i, j) and the (j, i) entries of M are both zero. However, $M + M^T$ is irreducible and hence $P(\mu)$ is irreducible for $\mu > 0$. Moreover, $P(\mu)$ is a nonnegative matrix and so, by the Perron-Frobenius Theorem, the spectral radius $r(\mu)$ of $P(\mu)$ is a simple eigenvalue of $P(\mu)$. It is clear that $r(\mu)$ is a continuous function of μ.

Since $E^T ME$ is an M-matrix, its real eigenvalues are all positive. Hence $r(0)$, the spectral radius of $P(0)$, is less than 1.

If $M + M^T$ is not positive definite, then $E(M + M^T)E$ has an eigenvalue that is not positive. Thus $2I - EM^T E - EME$ has an eigenvalue greater than or equal to 2 and $r(1) \geq 2$. By continuity, there exists μ with $0 < \mu \leq 1$ and $r(\mu) = 1 + \mu$. Now

$$|-EM^T E - \mu EME| = 0.$$

Taking the transpose gives

$$|-EME - \mu EM^T E| = 0.$$

Thus $|-\mu^{-1}I - M^{-1}M^T| = 0$, and $\lambda = \mu^{-1}$ is an eigenvalue of $-M^{-1}M^T$ with $\lambda \geq 1$. This proves (i).

If $M + M^T$ is not positive semidefinite, then $E(M + M^T)E$ has a negative eigenvalue; therefore, in the above argument, $\mu < 1$. Thus $\lambda > 1$, and (ii) is proved.

Now let $M + M^T$ be positive semidefinite but not positive definite. Thus $M + M^T$ is singular and hence 1 is an eigenvalue of $-M^{-1}M^T$. Let

$$q(x) = |xI + M^{-1}M^T|.$$

Then

$$q\left(\frac{1}{x}\right) = \left|\frac{1}{x}I + M^{-1}M^T\right|$$

$$= \frac{1}{x^n}|I + xM^{-1}M^T|$$

$$= \frac{1}{x^n}|M||I + xM^{-1}M^T||(M^T)|^{-1}$$

$$= \frac{1}{x^n}|M(M^T)^{-1} + xI|$$

$$= \frac{1}{x^n}|M^{-1}M^T + xI|$$

$$= \frac{1}{x^n}q(x).$$

Therefore, $q(x)$ has the form

$$(1 + x^n) + a_1(x + x^{n-1}) + a_2(x^2 + x^{n-2}) + \cdots \qquad (5.4.2)$$

for some $a_1, a_2, \ldots$.

We also need the elementary fact that

$$(x^i - 1)(x^{n-i} - 1) = 0 \bmod (x - 1)^2, \quad 0 \le i \le n$$

and hence

$$x^i + x^{n-i} = 1 + x^n \bmod (x - 1)^2, \quad 0 \le i \le n. \qquad (5.4.3)$$

Using (5.4.2) and (5.4.3) we have

$$q(x) = (1 + a_1 + a_2 + \cdots)(1 + x^n) \bmod (x - 1)^2. \qquad (5.4.4)$$

Let $\theta = (1 + a_1 + a_2 + \cdots)$. Then from (5.4.4) we get

$$q(x) - \theta(1 + x^n) = t(x)(x - 1)^2$$

for some $t(x)$. Therefore

$$q(x) - 2\theta - \theta(x^n - 1) = t(x)(x - 1)^2$$

and hence

$$q(x) = 2\theta \bmod (x - 1). \qquad (5.4.5)$$

It follows from (5.4.3), (5.4.4), and (5.4.5) that $\theta = 0$ and thus $(x - 1)^2$ divides $q(x)$. So 1 is an eigenvalue of $-M^{-1}M^T$ with algebraic multiplicity at least 2.

Since $r(1) = 2$ is a simple eigenvalue of $2I - EM^TE - EME$, the null space of $-EM^TE - EME$ has dimension 1. Thus the null space of $-M^{-1}M^T - I$ has dimension 1, which is the geometric multiplicity of 1. Since this is less than the algebraic multiplicity, by Lemma 5.4.5, $-M^{-1}M^T$ is not diagonalizable. This completes the proof of (*iii*). ∎

We are now ready to prove the following theorem:

Theorem 5.4.7. *Let G be a connected graph with W as the associated Coxeter group. Then W is finite if and only if G is in C_1.*

Proof. Let A be the adjacency matrix and C the Cartan matrix of G. Since G is connected, by Lemma 5.4.1, A and C are irreducible. First suppose that G is not in C_1. Then by Theorem 5.4.3, $\rho(G) \geq 2$ and hence C is not positive definite. Note that $C = U + U^T$, where U, defined in Lemma 5.4.4, is clearly an M-matrix. Let $D = -U^{-1}U^T$. It follows by Lemma 5.4.6 that either D has an eigenvalue greater than 1 or that D is not diagonalizable. Thus, by Lemma 5.4.5, there does not exist a positive integer N such that $D^N = I$. Since, by Lemma 5.4.4, $D = B_1 B_2 \cdots B_n$, then it follows that the element $B_1 B_2 \cdots B_n$ in W does not have finite order. Therefore, W is infinite.

Conversely, if G is in C_1, we must show that W is finite. We only offer some remarks concerning the proof of this fact and refer to Humphreys (1990) for complete details. We have already observed that the Coxeter group of A_n is S_{n+1}. A similar reasoning shows that the Coxeter group of $D_n, n \geq 4$, is the *sign change group* of order n, i.e., the group of $n \times n$ matrices in which each row and column has precisely one nonzero entry that is either 1 or -1. The Coxeter groups of E_6, E_7, and E_8 are also finite. The order of the group in each case is $\prod_j (m_j + 1)$, where m_j are the *exponents,* i.e., the numbers that occur in the eigenvalues of the adjacency matrices of the graphs. As an example, the order of the Coxeter group of E_6 is $2 \times 5 \times 6 \times 8 \times 9 \times 12 = 51840$. ∎

For an introduction to Coxeter groups and related topics see Humphreys (1990), Grove and Benson (1985), and Coxeter and Moser (1980). The graphs in C_1 and C_2 arise, sometimes unexpectedly, in many diverse areas. For some other characterizations of these graphs, see Deodhar (1982). A characterization based on the notion of "path-positivity" is given in Bapat and Lal (1991, 1993a). The main result there can be described as follows. For any positive integer k, let $P_k(\lambda)$ denote the characteristic polynomial of A_k, the path on k vertices. Let G be an arbitrary graph and consider the sequence of matrices $P_k(A(G))$, $k = 1, 2, \ldots$ obtained by evaluating the polynomial $P_k(\lambda)$ at the matrix $A(G)$. Then for any graph G not contained in C_1, each $P_k(A(G))$ is a nonnegative matrix. Furthermore, $P_k(A(G)) = 0$ for some k if and only if G is in C_2. This last statement has also been proved in Seidel (1992).

The eigenvalues of graphs in C_1 can be found, for example, in Mehta (1989). The area of spectra of graphs is vast and has many applications [see Cvetković, Doob, and Sachs (1979), Seidel (1989), Merris (1993), and the references contained therein]. The proof of Theorem 5.4.7 is based on that of Howlett (1982).

5.5. Matrices over the max algebra

In this section we describe an algebraic system called the "max algebra." We show that, for matrices over this general structure, the basic assertions of the Perron-Frobenius Theorem, such as the existence of a positive eigenvalue with a positive eigenvector, continue to hold. Max algebra is useful in several areas, including transportation networks, machine scheduling, and parallel computation. It can be used to represent, in a linear fashion, phenomena that are nonlinear in the conventional algebra.

For convenience, we work with a special instance of max algebra. Thus, for our purpose, the *max algebra* consists of the set $\mathcal{M} = R \cup \{-\infty\}$ (where R is the set of real numbers), equipped with two binary operations, addition and multiplication, denoted by $\oplus$ and $\otimes$, respectively. (In this section $\otimes$ does not denote the Kronecker product.) The operations are defined as follows: $a \oplus b = \max(a, b)$, the maximum of a, b and $a \otimes b = a + b$. Observe that $-\infty$ and 0 serve as identity elements for the operations $\oplus$ and $\otimes$, respectively. We denote $x_1 \oplus \cdots \oplus x_n$ by $\sum_{i=1}^{n} \oplus$, or simply by $\sum_{\oplus} x_i$, where the range of summation is to be understood from the context.

We work with vectors and matrices over the max algebra. Basic operations on matrices are defined in the natural way. Thus if $A = (a_{ij})$, $B = (b_{ij})$ are $m \times n$ matrices over $\mathcal{M}$, then $A \oplus B$ is the $m \times n$ matrix with (i, j)-th entry $a_{ij} \oplus b_{ij}$. If $c \in \mathcal{M}$, then $c \otimes A$ is the matrix $(c \otimes a_{ij}) = (c + a_{ij})$. If A is $m \times n$ and B is $n \times p$, then $A \otimes B$ is the $m \times p$ matrix with (i, j)-th entry

$$\sum_{k=1}^{n} \oplus a_{ik} \otimes b_{kj} = \max_k (a_{ik} + b_{kj}).$$

Thus, if

$$A = \begin{bmatrix} 2 & -\infty & 1 \\ -4 & 1 & -\infty \\ 1 & -\infty & 0 \end{bmatrix}, \qquad B = \begin{bmatrix} 0 & -\infty & 3 \\ -1 & 2 & -\infty \\ 1 & 5 & -\infty \end{bmatrix},$$

then $A \oplus B$ and $A \otimes B$ are given by

$$\begin{bmatrix} 2 & -\infty & 3 \\ -1 & 2 & -\infty \\ 3 & 5 & 0 \end{bmatrix} \quad \text{and} \quad \begin{bmatrix} 2 & 6 & 5 \\ 0 & 3 & -1 \\ 1 & 5 & 4 \end{bmatrix},$$

respectively.

It is easily seen that matrix multiplication is associative and that it distributes over matrix addition. If A is $n \times n$, we denote $A \otimes \cdots \otimes A$ taken k times by A^k.

The $n \times n$ matrix with each diagonal entry zero and each off-diagonal entry $-\infty$ is the *identity matrix* over the max algebra; if we permute its rows (or columns) we obtain a *permutation matrix* over the max algebra.

A vector over $\mathcal{M}$ is be said to be *finite* if each component of the vector is finite. A vector is called *partly infinite* if it has a finite component as well as an infinite component. A matrix or vector with each component $-\infty$ is called *infinite*, and we denote it by $-\infty$ as well.

Let A be an $n \times n$ matrix over $\mathcal{M}$. We associate a directed graph $H(A)$ with A as follows. The vertices of $H(A)$ are $1, 2, \ldots, n$. There is an edge from vertex i to vertex j, denoted by (i, j), if a_{ij} is finite. In this case we say that a_{ij} is the weight of the edge (i, j). Note that $B = (e^{a_{ij}})$ is a nonnegative matrix and if we ignore weights, then the graph $H(A)$ is identical with the graph $G(B)$, associated with B, which was defined in Section 1.1.

The *weight* of a path in $H(A)$ is the sum of the weights of the edges in the path; the corresponding average is called the *average weight* of the path.

A *circuit* of length ℓ in $H(A)$ is defined to be a closed path $i_1 \rightarrow i_2 \rightarrow \cdots \rightarrow i_\ell \rightarrow i_1$, where $i_1, i_2, \ldots, i_\ell$ are distinct. A circuit of length one is called a *loop*. We denote the set of circuits in $H(A)$ by $C(A)$. If $\tau \in C(A)$, then the average weight of τ is called the *mean* of the circuit τ, denoted by $M_A(\tau)$. We define the maximal circuit mean of A, denoted by $\mu(A)$, as

$$\mu(A) = \max_{\tau \in C(A)} M_A(\tau)$$

if $C(A)$ is nonempty, and we set $\mu(A) = -\infty$ otherwise.

We say that A is *irreducible* over the max algebra if $H(A)$ is strongly connected, or equivalently, if $B = (e^{a_{ij}})$ is irreducible. In this section, by irreducible we always mean irreducible over the max algebra. A matrix is *reducible* (over the max algebra) if it is not irreducible. Note that A is reducible if and only if either A is 1×1 containing $-\infty$ or there exists a permutation matrix Q (over the max algebra) such that

$$Q \otimes A \otimes Q^T = \begin{bmatrix} A_{11} & -\infty \\ A_{21} & A_{22} \end{bmatrix},$$

where A_{11}, A_{22} are square matrices of order at least one.

We now define the concepts of eigenvalues and eigenvectors over the max algebra. Let A be an $n \times n$ matrix over $\mathcal{M}$. We say that λ is an *eigenvalue* of A if there exists a vector $x \neq -\infty$ such that $A \otimes x = \lambda \otimes x$. In this case we say that x is an *eigenvector* of A corresponding to the eigenvalue λ. Furthermore, we call (λ, x) an *eigenpair* of A. Note that (λ, x) is an eigenpair of A if and only if $x \neq -\infty$ and $\max_j(a_{ij} + x_j) = \lambda + x_i, i = 1, 2, \ldots, n$. For example, if

$$A = \begin{bmatrix} 3 & -\infty \\ 2 & 4 \end{bmatrix},$$

then

$$A \otimes \begin{bmatrix} -\infty \\ 0 \end{bmatrix} = 4 \otimes \begin{bmatrix} -\infty \\ 0 \end{bmatrix},$$

and thus 4 is an eigenvalue of A with $[-\infty, 0]^T$ as a corresponding eigenvector. It is easily verified that 4 is the only eigenvalue of A. Further, A^T has both 3 and 4 as eigenvalues.

Here is an example indicating how the eigenproblem may arise in a real life problem. Consider an industrial process that runs in cycles and uses four machines. The machines do not run independently and the output of one machine may be the input of other machines. For this and other technical reasons, we may be faced with a constraint where the $(k + 1)$-th cycle of any machine can begin only after the k-th cycle of certain other machines is completed and a certain time period has elapsed. The following matrix A summarizes this information:

$$A = \begin{bmatrix} 0 & 3 & 2 & -\infty \\ 1 & 1 & -\infty & 2 \\ 4 & 0 & 1 & -\infty \\ 1 & -\infty & 2 & 2 \end{bmatrix}.$$

The (i, j)-th entry of the matrix indicates the waiting time for the $(k + 1)$-th cycle of the i-th machine to begin after the k-th cycle of the j-th machine has been completed. An entry equal to $-\infty$ indicates that the cycles of the two machines have no direct relation.

If the components of the vector x denote the starting time of the initial cycle of the four machines, then observe that $A \otimes x, A^2 \otimes x, A^3 \otimes x, \ldots$ give the starting times of the machines in the successive cycles. Thus if we choose x to be an eigenvector of A with λ as the corresponding eigenvalue, then the system will run in such a way that the time interval between the beginnings of consecutive cycles on any machine is λ. In our example it can be verified that $x = [1, 0, 2, 1]^T$ is an eigenvector of A with 3 as the eigenvalue. Thus if we start the second machine first, then machines 1 and 4 a unit time later, and machine 3 two units of time later, then the system will continue to operate so that the consecutive cycles of any machine are started after a fixed time interval of 3 units.

We continue our discussion of the theoretical aspects of the eigenproblem. If Q is a permutation matrix over the max algebra and $\lambda \in \mathcal{M}$, then (λ, x) is an eigenpair of A if and only if $(\lambda, Q \otimes x)$ is an eigenpair of $Q \otimes A \otimes Q^T$. In particular, A and $Q \otimes A \otimes Q^T$ have the same eigenvalues. In view of these observations, we often find it convenient to deal with $Q \otimes A \otimes Q^T$ for a suitable permutation matrix Q.

The next few results are concerned with the occurrence of $-\infty$ as an eigenvalue.

Lemma 5.5.1. *Let A be an $n \times n$ matrix over $\mathcal{M}$ and suppose $C(A) = \phi$. Then $-\infty$ is the unique eigenvalue of A.*

Proof. Since $C(A)$ has no circuit, we can relabel the vertices of $H(A)$ such that there is no edge from i to j if $i \leq j$ [see, for example, Carré (1979), p. 50]. Thus we may assume, without loss of generality, that

$$A = \begin{bmatrix} -\infty & \cdots & \cdots & -\infty \\ a_{21} & -\infty & \cdots & -\infty \\ \vdots & \vdots & \ddots & \vdots \\ a_{n1} & a_{n2} & \cdots & a_{nn} \end{bmatrix}.$$

The vector $x = [-\infty, \ldots, -\infty, 0]^T$ satisfies $A \otimes x = -\infty \otimes x$ and hence $-\infty$ is an eigenvalue of A. Now suppose (λ, y) is an eigenpair of A, so that

$$A \otimes y = \lambda \otimes y. \tag{5.5.1}$$

Suppose $\lambda \neq -\infty$. Then the first equation in the system (5.5.1) gives $y_1 = -\infty$. The second equation in (5.5.1) is $a_{21} + y_1 = \lambda + y_2$, and since $y_1 = -\infty$, we have $y_2 = -\infty$. Continuing this way we conclude that all components of y are $-\infty$, which is a contradiction. Thus $\lambda = -\infty$ and the proof is complete. ∎

Lemma 5.5.2. *Let A be an $n \times n$ matrix over $\mathcal{M}$. Then $-\infty$ is an eigenvalue of A if and only if A has an infinite column.*

Proof. If $(-\infty, x)$ is an eigenpair of A and if x_j is finite, then it follows from the eigenvalue-eigenvector equation that the j-th column of A is $-\infty$. Conversely, if the j-th column of A is $-\infty$, then a vector with all components $-\infty$ except the j-th component serves as an eigenvector of A with $-\infty$ as the eigenvalue. ∎

The next result follows immediately from Lemma 5.5.2.

Corollary 5.5.3. *If A is an irreducible matrix, then any eigenvalue of A is finite.*

We will see later that if A is irreducible, then $\mu(A)$ is the only eigenvalue of A.

Lemma 5.5.4. *Let A be an $n \times n$ matrix over $\mathcal{M}$. If A has a partly infinite eigenvector, then A is reducible.*

Proof. Let (λ, x) be an eigenpair of A with x partly infinite. If $\lambda = -\infty$, then by Lemma 5.5.2 we conclude that A has an infinite column and hence A is reducible. So suppose λ is finite. We assume, without loss of generality, that $x = [-\infty, y]^T$, where y is finite. Partition A conformally so that we have

$$\begin{bmatrix} A_{11} & A_{12} \\ A_{21} & A_{22} \end{bmatrix} \otimes \begin{bmatrix} -\infty \\ y \end{bmatrix} = \lambda \otimes \begin{bmatrix} -\infty \\ y \end{bmatrix}.$$

Then $A_{12} \otimes y = \lambda \otimes -\infty$. Since y is finite, $A_{12} = -\infty$ and hence A is reducible. ∎

Corollary 5.5.5. *If A is an $n \times n$ irreducible matrix over $\mathcal{M}$, then any eigenvector of A is finite.*

The following is a simple fact about directed graphs.

Lemma 5.5.6. *Let G be a directed graph in which every vertex has out-degree at least one. Then G has a circuit.*

Proof. If G has no circuit, then as noted in the proof of Lemma 5.5.1, we can relabel the vertices of G so that there is no edge from i to j if $i \leq j$. But then vertex 1 must have out-degree zero, which is a contradiction. Therefore G must have a circuit. ∎

Theorem 5.5.7. *Let A be an $n \times n$ matrix over $\mathcal{M}$, let (λ, x) be an eigenpair of A, and suppose x is finite. Then $\lambda = \mu(A)$.*

Proof. If $\lambda = -\infty$, then, as noted in the proof of Lemma 5.5.2, every column of A must be $-\infty$ and hence $\lambda = \mu(A) = -\infty$. So we assume that λ is finite. We have

$$\max_j (a_{ij} + x_j) = \lambda + x_i, \quad i = 1, 2, \ldots, n \tag{5.5.2}$$

and hence, in particular,

$$a_{ij} + x_j - x_i \leq \lambda, \quad i, j = 1, 2, \ldots, n. \tag{5.5.3}$$

Since λ is finite, we must have $C(A) \neq \phi$ by Lemma 5.5.1. Let $\tau \in C(A)$ be the circuit formed by the edges $(i_1, i_2), (i_2, i_3), \ldots, (i_k, i_1)$. Apply (5.5.3) to the entries of A in positions $(i_1, i_2), (i_2, i_3), \ldots, (i_k, i_1)$ and add the resulting inequalities to get $M_A(\tau) \leq \lambda$.

Now construct a directed graph K with vertices $1, 2, \ldots, n$ and an edge (i, j) if and only if there is equality in (5.5.3), i.e., $a_{ij} + x_j - x_i = \lambda$. By (5.5.2) each vertex in K has out-degree at least one and hence, by Lemma 5.5.6, K has a circuit, say, τ. Since λ is finite, K is clearly a subgraph of $H(A)$. Thus $\tau \in C(A)$, and furthermore, $M_A(\tau) = \lambda$. Therefore, $\lambda = \mu(A)$, and the proof is complete. ∎

Theorem 5.5.8. *Let A be an irreducible $n \times n$ matrix over $\mathcal{M}$. Then $\mu(A)$ is the only eigenvalue of A and every eigenvector corresponding to $\mu(A)$ is finite.*

Proof. Since A is irreducible, using an argument as in the proof of Lemma 5.5.1 we see that $C(A) \neq \phi$ and hence $\mu(A)$ is finite. Let B be the $n \times n$ matrix with $b_{ij} = a_{ij} - \mu(A)$ for all i, j, i.e., $A = \mu(A) \otimes B$. Define $\Gamma(B)$ by

$$\Gamma(B) = B \oplus B^2 \oplus \cdots \oplus B^n. \qquad (5.5.4)$$

The (i, j)-th entry of B^k is precisely the maximal weight of a path of length k from vertex i to vertex j in $H(B)$. Any path in $H(B)$ of length more than n must contain a circuit and because any circuit mean in $C(B)$ is at most zero, we have

$$\Gamma(B) \geq B^k, \quad k \geq n + 1. \qquad (5.5.5)$$

It follows from (5.5.4) and (5.5.5) that the (i, j)-th entry of $\Gamma(B)$, denoted by $\Gamma_{ij}(B)$, is the maximal weight of a path (of any length) from i to j in $H(B)$. Let $\tau \in C(A)$ be such that $M_A(\tau) = \mu(A)$ and let j be any vertex that belongs to τ. Any path from j to itself in $H(B)$ must have weight at most zero, and because j is in τ, there is at least one path from j to itself in $H(B)$ with weight zero. Thus $\Gamma_{jj} = 0$ and, in particular, $\Gamma_j(B)$, the j-th column of $\Gamma(B)$, is not $-\infty$.

For any i, the maximal weight of a path in $H(B)$ (of any length) from i to j may also be expressed as

$$\sum_{\oplus} b_{i\ell} \otimes \Gamma_{\ell j} \qquad (5.5.6)$$

since there exists such a path (in which a vertex may occur more than once) of length at least two, in view of the fact that j is in τ. However, (5.5.6) is precisely the i-th entry of $B \otimes \Gamma_j(B)$. Therefore

$$B \otimes \Gamma_j(B) = \Gamma_j(B) = 0 \otimes \Gamma_j(B)$$

and hence $(0, \Gamma_j(B))$ is an eigenpair of B. Now

$$A \otimes \Gamma_j(B) = \mu(A) \otimes B \otimes \Gamma_j(B) = \mu(A) \otimes \Gamma_j(B)$$

and therefore $\mu(A)$ is an eigenvalue of A.

Now suppose (λ, x) is an eigenpair of A. Since A is irreducible, by Corollary 5.5.3, $\lambda \neq -\infty$. By Corollary 5.5.5, x is finite. It then follows from Theorem 5.5.7 that $\lambda = \mu(A)$. By Corollary 5.5.5 any eigenvector corresponding to $\mu(A)$ is finite, and the proof is complete. ∎

A very thorough discussion of all aspects of matrices over max algebra is contained in Cuninghame-Green (1979). Some related references are Carré (1979, Chapter 6) and Gondran and Minoux (1984). For recent work on max algebra systems, see Baccelli et al (1992), Braker and Olsder (1993), Cohen et al (1983, 1985), Dudnikov and Samborskii (1992), Gaubert (1992), and Olsder and Roos (1988). Bapat, Stanford, and Van den Driessche (1993) contains properties of eigenvalues and eigenvectors based on the pattern of finite and infinite entries in the matrix as well as inequalities similar to those for the Perron root.

5.6. Boolean matrices

The binary Boolean algebra $\mathcal{B}$ consists of the set $\{0, 1\}$, equipped with the operations of addition and multiplication, defined as follows:

+	0	1		·	0	1
0	0	1		0	0	0
1	1	1		1	0	1

By a Boolean matrix we mean a matrix over $\mathcal{B}$. In this section we confine ourselves to Boolean matrices. The operations of matrix addition, scalar multiplication and matrix multiplication are defined in the usual way. A Boolean matrix may be regarded as a matrix over the max algebra with entries 0 and $-\infty$. This is achieved by identifying 0 and $-\infty$ in the max algebra with 1 and 0 in the Boolean algebra, respectively.

Example 5.5. Let

$$A = \begin{bmatrix} 0 & 1 & 1 \\ 1 & 0 & 1 \\ 0 & 0 & 1 \end{bmatrix}, \quad B = \begin{bmatrix} 0 & 1 & 0 \\ 1 & 1 & 0 \\ 0 & 1 & 1 \end{bmatrix}.$$

Then

$$A + B = \begin{bmatrix} 0 & 1 & 1 \\ 1 & 1 & 1 \\ 0 & 1 & 1 \end{bmatrix} \quad \text{and} \quad AB = \begin{bmatrix} 1 & 1 & 1 \\ 0 & 1 & 1 \\ 0 & 1 & 1 \end{bmatrix}.$$

As usual, the square matrix with ones on the diagonal and zeros elsewhere is the *identity matrix*. The transpose of the matrix A is denoted by A^T.

We denote by B^n, the set of $n \times 1$ column vectors over B. Occasionally, we will use B^n to denote the set of $1 \times n$ row vectors over B but this should not cause any confusion. A set of vectors $x_1, \ldots, x_n$ is said to be *linearly independent* if the zero vector is not in the set and no vector is the sum of some of the remaining vectors. The *linear span* of the vectors $x_1, \ldots, x_n$, denoted by $S(x_1, \ldots, x_n)$, is the set of all linear combinations $c_1 x_1 + \cdots + c_n x_n$, each $c_i \in B$. A vector y is said to be *dependent* on the vectors $x_1, \ldots, x_n$ if $y \in S(x_1, \ldots, x_n)$.

If A, B are matrices over B of the same order, then we say $A \geq B$ whenever $a_{ij} \geq b_{ij}$ for all i, j. A similar notation applies to vectors.

A set $T \in B^n$ is a *vector space* if the set contains the zero vector and is closed under addition. If T_1, T_2 are vector spaces and $T_1 \subset T_2$, then T_1 is a *subspace* of T_2.

Example 5.6. Let $x_i, i = 1, \ldots, 5$ be given by

$$
x_1 = \begin{bmatrix} 0 \\ 0 \\ 0 \\ 0 \end{bmatrix}, x_2 = \begin{bmatrix} 0 \\ 1 \\ 1 \\ 0 \end{bmatrix}, x_3 = \begin{bmatrix} 1 \\ 0 \\ 1 \\ 1 \end{bmatrix}, x_4 = \begin{bmatrix} 1 \\ 1 \\ 1 \\ 1 \end{bmatrix}, x_5 = \begin{bmatrix} 0 \\ 1 \\ 0 \\ 0 \end{bmatrix}.
$$

Then $\{x_2, x_3, x_5\}$ is a linearly independent set, x_4 depends on x_2 and x_3 since $x_4 = x_2 + x_3$, and $\{x_1, x_2, x_3, x_4, x_5\}$ form a vector space.

Let T be a vector space and let $S \subset T$. Then S is said to be a *basis* for T if S is linearly independent and the linear span $S(S)$ equals T.

The next result is the first example of some peculiar things that can occur when we look at vector spaces over the Boolean algebra.

Theorem 5.6.1. *Let $T \subset B_n$ be a vector space. Then T admits a basis, and the basis is unique.*

Proof. Let W be the set of all nonzero vectors in T that cannot be expressed as sum of some other vectors in T. We claim that W is the unique basis for T. It is clear that W is linearly independent. Let V be the linear span of W and suppose $V \neq T$. Let v be a vector in $T \setminus V$ with the least number of ones. Since $v \notin W$, it can be expressed as a linear combination of other vectors and these vectors must be in V, in view of the minimality of v. However, then v must also belong to V, and this is a contradiction. Thus $V = T$, and W is a basis for T. Let, if possible, W' be another basis for T. It is clear that $W \subset W'$. Let, if possible, $W \neq W'$, and let $z \in W' \setminus W$. Since W is a basis, z must be a sum of vectors in W. However, then W' is not a linearly independent set, which is a contradiction. Hence W is the unique basis for T. ∎

Observe that $\{x_2, x_3, x_5\}$ is a basis for the vector space $\{x_1, x_2, x_3, x_4, x_5\}$ given in Example 5.6.

The *dimension* of a vector space is defined as the cardinality of its (unique) basis. Let A be an $m \times n$ matrix. The *row space* of A, $\mathcal{R}(A)$, is the vector space spanned by the row vectors of A; the column space of A, $\mathcal{C}(A)$, is the vector space spanned by the columns of A. The *row rank* of A is the dimension of $\mathcal{R}(A)$, whereas the *column rank* of A is the dimension of $\mathcal{C}(A)$.

The next few examples illustrate some of the pathologies that can occur.

Example 5.7. Let T be a vector space, and let $W \subset T$ be linearly independent. Then W does not necessarily extend to a basis. Consider the vector space $\mathcal{B}_3$, which has the unique basis

$$\begin{bmatrix} 1 \\ 0 \\ 0 \end{bmatrix}, \begin{bmatrix} 0 \\ 1 \\ 0 \end{bmatrix}, \begin{bmatrix} 0 \\ 0 \\ 1 \end{bmatrix}.$$

Although the set consisting of the vectors

$$\begin{bmatrix} 1 \\ 1 \\ 1 \end{bmatrix}, \begin{bmatrix} 1 \\ 1 \\ 0 \end{bmatrix}$$

is linearly independent, it does not extend to a basis.

Example 5.8. If T_1 is a subspace of T_2, it is not necessarily true that the dimension of T_1 is at most equal to that of T_2. Let W be the set of all vectors in $\mathcal{B}_4$ with at least two coordinates equal to 1, together with the zero vector. Then the dimension of W is 6 since

$$\begin{bmatrix} 1 \\ 1 \\ 0 \\ 0 \end{bmatrix}, \begin{bmatrix} 1 \\ 0 \\ 1 \\ 0 \end{bmatrix}, \begin{bmatrix} 1 \\ 0 \\ 0 \\ 1 \end{bmatrix}, \begin{bmatrix} 0 \\ 1 \\ 1 \\ 0 \end{bmatrix}, \begin{bmatrix} 0 \\ 1 \\ 0 \\ 1 \end{bmatrix}, \begin{bmatrix} 0 \\ 0 \\ 1 \\ 1 \end{bmatrix}$$

form a basis for W. Note that W is a subspace of $\mathcal{B}_4$, which has dimension 4.

Example 5.9. The row rank of a matrix need not equal its column rank. Let

$$A = \begin{bmatrix} 1 & 0 & 0 & 1 \\ 0 & 1 & 0 & 1 \\ 1 & 0 & 1 & 0 \\ 0 & 1 & 0 & 1 \end{bmatrix}.$$

Then the row rank of A is 3 whereas the column rank is 4.

Although the row rank of a matrix need not equal its column rank, the row space and the column space have the same cardinality. This is proved next.

Theorem 5.6.2. *Let A be an $m \times n$ matrix. Then card $\mathcal{C}(A) = $ card $\mathcal{R}(A)$.*

Proof. Let $x_1, \ldots, x_n$ denote the columns, and let $y_1^T, \ldots, y_m^T$ denote the rows of A, respectively. We define a map $f : \mathcal{C}(A) \to \mathcal{R}(A)$ as follows. Suppose $v \in \mathcal{C}(A)$. Set

$$f(v) = \sum_{i:v_i=0} y_i^T.$$

We claim that f is one-to-one. To see this, suppose $v, w \in \mathcal{C}(A)$ and that $f(v) = f(w)$. Thus

$$\sum_{i:v_i=0} y_i^T = \sum_{i:w_i=0} y_i^T.$$

If $v \neq w$, we may assume, without loss of generality, that there exists k with $v_k = 0, w_k = 1$. Since $w \in \mathcal{C}(A)$, there exists ℓ such that $a_{k\ell} = 1$ and $x_\ell \leq w$. Since $v_k = 0$, we have $f(v) \geq y_k^T$ and further, since $a_{k\ell} = 1$, we have $(f(v))_k = 1$. Thus $(f(w))_k = 1$ and, therefore there exists r such that $w_r = 0$ and $a_{r\ell} = 1$. Now since $y_\ell \leq w$, we have $w_r = 1$, which is a contradiction. Thus $f(v) \neq f(w)$, and f must be one-to-one. Hence the claim is proved. It follows that card $\mathcal{C}(A) \leq$ card $\mathcal{R}(A)$. We may similarly show that card $\mathcal{R}(A) \leq$ card $\mathcal{C}(A)$ and, therefore, card $\mathcal{C}(A) = $ card $\mathcal{R}(A)$. That completes the proof. ∎

Example 5.10. Let

$$A = \begin{bmatrix} 1 & 1 & 0 & 0 \\ 0 & 1 & 1 & 0 \\ 0 & 0 & 1 & 1 \end{bmatrix}.$$

Then $\mathcal{C}(A)$ consists of the vectors

$$\begin{bmatrix} 0 \\ 0 \\ 0 \end{bmatrix}, \begin{bmatrix} 1 \\ 0 \\ 0 \end{bmatrix}, \begin{bmatrix} 1 \\ 1 \\ 0 \end{bmatrix}, \begin{bmatrix} 0 \\ 1 \\ 1 \end{bmatrix}, \begin{bmatrix} 0 \\ 0 \\ 1 \end{bmatrix}, \begin{bmatrix} 1 \\ 0 \\ 1 \end{bmatrix}, \begin{bmatrix} 1 \\ 1 \\ 1 \end{bmatrix},$$

whereas $\mathcal{R}(A)$ consists of the vectors

$$[0 \ 0 \ 0 \ 0], [1 \ 1 \ 0 \ 0], [0 \ 1 \ 1 \ 0], [0 \ 0 \ 1 \ 1],$$

$$[1 \ 1 \ 1 \ 0], [0 \ 1 \ 1 \ 1], [1 \ 1 \ 1 \ 1].$$

Thus card $\mathcal{C}(A) = $ card $\mathcal{R}(A) = 8$.

We now introduce some definitions. Let A be an $m \times n$ matrix. An $n \times m$ matrix G is called a *generalized inverse* of A if $AGA = A$. The *Moore-Penrose inverse* of A is a matrix G of order $n \times m$ that satisfies the following equations:

$$AGA = A, \quad GAG = G, \quad (AG)^T = AG, \quad (GA)^T = GA. \quad (5.6.1)$$

The Moore-Penrose inverse, when it exists, must be unique. This is proved in the next result.

Theorem 5.6.3. *Let A be an $m \times n$ matrix, and suppose G_1 and G_2 are $n \times m$ matrices satisfying (5.6.1). Then $G_1 = G_2$.*

Proof. We have

$$
\begin{aligned}
G_1 &= G_1 A G_1 \\
&= G_1 G_1^T A^T \\
&= G_1 G_1^T A^T G_2^T A^T \\
&= G_1 G_1^T A^T A G_2 \\
&= G_1 A G_1 A G_2 \\
&= G_1 A G_2 \\
&= G_1 A G_2 A G_2 \\
&= G_1 A A^T G_2^T G_2 \\
&= A^T G_1^T A^T G_2^T G_2 \\
&= A^T G_2^T G_2 \\
&= G_2 A G_2 \\
&= G_2.
\end{aligned}
$$

That completes the proof. ∎

In the remainder of this section our main objective is to characterize Boolean matrices that admit the Moore-Penrose inverse. In the process we obtain some results concerning generalized inverse, rank, and idempotent matrices.

An $n \times n$ matrix is said to be *idempotent* if $A^2 = A$. As usual, $A(i, j)$ will denote the submatrix obtained by deleting row i and column j of A.

Lemma 5.6.4. *Let A be an $n \times n$ idempotent matrix and suppose $a_{ii} = 0$. Then the i-th column (row) of A depends on the remaining columns (rows). Furthermore, $A(i, i)$ is idempotent.*

Proof. We assume, without loss of generality, that $a_{nn} = 0$, and let

$$A = \begin{bmatrix} B & x \\ y^T & 0 \end{bmatrix}.$$

Since $A^2 = A$, we have

$$B^2 + xy^T = B, \quad Bx = x, \quad y^T B = y^T, \quad \text{and} \quad y^T x = 0.$$

Thus

$$\begin{bmatrix} x \\ 0 \end{bmatrix} = \begin{bmatrix} B \\ y^T \end{bmatrix} x, \quad [y^T, 0] = y^T [B, x],$$

and the first part of the result is proved. Now $B^2 + xy^T = B$ implies $xy^T \leq B$. Also, $Bx = x$ implies $Bxy^T = xy^T$. Thus $xy^T = Bxy^T \leq B^2$ and so $B = B^2 + xy^T = B^2$. That completes the proof. ∎

Lemma 5.6.5. *Let A be an $n \times n$ idempotent matrix and suppose $a_{ii} = 1$ for all i. Then the i-th column of A depends on the remaining columns if and only if the i-th row depends on the remaining rows. Furthermore, $A(i, i)$ is idempotent.*

Proof. We assume, without loss of generality, that the n-th row depends on the remaining rows, and let

$$A = \begin{bmatrix} B & x \\ y^T & 1 \end{bmatrix}.$$

Since $A^2 = A$, we have

$$B^2 + xy^T = B, \quad Bx + x = x, \quad \text{and} \quad y^T B + y^T = y^T.$$

Since $[y^T, 1]$ depends on the remaining rows, there exists a vector z such that

$$[y^T, 1] = z^T [B, x]. \tag{5.6.2}$$

Thus

$$y^T x = z^T Bx. \tag{5.6.3}$$

Since $Bx + x = x$, we have $Bx \leq x$. However, $B \geq I$ implies that $Bx \geq x$ and therefore $Bx = x$. It follows from (5.6.2) and (5.6.3) that $z^T Bx = z^T x = 1$. Thus

$$\begin{bmatrix} x \\ 1 \end{bmatrix} = \begin{bmatrix} B \\ y^T \end{bmatrix} x$$

and therefore the n-th column depends on the remaining columns. Now $B^2 + xy^T = B$ implies that $B^2 \leq B$, and since $B \geq I$, we have $B^2 \geq B$. It follows that $B^2 = B$, and the proof is complete. ∎

Corollary 5.6.6. *If A is an idempotent matrix, then the row rank of A equals its column rank.*

Proof. If $a_{ii} = 0$ for some i, then by Lemma 5.6.4 the i-th column (row) of A depends on the remaining columns (rows). Also, the matrix $A(i, i)$ is idempotent. Thus we may delete the rows and columns corresponding to the zero diagonal entries and thus assume, without loss of generality, that $a_{ii} = 1$ for all i. If the rows of A are linearly independent, then by Lemma 5.6.5 the columns must be linearly independent, in which case the row rank and the column rank both equal n. If some row of A depends on the remaining rows, then again by Lemma 5.6.5, the corresponding column must depend on the remaining columns. We may then delete that row and the corresponding column without affecting either the row rank or the column rank. The resulting matrix is again idempotent by Lemma 5.6.5, and the result is proved by induction. ∎

We observed that the row rank of a matrix need not equal its column rank. However, if A admits a generalized inverse, then the two ranks are equal. This is proved in the next result.

Theorem 5.6.7. *Let A be an $m \times n$ matrix and suppose A admits a generalized inverse. Then the row rank of A equals its column rank.*

Proof. Let G be a generalized inverse of A. Then $AGAG = AG$; thus AG is idempotent. By Corollary 5.6.6 the row rank of AG equals its column rank, which we denote by r. In view of Lemmas 5.6.4 and 5.6.5, we assume, without loss of generality, that the rows (columns) $r + 1, \ldots, n$ depend on the first r rows (columns). Thus AG can be partitioned as

$$AG = \begin{bmatrix} B & BC \\ DB & DBC \end{bmatrix},$$

where B is an $r \times r$ idempotent matrix with rank r. Note that the first r columns of AG form a basis for $\mathcal{C}(AG)$. Since $A = (AG)A$, we have $\mathcal{C}(A) \subset \mathcal{C}(AG)$. Also, $\mathcal{C}(AG) \subset \mathcal{C}(A)$; and thus $\mathcal{C}(A) = \mathcal{C}(AG)$. Thus the first r columns of AG form a basis for $\mathcal{C}(A)$. It follows, by the uniqueness of basis (Theorem 5.6.1), that the first r columns of AG are in fact columns of A as well. Therefore we may assume, without loss of generality, that

$$A = \begin{bmatrix} B & BX \\ DB & DBX \end{bmatrix}.$$

Because the rows of B are linearly independent, the row rank of A is r, and the proof is complete. ∎

Example 5.11. If the row rank of A equals its column rank, it is not necessarily true that A admits a generalized inverse. Let

$$A = \begin{bmatrix} 1 & 0 & 0 & 1 \\ 0 & 1 & 1 & 0 \\ 1 & 0 & 1 & 0 \\ 0 & 1 & 0 & 1 \end{bmatrix}.$$

Then the row rank as well as the column rank of A is 4. However, it can be verified that A has no generalized inverse.

Lemma 5.6.8. *Let A be an $m \times n$ matrix. Then $A \le A A^T A$.*

Proof. Let $B = A A^T A$. We must show that $a_{ij} \le b_{ij}$ for all i, j. This is obvious if $a_{ij} = 0$. Now assume that $a_{ij} = 1$. We have

$$b_{ij} = \sum_{k=1}^{n} \sum_{\ell=1}^{n} a_{ik} a_{\ell k} a_{\ell j}. \tag{5.6.4}$$

If we set $k = j$, $\ell = i$, then $a_{ik} a_{\ell k} a_{\ell j} = a_{ij}^3 = a_{ij}$. It follows from (5.6.4) that $b_{ij} = 1$, and the proof is complete. ∎

Lemma 5.6.9. *Let A be an $m \times n$ matrix and suppose A admits a Moore-Penrose inverse. Then $A A^T A \le A$.*

Proof. Let $B = A^T A$. Since there are finitely many Boolean matrices of order n, there must exist integers $k \ge 1$, $t \ge 1$ such that $B^k = B^{k+t}$. We assume, without loss of generality, that k is the least integer for which $B^k = B^{k+t}$ for some integer $t \ge 1$.

We now claim that $k = 1$. Let, if possible, $k \ge 2$. Let G be the Moore-Penrose inverse of A. Then

$$G^T A^T A B^{k-1} = G^T A^T A B^{k+t-1}. \tag{5.6.5}$$

Since $G^T A^T = A G$, it follows from (5.6.5) that $A B^{k-1} = A B^{k+t-1}$. Thus $A B B^{k-2} = A B B^{k+t-2}$ and hence

$$G A A^T A B^{k-2} = G A A^T A B^{k+t-2}. \tag{5.6.6}$$

Again, since $G A = A^T G^T$, it follows from (5.6.6) that $B^{k-1} = B^{k+t-1}$, contradicting the minimality of k. Thus $k = 1$ and the claim is proved. We therefore have

$$B = B^{1+t}. \tag{5.6.7}$$

Observe that if $t = 1$ then B is idempotent, in which case, $B = B^{1+\ell}$ for all $\ell \geq 1$. Thus we may assume, without loss of generality, that $t \geq 2$. Premultiply both sides of (5.6.7) by G^T to get $G^T A^T A = G^T A^T A B^t$. Since $G^T A^T = AG$, it follows that

$$A = AB^t = A(A^T A)^t. \tag{5.6.8}$$

By Lemma 5.6.8, $A \leq AA^T A$ and hence $AA^T A \leq A(A^T A)^2$. By a repeated application of Lemma 5.6.8 we get

$$AA^T A \leq A(A^T A)^t. \tag{5.6.9}$$

The result follows from (5.6.8) and (5.6.9). ∎

In the next result we present several characterizations of matrices that admit the Moore-Penrose inverse.

Theorem 5.6.10. *Let A be an $m \times n$ matrix. Then the following assertions are equivalent:*

(i) *The Moore-Penrose inverse of A exists.*

(ii) *The Moore-Penrose inverse of A exists and equals A^T.*

(iii) *$AA^T A = A$.*

(iv) *$AA^T A \leq A$.*

(v) *Any two rows of A are either identical or disjoint (i.e., there is no column with a 1 in both the rows.)*

(vi) *Any two columns of A are either identical or disjoint.*

(vii) *The number of ones in any 2×2 submatrix of A is not 3.*

(viii) *Any 2×2 submatrix of A admits a Moore-Penrose inverse.*

(ix) *There exist permutation matrices P and Q such that*

$$PAQ = \begin{bmatrix} J_1 & 0 & \cdots & 0 & 0 \\ 0 & J_2 & \cdots & 0 & 0 \\ \vdots & \vdots & \ddots & \vdots & \vdots \\ 0 & 0 & \cdots & J_t & 0 \\ 0 & 0 & \cdots & 0 & 0 \end{bmatrix},$$

where $J_1, \ldots, J_t$ are matrices (not necessarily square) of all ones.

(x) *There exist permutation matrices P, and Q such that*

$$PAQ = \begin{bmatrix} I & C \\ D & DC \end{bmatrix},$$

where C and D satisfy $CC^T \leq I$, $D^T D \leq I$.

(xi) *There exists a matrix G such that $GAA^T = A^T$ and $A^T AG = A^T$.*

Proof. We first observe that if A is a row vector or a column vector, then A^T is the Moore-Penrose inverse of A. We therefore assume that $m \geq 2, n \geq 2$.

(i) $\Rightarrow$ (ii): Suppose A admits a Moore-Penrose inverse. Then by Lemma 5.6.9, $AA^T A \leq A$. Also, by Lemma 5.6.8, $A \leq AA^T A$ and therefore $AA^T A = A$. It then follows from the definition that A^T is the Moore-Penrose inverse of A.

The proof of (ii) $\Rightarrow$ (iii) and (iii) $\Rightarrow$ (iv) is trivial, whereas the proof of (iv) $\Rightarrow$ (i) is essentially contained in our proof of (i) $\Rightarrow$ (ii). We have thus shown the equivalence of (i)–(iv) and now we must show that these are equivalent to the remaining assertions.

We now show (iii) $\Rightarrow$ (v). If (iii) holds, then we have

$$a_{ij} = \sum_{k=1}^{n} \sum_{\ell=1}^{n} a_{ik} a_{\ell k} a_{\ell j}$$

for all i, j and therefore

$$a_{ij} \geq a_{ik} a_{\ell k} a_{\ell j}. \tag{5.6.10}$$

for all i, j, k, ℓ. Consider rows i and ℓ of A and suppose these are not disjoint. Then there exists k such that $a_{ik} = a_{\ell k} = 1$. We must show that the rows i and ℓ are identical. Suppose $a_{\ell j} = 1$. Then it follows from (5.6.10) that $a_{ij} = 1$. We may similarly show that if $a_{ij} = 1$, then $a_{\ell j} = 1$. Thus the rows i and ℓ are identical and (v) is proved.

The proof of (iii) $\Rightarrow$ (vi) is similar. The equivalence of (v) and (vii) is easy to prove and so is the implication (v) $\Rightarrow$ ($viii$). Now if ($viii$) holds, then by the implication (i) $\Rightarrow$ (v), the two rows of any 2×2 submatrix of A are either identical or disjoint and hence (v) holds. The implications (v) $\Rightarrow$ (ix) and (ix) $\Rightarrow$ (iii) are easy to prove. Thus we have shown that assertions (i)–(ix) are equivalent.

It is left as an exercise to see that (ix) $\Rightarrow$ (x). If (x) holds, then it can be verified that A^T is the Moore-Penrose inverse of A and thus (i) holds.

We finally show the equivalence of (xi) with the remaining conditions. If (i) holds, then (iii) holds, and setting $G = A^T$ we see that (xi) holds as well. Conversely, suppose (xi) is true. Then $GAA^T G^T = A^T G^T = (GA)^T$ and thus GA is symmetric. From $GAA^T = A^T$ it follows that $AGA = A$. Similarly, using $A^T AG = A^T$ we conclude that AG is symmetric. It can now be verified that GAG must be the Moore-Penrose inverse of A, and thus (i) holds. That completes the proof. ∎

A comprehensive reference on Boolean matrices is provided by Kim (1982). Results concerning nonnegative matrices that are limited to matrix addition and multiplication or that merely involve the zero-nonzero pattern of the entries really belong to the theory of Boolean matrices. We refer to Berman and Plemmons (1994, Chapter 4), for several such results. Our treatment is mainly based on Prasada Rao and Bhaskara Rao (1975) and Kim (1982), although the formulation and the proof of Theorem 5.6.10 have some novelty. The result is based on the work of Bapat, Jain, and Pati (1995), who consider the weighted Moore-Penrose inverse as well. Plemmons (1971) also contains some related results. Many of the results presented in this section can be extended to matrices over an arbitrary Boolean algebra using a standard homomorphism technique; see, for example, Prasada Rao and Bhaskara Rao (1975) and Kim (1982). Extensions to even more general structures are possible; see Cao, Kim, and Roush (1984). For some recent work on Boolean matrices, see Kirkland and Pullman (1992) and Gregory, Kirkland, and Pullman (1993).

Exercises

1. Show, using Lemma 5.2.1, that the mixed discriminant of a set of positive definite matrices is positive.

2. Let $A^k, B^k, k = 1, 2, \ldots, n$ be $n \times n$ matrices and suppose $A^k \succeq B^k, k = 1, 2, \ldots, n$. Show that

$$D(A^1, \ldots, A^n) \geq D(B^1, \ldots, B^n).$$

3. If $x_i, y_i; i = 1, 2, \ldots, n$ are vectors in R^n, show that

$$D\left(x_1 y_1^T, \ldots, x_n y_n^T\right) = \frac{1}{n!} |(x_1, \ldots, x_n)| |(y_1, \ldots, y_n)|.$$

Hence derive a formula, as in Lemma 5.2.1, for $D(A^1, \ldots, A^n)$, where $A^k = X_k Y_k^T$ for each k. This is the *Cauchy-Binet formula* for mixed discriminants.

4. Use the Cauchy-Binet formula, derived in Exercise 3, and the Cauchy-Schwarz Inequality to show that if $A^k, k = 1, 2, \ldots, n$ are positive definite matrices and if $B^k = (A^k)^{-1}$ for each k, then

$$D(A^1, \ldots, A^n) \, D(B^1, \ldots, B^n) \geq 1.$$

5. If $\mathbf{A} = (A^1, \ldots, A^n)$ is an element of $\mathcal{D}_n$ and if each A_i is of rank 1, then show that $\mathbf{A}$ is an extreme point of $\mathcal{D}_n$.

6. Show that the extreme points of $\mathcal{D}_2$ are precisely the elements $\mathbf{A} = (A^1, A^2)$ such that A^1 and A^2 are both of rank 1.

7. Let G be a graph with n vertices, and suppose that each edge of G is assigned a color out of m ($\geq n - 1$) possible colors, $1, 2, \ldots, m$. Show that G has a spanning tree in which all edges have distinct colors if and only if the following condition holds: For any $T \subset \{1, 2, \ldots, m\}$, the subgraph of G consisting of edges colored by colors in T has at most $n - \text{card}(T)$ components.

8. Let A be an irreducible, stochastic $n \times n$ matrix and let $G(A)$ be the directed graph associated with A, as in Section 1.1. The *weight* of a spanning tree in $G(A)$ is defined to be the product of the weights of its edges. Let X be the adjoint matrix of $I - A$. Show that the j-th column of X has all entries equal, say α_j, $j = 1, 2, \ldots, n$. Furthermore, show that α_j is the sum of the weights of the spanning trees in $G(A)$ in which all edges are directed toward the vertex j, $j = 1, 2, \ldots, n$. This is a weighted version of the Matrix-Tree Theorem. Observe that any row of X is a left eigenvector of A, and using the graph theoretic interpretation just derived, it follows that $X \geq 0$. Thus, this exercise points toward a relationship between the Perron-Frobenius Theorem and the Matrix-Tree Theorem. For similar generalizations of the Matrix-Tree Theorem, see Chaiken and Kleitman (1978) and Chaiken (1982).

9. Let A^k, $k = 1, 2, \ldots, n$ be $n \times n$ positive semidefinite matrices. Show that

$$D(A^1, \ldots, A^n) \geq \left\{ \prod_{k=1}^{n} |A^k| \right\}^{\frac{1}{n}}.$$

(Hint: Use the Alexandroff Inequality.)

10. Let A be a nonnegative $m \times n$ matrix and let α_k, $k = 1, 2, \ldots, \min(m, n)$, be the sum of all $k \times k$ subpermanents of A. (We set $\alpha_0 = 1$.) Show that the polynomial

$$\alpha_0 + \alpha_1 x + \alpha_2 x^2 + \cdots + \alpha_k x^k \tag{5.6.11}$$

has only real roots.
[See Nijenhuis (1976) for a proof similar to our proof of Theorem 5.3.3; both are based on interlacing. When A is a $(0, 1)$-matrix, α_k is the number of ways in which k nonattacking rooks can be placed on an $m \times n$ chessboard, using only positions corresponding to nonzero entries in A. In this case, (5.7.1) is called a *rook polynomial*. For a more general result, see Heilman and Lieb (1972).]

11. Let $a^1, a^2, \ldots, a^n, b$, and c be positive vectors in R^n. For $r = 0, 1, \ldots, n$, consider the polynomial

$$g_r(x) = \text{per}(a^1, a^2, \ldots, a^{n-r}, \underbrace{b + xc, \ldots, b + xc}_{r}).$$

Thus $g_r(x)$ is a polynomial of degree r in the real variable x. Show that $g_r(x)$ has only real roots and that they interlace the roots of $g_{r-1}(x)$, $r = 1, 2, \ldots, n$.

12. Let M be an $n \times n$ matrix such that $M + M^T$ is positive definite. Show that M is nonsingular and that the eigenvalues of $M^{-1}M^T$ are all of modulus one.

13. Show that the Coxeter group of D_n, $n \geq 4$, is the sign change group of order n.

14. Let A be an irreducible $n \times n$ matrix over the max algebra $\mathcal{M}$. Consider the linear programming problem

$$\text{minimize } \tau, \text{ subject to } a_{ij} + x_j - x_i \leq \tau \text{ for all } i, j.$$

Write the dual problem and show that both problems have optimal solutions. Hence show that there exists a finite vector x such that $A \otimes x = \lambda \otimes x$ for some λ.

15. Let B be a positive $n \times n$ matrix and let $A = (\log b_{ij})$. For a positive number r, recall that the matrix $B^{(r)}$ is defined to be (b_{ij}^r). Let ρ_r denote the Perron root of $B^{(r)}$ and let $y_{(r)}$ be the Perron vector, normalized so that its components add up to 1. Show that as $r \to \infty$, $\rho_r^{\frac{1}{r}} \to \lambda$ and $y_{(r)}^{(\frac{1}{r})} \to y$, where $A \otimes y = \lambda \otimes y$. (Note the connection of this exercise with the results in Section 3.5.)

16. Let A be an $n \times n$ matrix over $\mathcal{M}$, which is lower triangular, i.e., satisfies $a_{ij} = -\infty$ if $i < j$. Show that any eigenvalue of A must be a diagonal entry of A. Conversely, describe exactly when a_{ii} is an eigenvalue of A. Now solve the problem of finding all the eigenvalues of an $n \times n$ matrix over $\mathcal{M}$, by first putting it in Frobenius Normal Form. Compare with Exercise 18 in Chapter 1.

17. Let S be a finite set equipped with two binary operations $\oplus$ and $\otimes$ such that both operations are commutative and admit identity elements, denoted by 0 and 1, respectively, and suppose the usual distributive laws hold. Such a structure is called a *semiring*. Now assume the following additional conditions: (i) $a \oplus a = a$, i.e., $\oplus$ is idempotent, and (ii) if $a \otimes b = a \otimes c$, $a \neq 0$, then $b = c$. Show that most of the results in Section 5.5 continue to hold under this general setup.

18. Let S be the set of all compact convex subsets of R^n, and for $A, B \in S$, consider the following operations: $A \oplus B$ is the convex hull of A, B and

$$A \otimes B = \{a + b : a \in A, b \in B\}.$$

Show that the hypotheses in the previous exercise are satisfied in this example.

19. Let A be an $n \times n$ matrix over $\mathcal{M}$. Show that

$$\min_i \max_j a_{ij} \leq \mu(A) \leq \max_{i,j} a_{ij}.$$

20. Let A and B be $n \times n$ matrices over $\mathcal{M}$. Show that

$$\mu(A + B) \leq \mu(A) + \mu(B) \quad \text{and} \quad \mu(A \oplus B) \geq \mu(A) \oplus \mu(B).$$

21. Let A be an $n \times n$ matrix over $\mathcal{M}$. The permanent of A over the max algebra, denoted by $\pi(A)$, is obtained by replacing the sum and the product in the definition of permanent by $\oplus$ and $\otimes$, respectively. Thus

$$\pi(A) = \max_{\sigma \in S_n} \sum_{i=1}^{n} a_{i\sigma(i)}.$$

Obtain the following analogue of the Alexandroff Inequality for the permanent over the max algebra: Let $A = (a_1, \ldots, a_n)$ be an $n \times n$ matrix over $\mathcal{M}$, where a_i denotes the i-th column of A. Then

$$2\pi(A) \geq \pi(a_1, \ldots, a_{n-2}, a_{n-1}, a_{n-1}) + \pi(a_1, \ldots, a_{n-2}, a_n, a_n).$$

[Hint: Apply the usual Alexandroff Inequality to the matrix $(t^{a_{ij}})$ and then take a suitable limit as $t \to \infty$; alternatively, use the Duality Theorem (See Bapat (1995).)]

22. If A is an $n \times n$ irreducible, idempotent, Boolean matrix then show that $A = J_{nn}$, the $n \times n$ matrix of all ones.

23. Let A be an $n \times n$ Boolean matrix and let P be a permutation matrix such that PAP^T is in the Frobenius Normal Form:

$$PAP^T = \begin{bmatrix} A_{11} & 0 & \cdots & 0 \\ A_{21} & A_{22} & \cdots & 0 \\ \vdots & \vdots & \ddots & \vdots \\ A_{k1} & A_{k2} & \cdots & A_{kk} \end{bmatrix}.$$

Thus each diagonal block A_{ii} is either irreducible or the 1×1 zero matrix. Now suppose that A is idempotent. Show that each A_{ij} is either the zero matrix or the matrix of all ones. In particular, obtain a complete description of symmetric, idempotent, Boolean matrices.

24. Let A be an $m \times n$ matrix of rank r and let $A^\dagger$ be the Moore-Penrose inverse of A. Show that for any generalized inverse G of A, $\sigma_i(G) \geq \sigma_i(A^\dagger)$, $i = 1, 2, \ldots, r$, where $\sigma_i(\cdot)$ denotes the i-th largest singular value. [See Bapat and Ben-Israel (1995).]

6

Scaling problems and
their applications

A variety of biological, statistical, and social science data come in the form of cross-classified tables of counts commonly known as *contingency tables*. Scaling the cell entries of such multidimensional matrices involves both mathematically and statistically well-posed problems of practical interest. In this chapter we first describe several situations where scaling can be useful. We then prove a very general theorem that demonstrates the existence of scaling factors. We also describe a natural scaling algorithm in the problem of scaling a nonnegative matrix to obtain prescribed row and column sums. In order to study the convergence properties of the algorithm it is convenient to work in terms of Hilbert's projective metric. Certain related concepts such as the contraction ratio of Birkhoff and the oscillation ratio of Hopf are introduced. In the last section we consider the problem of maximum likelihood estimation in contingency tables. This area of statistics, which forms part of discrete multivariate analysis, is of considerable interest to research workers at present.

6.1. Practical examples of scaling problems

Before we take up the mathematical problem of scaling, we illustrate some practical situations where scaling is useful.

6.1.1. Budget allocation problem

The Air Force, the Army, and the Navy have received their budget for the next fiscal year measured in some units to be allocated among technical, administrative, and research categories. The federal government has fixed some amount to be allocated to each of these three categories. Given the allocation for each defense division for the above three categories for the current year, the problem is to somehow modify the current budget allocations to a new set of allocations that will fulfill the federal government stipulations.

239

For example, the matrix A summarizes the current budget:

Matrix A

	technical	non-tech	research
Air Force	200	75	15
Army	135	165	12
Navy	175	160	25

The cells of matrix B are to be filled in to meet the federal demands specified in terms of the row and column sums, the specified budget for next year:

Matrix B

	technical	nontech.	research	budget
Air Force	?	?	?	400
Army	?	?	?	350
Navy	?	?	?	370
Total	700	320	100	1120

Even though one could find many matrices with the newly specified marginals, many of them may be quite unrelated to the original matrix A. For example, the matrix C

Matrix C

	technical	nontech.	research
Air Force	0	320	80
Army	330	0	20
Navy	370	0	0

satisfies the new constraints. However, such a solution stipulates no technical staff for the Air Force, no nontechnical staff for the Army, and no research division for the Navy! Such a decision would naturally involve several problems, including the sudden social problems that could be created by dislocating personnel. We need to find a matrix that is somewhat closer to the original matrix A in some suitable measure of distance.

6.1.2. Markov chains

Suppose we know that a stochastic process $\{X_t\}$ is a discrete time stationary Markov chain with finitely many states with a doubly stochastic transition probability matrix $P = (p_{ij})$. The problem is to estimate the probabilities p_{ij} based on observing the process for a finite time. The maximum likelihood estimates based on the observed data are quite hard to find. (Roughly speaking, the maximum likelihood estimates of the parameters are values of the parameters that lead to the maximum probability of getting the sample that has been observed.) Based on the data, we can find the frequency of visits from any state to any other state in unit time. For example, if the stationary Markov chain has three states and the observations are

$$x_1 = 1, \quad x_2 = 1, \quad x_3 = 2, \quad x_4 = 1, \quad x_5 = 1, \quad x_6 = 3,$$
$$x_7 = 2, \quad x_8 = 3, \quad x_9 = 1, \quad x_{10} = 2, \quad x_{11} = 3, \quad x_{12} = 3, ,$$
$$x_{13} = 1, \quad x_{14} = 2, \quad x_{15} = 2, \quad x_{16} = 3, \quad x_{17} = 3, \quad x_{18} = 1$$

then summarizing the frequency of transitions, we get

$$\text{from state } i = \begin{array}{c} \\ 1 \\ 2 \\ 3 \end{array} \begin{array}{ccc} & \text{To} \quad \text{state} \quad j & \\ 1 & 2 & 3 \\ \left(\begin{array}{ccc} 2 & 3 & 1 \\ 1 & 1 & 3 \\ 3 & 1 & 2 \end{array}\right). \end{array}$$

Since half of the time the system moved from state 1 to state 2 in the sample, an estimate of p_{12} is 0.5. By our second assumption, the matrix is doubly stochastic and thus the proportions of visits to state 2 from states 1, 2, and 3 must add up to 1. Thus we get a second estimate 0.6 for p_{12} based on our sample. Because we basically scale the rows to arrive at one estimate, and scale the columns to arrive at the second estimate, a heuristic approach would be to alternately scale the rows and columns of the matrix repeatedly and hope to arrive at a doubly stochastic matrix. Because each such row normalization corresponds to premultiplying the original matrix by a diagonal matrix with positive diagonal entries, and because each such column normalization corresponds to post-multiplying the original matrix by a diagonal matrix with positive diagonal entries, we are led to the following question: Given an $n \times n$ matrix A with positive entries, can we find $x_i > 0, y_j > 0, i, j = 1, \ldots, n$ such that the matrix $C = (c_{ij}) = (a_{ij} x_i y_j)$ is doubly stochastic? An affirmative answer to this question was given in Chapter 2 (Theorem 2.7.7).

6.1.3. Scaling in Gaussian elimination

The following example is contained in Chvátal (1983, p. 75.) Consider the solution of the simultaneous equations

$$x_1 + 10000x_2 = 10000$$
$$0.5x_1 + 0.5x_2 = 1.$$

By using Gaussian elimination with partial pivoting, and working up to three decimal places, we obtain

$$x_1 = 10000 - 10000x_2$$
$$-4999.5x_2 = -4999.$$

The solution, rounded to three decimal places, gives $x_2 = 1, x_1 = 0$. This is quite unsatisfactory even from the point of view of backward error analysis as described below. Let

$$\tilde{a}_{11}x_1 + \tilde{a}_{12}x_2 = 10000$$

$$\tilde{a}_{21}x_1 + \tilde{a}_{22}x_2 = 1$$

be a system that is close to the original system. The backward error analysis demands that the current solution $(\tilde{x}_1, \tilde{x}_2) = (0, 1)$ be an exact solution to one such system. However, this is never possible in our case. In fact, $\tilde{a}_{22} = 1$ is the requirement for the new coefficients. Thus the new matrix is never near the given matrix and our computational inaccuracies cannot be overcome with the given system. Fortunately, there is a way out for such errors. The key idea is that, in practice, the individual rounding errors are less significant when the nonzero entries are roughly of the same magnitude. Such systems are called well-scaled systems. It is possible to choose positive scale factors r_i and s_j such that, instead of solving

$$\sum_{j=1}^{n} a_{ij}x_j = b_i, \quad i = 1, \ldots, n,$$

one solves

$$\sum_{j=1}^{n} a_{ij}r_i y_j s_j = b_i r_i, \quad i = 1, \ldots, n,$$

where we identify $x_j = y_j s_j$. The new matrix $\hat{a}_{ij} = a_{ij}r_i s_j$ has to be so chosen that $0.1 < \max_j |\hat{a}_{ij}| < 1$ for all i, j. If we can find such a matrix, the rounding errors can be minimized. For example, in the present problem with $r_1 = 10^{-4}, r_2 = 1, s_1 = 1$, and $s_2 = 1$ we get

$$\tilde{x}_1 + 9999\tilde{x}_2 = 10000$$

$$.5\tilde{x}_1 + .5\tilde{x}_2 = 1$$

with $\tilde{x}_1 = 1, \tilde{x}_2 = 1$. From the point of view of backward error analysis the above scaling is good. However, not every scaling gives a satisfactory solution, even when the coefficients are close to each other. For example, the scaling $r_1 = 1, r_2 = 1, s_1 = 1$, and $s_2 = 10^{-4}$ is quite unsatisfactory. The first scaling was chosen in an ad hoc fashion.

6.1.4. Scaling in Leontief input-output systems

The economy of a nation can be modeled after an inter-industry system in which the economy is quantified in terms of the consumptions and outputs of some major industries for each period. More specifically, let $A = (a_{ij})$ be the input-output matrix, where a_{ij} is the amount of the i-th good needed as input in a unit production of the j-th industrial output. Over a period of time when the technology improves and the demand pattern changes, the matrix $A = (a_{ij})$ changes and one may have to replace A with a new matrix B. However, B will in general be unknown; one natural procedure is to scale suitably the matrix A by scale factors x_i, y_j for all i, j and use the new matrix $C = (c_{ij}) = (a_{ij}x_iy_j)$ for our purpose. Here, one could interpret x_i as the substitution effect of factor i for other factors and y_j as the extent to which intermediate inputs have uniformly assumed increased weights in the input structure of the j-th commodity. Often, intermediate data about the progress of the economy is used to estimate the scale factors.

6.1.5. Scaling in contingency tables

Even before the advent of any sound mathematical theory of scaling, the General Register Offices in England and Wales have been using many heuristic approaches to the problem of estimating cell entries in contingency tables for the future, based on the past data. Friedlander (1961) gives the distribution according to marital status and age for a certain county for the year 1957. The problem is to estimate the same frequencies for the entire year 1958 in the middle of the year 1958. In Table 6.1, from Friedlander (1961), we have only shown some entries for illustration.

In mid-1958, though exact data for the age distribution was available, only estimates on marital status were available. The problem was to estimate the entire contingency table for 1958. By alternately adjusting the row and column totals several times, the following cell estimates were arrived at: as shown in Table 6.2.

Table 6.1. *Age and marital status data*

Age	Single	Married	Widowed or divorced
15–19	—	—	—
20–24	619	765	3
25–29	263	1194	9
30–34	—	—	—
35–39	—	—	—
40–44	159	1372	81
45–49	—	—	—
50	1116	4100	2329

Table 6.2. *Mid-1958 estimates*

Age	Single	Married	Widowed or divorced	Total
15–19	—	—	—	—
20–24	614	785	3	1402
25–29	254	1187	9	1450
30–34	—	—	—	—
35–39	—	—	—	—
40–44	147	1309	76	1532
45–49	—	—	—	—
50–	1108	4177	2359	7644
total	3988	11702	2634	18324

As we mentioned above, the row totals are generally known whereas column totals are generally estimated. As in the Markov chain example, the alternate scaling algorithm can be implemented as a practical solution.

From the point of view of applications, the need for efficient algorithms is clear and the algorithm of alternate scaling is known to be too slow, even when it converges to a matrix with prescribed row and column sums. The pitfalls of the algorithm become clear as soon as there are some zero entries or some small entries in the matrix. Suppose we are given the matrix

$$A = \begin{bmatrix} 0 & 0 & 1 \\ 0 & 0 & 1 \\ 1 & 1 & 0 \end{bmatrix}$$

and we want it scaled to a doubly stochastic matrix. The successive iterations produce, alternately,

$$A^{(n)} = \begin{bmatrix} 0 & 0 & 1 \\ 0 & 0 & 1 \\ .5 & .5 & 0 \end{bmatrix} \text{ and } \begin{bmatrix} 0 & 0 & .5 \\ 0 & 0 & .5 \\ 1 & 1 & 0 \end{bmatrix}.$$

Thus the algorithm does not converge. The reason for the failure is that the matrix A does not have doubly stochastic pattern. Since scaling does not affect the zero pattern of a given matrix, we cannot reach a doubly stochastic matrix from the given matrix by scaling the rows and columns. Thus with zero entries present, it is even more important to resolve the existence of scale factors. We close with a final example for scaling in three dimensions.

6.1.6. Transportation planning

A city is divided into three regions. Passengers can be classified according to such factors as their travel regions, professional groups or types, and time of travel. For simplicity, let there be three regions and three types of passengers (blue and white collar workers, senior citizens, and students). Let there be three travel time periods, say 5 A.M. to 10 A.M., 10 A.M. to 3 P.M., and 3 P.M. to 8 P.M. Suppose we know the data regarding the average number of passengers of various types and their travel patterns, such as how many travel in the morning between 5 A.M. and 10 A.M. from region I to region II and how many among them are senior citizens, etc. The city must plan for future expansion of road services, increased transportation facilities based on projected estimates of the job patterns, changes in the regions, and so on. It may be necessary to base the estimates using the current data and some information on dynamics of the marginals in terms of regions, population patterns, and types of jobs.

Thus, given the (multidimensional) matrix $A = (a_{ijk})$ and an unknown matrix $B = (b_{ijk})$ with *known* one-dimensional marginals $r_i = \sum_{j,k} b_{ijk}$, $c_j = \sum_{i,k} b_{ijk}$, and $t_k = \sum_{i,j} b_{ijk}$ for all i, j, k, the problem is to find a matrix $C = (c_{ijk})$ that has the same one-dimensional marginals as B and further for some scale factors x_i, y_j, z_k, the matrix $(c_{ijk}) = (a_{ijk} x_i y_j z_k)$. The scaling is used to reflect C as an estimate evolved from the matrix A.

Here, for example, i could refer to the region, j to the type of passenger, and k to the time of travel. Then a_{ijk} might represent the average number of members in region i, belonging to type j, traveling at time period k in a day.

Such a scaling model has other interpretations in the transportation literature and is called a *gravity model*. The following law propounded by Sir Isaac Newton in 1696 is well known: Given two bodies with masses m and n, and given the distance d between the two bodies, the force between them is given by

$$F = \frac{Gmn}{d^2},$$

where G is the gravitational constant. Mimicking this law, transportation researchers claimed [Stopher and Meyburg (1975)] that given a residential area with population P and a shopping center with attraction Q, measured in terms of number of shops, and given the distance d between them, the number of shopping visits F between the two behaves the same way as Newton's gravitational law. Generalizing the above force equation for many bodies, and mimicking the same for many zones, the expected number t_{ijk} of passengers who travel from zone i to zone j, of population type k, can be written as

$t_{ijk} = p_i q_j r_k f_{ijk}$. The problem is to estimate the parameters p_i, q_j, r_k, and f_{ijk} based on the given data. The transportation planner essentially has a model where $\log t_{ijk} = \log f_{ijk} + \log p_i + \log q_j + \log r_k$.

As indicated by these examples, scaling of matrix entries is one of the intuitive techniques that finds its applications in several areas. The earliest reference to the technique of alternately scaling rows and columns is perhaps Deming and Stephan (1940). Brown (1959) used the alternate scaling procedure to approximate the joint density of discrete vector random variables, with a prescribed subset of marginal densities. The problem of estimating the transition probabilities of a stationary Markov chain via alternate scaling was first discussed by Sinkhorn (1964). Independently, in his Ph.D. thesis, Bacharach (1965, 1970) dealt with scaling a given nonnegative matrix to achieve fixed row sums and column sums and used it to estimate the changing coefficients of a Leontief input-output matrix. For an example of a Markov chain, where the transition matrix happens to be doubly stochastic, see Feller (1968, p. 406). Some other references dealing with matrix scalings are Bapat (1982), Brualdi, Parter and Schneider (1966), Menon (1968), and Raghavan (1984a,b). Additional references are contained in the remaining sections.

As early as in 1943, Hotelling showed that errors in solving systems of linear equations can reach such high proportions that for a 78×78 symmetric matrix A, the matrix equation $Ax = b$, where each coefficient has a level of precision up to 46 decimal places, can still end up with a solution that is not even accurate up to the first decimal when pivoting is done systematically using, say, the Doolittle method [see Dwyer (1941)]. One of the main reasons that errors creep in is the improper choice of pivots in the Doolittle method, in which one tries to eliminate variable x_p at step p from equations $p+1, p+2, \ldots, n$, where n is the number of equations. The need for proper choice of pivots in Gaussian elimination has led to several heuristic procedures [see Wilkinson (1963) and Duff, Erisman, and Reid (1986)]. The choice of pivot is often based on its relative size among potential pivots and, in turn, the size of a pivot can be easily changed by simply scaling the matrix along rows or columns. The rows of a matrix are said to be equilibrated at a fixed norm [van der Sluis (1970)] if all the rows have the same norm. However, Skeel (1981) showed that equilibrated rows with column pivoting has the same error bound as row pivoting without equilibrated rows. Another technique for square matrices, called balancing [Parlett and Reinsch (1969)], entails scaling the rows by a set of scale factors and then scaling the columns by the inverses of these scale factors.

The model for transportation planning is a particular case of the well-known gravity model in the transportation literature. Quandt and Baumol (1966) proposed an exponential model for estimating the travel distribution between 16 cities in the state of California. Their model is very close to the log-linear model for densities considered in the theorem of Darroch and Ratcliff in the next section. As an example, if there are four ways of traveling from city i to city j, the problem is to predict the current travel pattern based on the population size, the average income at the origin, and the destination, travel time, travel cost, frequency of departure, and the performance of the specific mode of transportation k such as plane, train, etc., relative to the best performance figures among all competing modes. The exponential model $t_{ijk} = a \prod_x u_x^{\alpha_x}$, where a is a constant, u_x are the above variables, and α_x are unknown constants with the known specifications $\sum_k t_{ijk} = t_{ij}$ for all i, j etc., has been effectively used for predicting the flow volumes.

We refer to Tilanus (1976) for further applications of scaling to budgeting problems.

6.2. Kronecker Index Theorem and scaling

The scaling problems considered in the previous section can all be subsumed as special cases of the following general result:

Theorem 6.2.1. *Let K be a nonempty, bounded polyhedron given by*

$$K = \{\pi \in R^n : \pi \geq 0, C\pi = b\},$$

where $C = (c_{ij})$ is an $m \times n$ matrix and $b \in R^m$ is a nonzero vector. Let $y \in K$. Then for any $x \geq 0$ with the same pattern as y, there exist $z_i > 0, i = 1, \ldots, m$ and there exists $\pi \in K$ such that

$$\pi_j = x_j \prod_{i=1}^{m} z_i^{c_{ij}}, \quad j = 1, \ldots, n. \tag{6.2.1}$$

Furthermore, any $\pi \in K$ of the above type is unique.

The proof of Theorem 6.2.1 will be given later in this section. We now show that the theorem applies to the scaling problems discussed in Section 6.1.

As an example, let $A = (a_{ij})$ and $B = (b_{ij})$ be 2×3 matrices with nonnegative entries and with the same pattern. Let r_1, r_2, c_1, c_2, and c_3 be the row sums and column sums of B. Suppose no row or column of B is the zero vector. Let K be the set of all nonnegative matrices with the same row sums and column sums as B. A typical element of K is a vector $(y_{11}, y_{12}, y_{13}, y_{21}, y_{22}, y_{23})^T$. Given the matrix

$$C = \begin{bmatrix} 1 & 1 & 1 & 0 & 0 & 0 \\ 0 & 0 & 0 & 1 & 1 & 1 \\ 1 & 0 & 0 & 1 & 0 & 0 \\ 0 & 1 & 0 & 0 & 1 & 0 \\ 0 & 0 & 1 & 0 & 0 & 1 \end{bmatrix} \quad \text{and } b = \begin{bmatrix} r_1 \\ r_2 \\ c_1 \\ c_2 \\ c_3 \end{bmatrix},$$

we have $K = \{y : y \geq 0, Cy = b\}$. We can associate variables $u_1, u_2, v_1, v_2,$ and v_3 with the five rows of the above matrix. The given matrix A can be identified with the vector x in Theorem 6.2.1. Identifying B with the vector y in Theorem 6.2.1, we have $z = (u_1, u_2, v_1, v_2, v_3)$ with $z > 0$ such that for some $\pi \in K$, we have, for example,

$$\pi_{13} = a_{13}u_1^1 u_2^0 v_1^0 v_2^0 v_3^1,$$

where we have used the column

$$\begin{bmatrix} 1 \\ 0 \\ 0 \\ 0 \\ 1 \end{bmatrix}$$

in C corresponding to the (13) coordinate of π. This is the same as $\pi_{13} = a_{13}u_1 v_3$. In general, we have $(\pi_{ij}) = (a_{ij}u_i v_j)$.

It is quite straightforward to write down such a $(0, 1)$-matrix for the three dimensional scaling problem corresponding to the transportation example. The $(0, 1)$-matrix C for the case of $m \times n \times k$ multidimensional matrices has $m+n+k$ rows and mnk columns. Theorem 6.2.1 extends, with trite notational changes, to the multidimensional case. In fact, the result can also be formulated for cases in which we specify not necessarily the one-dimensional marginals, but perhaps all two-dimensional marginals. The next result is an extension of Lemma 2.7.3 to set-valued maps.

Theorem 6.2.2. *Let $K = \{x \in R^n : x \geq 0, Ax = b\}$ be a nonempty, bounded polyhedron. Let $\Phi : K \to 2^K$ be a point-to-set map satisfying the following conditions:*

(i) *For each $x \in K$, $\Phi(x)$ is a nonempty, closed, bounded, convex subset of K.*

(ii) *The map Φ is upper semicontinuous, i.e., $x^{(n)} \to x$, $y^{(n)} \in \Phi(x^{(n)})$, and $y^{(n)} \to y \Rightarrow y \in \Phi(x)$.*

(iii) *If the j-th coordinate of x is zero, then for any $y \in \Phi(x)$ the j-th coordinate of y is also zero.*

Then for any $u \in K$, there exists $z \in K$ such that $u \in \Phi(z)$.

Proof. Let $\mathbf{P}_k$ be a fine mesh of simplicial partitions $k = 1, 2, \ldots$ of K defined by:

(i) If $T \in \mathbf{P}_k$, then T is a simplex.

(ii) For each k the number of elements in $\mathbf{P}_k$ is finite.

(iii) If $T_1, T_2 \in \mathbf{P}_k$, then either $T_1 \cap T_2$ is a face of T_1 and T_2 or $T_1 \cap T_2 = \emptyset$.

(iv) If $T \in \mathbf{P}_k$ and F is a face of T, then $F \in \mathbf{P}_k$.

(v) $\bigcup_{T \in \mathbf{P}_k} T = K$.

(vi) If $T \in \mathbf{P}_k$ and $x, y \in T$, then $\|x - y\| < \frac{1}{k}$.

(vii) If $X^{(k)}$ is the collection of all vertices of the simplexes in $\mathbf{P}_k$, then
$$X^{(k)} \subset X^{(k+1)}, k = 1, 2, \ldots.$$

(viii) $X^{(1)}$ contains all the extreme points of K.

The existence of such a fine mesh of partitions for any polyhedron is well known [see Stillwell (1980)]. For each $x \in X^{(k)}$, choose any $y \in \Phi(x)$ and define the single-valued function $f_k(x) = y$ if $x \in X^{(k)}$. For any general

$$x = \sum_{i=1}^{r} \lambda_i x_i^{(k)}, \quad \lambda_i > 0, \quad \sum_{i=1}^{r} \lambda_i = 1, \quad x_i^{(k)} \in X^{(k)},$$

define $f_k(x) = \sum_{i=1}^{r} \lambda_i f_k(x_i^{(k)})$. By assumption (iv), $f_k(x)$ is well defined and is continuous on K. We check that for $u \in K$, if $u_j = 0$, then $f_k(u_j) = 0$. Let $u = \sum_{i=1}^{r} \lambda_i x_i^{(k)}$, where $\lambda_i > 0$, and let $x_1^{(k)}, \ldots, x_r^{(k)}$ be the vertices of the smallest dimensional simplex containing u. Since $u_j = 0$ and $x_i^{(k)} \geq 0$, the j-th coordinate of each vector $x_i^{(k)}$ is zero. By the conditions imposed on the map $\Phi(x)$, the j-th coordinate of any $y \in \Phi(x_i^{(k)})$, and in particular of $f_k(x_i^{(k)})$, is zero. Thus for $u = \sum_{i=1}^{r} \lambda_i x_i^{(k)}$,

$$\left(f_k \left(\sum_{i=1}^{r} \lambda_i x_i^{(k)} \right) \right)_j = \sum_{i=1}^{r} \lambda_i \left(f_k(x_i^{(k)}) \right)_j = 0.$$

This shows that $f_k : \partial K \to \partial K$. By Lemma 2.7.3 applied to f_k, we conclude that for any $u \in K$, there exists $z^{(k)} \in K$ such that $u \in f_k(z^{(k)})$. Without loss of generality (by passing to a subsequence if necessary), $z^{(k)} \to z$ for some z. By the upper semicontinuity of Φ, $u \in \Phi(z^*)$, and the proof is complete. ∎

Proof of Theorem 6.2.1. Let S be the subset of K consisting of all vectors $\pi \in K$ with the same pattern as y. Clearly, since $y \in S$, S is nonempty. Also, by assumption, S is bounded. For any $h \in S$, let

$$\begin{cases} p_j(h) = \log\left(\frac{x_j}{h_j}\right) & \text{if } x_j > 0, h_j > 0 \\ 0 & \text{if } x_j = 0 \\ \infty & \text{if } x_j > 0, h_j = 0. \end{cases}$$

Define $0 \log 0 = 0$, $0 \log \infty = 0$, and $0 + \infty = \infty$. Let

$$\phi(h) = \left\{ \pi^* : (p, \pi^*) = \min_{\pi \in S}(p, \pi) \right\}.$$

It can be shown that ϕ is well defined and satisfies the hypotheses of Theorem 6.2.2. The proof of this fact is somewhat technical so is omitted [see Bapat and Raghavan (1989)]. Now by Theorem 6.2.2, there exists $h^* \in S$ such that $y \in \phi(h^*)$. Let $p^* = p(h^*)$. We have y optimal to the linear program:

$$\min \sum_{j=1}^{n} p_j^* \pi_j$$

subject to

$$\sum_{j=1}^{n} c_{ij} \pi_j = b_i, \quad i = 1, \ldots, m,$$

$$\sum_{j=1}^{n} c_{(m+1), j} \pi_j = 0,$$

$$\pi_j \geq 0, \quad j = 1, 2, \ldots, n.$$

(Here $c_{(m+1), j} = 0$ if $y_j > 0$ and $c_{(m+1), j} = 1$ if $y_j = 0$.)

By the Duality Theorem, we have an optimal solution to the dual program, satisfying the complementary slackness conditions. Thus there is an optimal solution to

$$\max \sum_{i=1}^{m} b_i w_i$$

subject to

$$\sum_{i=1}^{m} c_i w_i + c_{(m+1), j} w_{m+1} \leq p_j^*, \quad j = 1, 2, \ldots, n.$$

Moreover, the dual inequalities are equalities at an optimal solution w^*, when $y_j > 0$. Also, when $y_j > 0$, $c_{(m+1), j} = 0$ and we have

$$\sum_{i=1}^{m} c_{ij} w_i^* = p_j^* = \log \frac{x_j}{h_j^*}.$$

Since $y \in \phi(h^*)$ when $h_j^* = 0$, then y_j must also be zero. Thus when $x_j > 0$, then $y_j > 0$ and hence $h_j^* > 0$. This shows that p is finite when $x_j > 0$. Let $z_i = \exp(-w_i^*)$, then from $\sum_{i=1}^{m} c_{ij} w_i^* = \log \frac{x_j}{h_j^*}$ we get $h_j^* = x_j \prod_{i=1}^{m} z_i^{c_{ij}}$ for all $x_j > 0$. When $x_j = 0$, the equality trivially holds.

For an alternate proof of the existence of π, as well as for a proof of the uniqueness of π we refer to Franklin and Lorenz (1989). This completes the proof of the theorem.   ∎

The proof of Theorem 6.2.1 is based on Bapat and Raghavan (1989). For alternate proofs based on Lagrangian arguments and on nonlinear programming, and for other related results, see Franklin and Lorenz (1989), Rothblum (1989), Rothblum and Schneider (1989), and Schneider (1989).

6.3. Hilbert's projective metric

Let A, B be a given pair of $m \times n$ positive matrices. By Theorem 6.2.1 there exists a unique matrix $C = XAY$, where X, Y are diagonal matrices with positive diagonal entries and C, B have the same row and column sums. The proof of Theorem 6.2.1 is nonconstructive. Also, we have not yet shown that the heuristic procedure of alternately scaling the rows and columns of A actually converges to a matrix C with the prescribed row and column sums.

A possible way out is to view the n-th iterate as a map $f_n: \mathcal{X} \to \mathcal{X}$, where $\mathcal{X}$ is the set of all matrices of the type XAY, for some diagonal matrices X, Y with positive diagonal entries, and then to show that this set has a suitable metric θ with respect to which $\mathcal{X}$ is a complete metric space and that the iteration maps f_n are contractions. Banach's Contraction Mapping Theorem would then guarantee a unique fixed point that could be identified with $C = XAY$. In fact, one can use this technique to prove the Perron-Frobenius Theorem in greater generality for some special cases; see Birkhoff (1957).

Recall that a closed set $K \subset R^n$ is called a *convex cone* if $x, y \in K, \lambda$, $\mu \geq 0 \Rightarrow \lambda x + \mu y \in K$. We assume that the cone K is *pointed*, that is, $x + y = 0, x, y \in K \Rightarrow x = y = 0$. We say $x \geq y$ (or $y \leq x$) if $x - y \in K$. Let $p \in K \setminus \{0\}$. For any $x \in R^n$ we define

$$\sup\left(\frac{x}{p}\right) = \inf\{\lambda : x \leq \lambda p, \lambda \text{ real}\},$$

$$\inf\left(\frac{x}{p}\right) = \sup\{\lambda : x \geq \mu p, \mu \text{ real}\}.$$

These two functions are extended real valued. For the empty set ϕ, we set $\inf \phi = \infty, \sup \phi = -\infty$.

Lemma 6.3.1. *Let $x \in R^n$, $p \in K \setminus \{0\}$, where K is a pointed convex cone. Then*

(i) $\inf(\frac{x}{p}) \leq \sup(\frac{x}{p})$.

(ii) $\sup(-\frac{x}{p}) = -\inf(\frac{x}{p})$; $\inf(-\frac{x}{p}) = -\sup(\frac{x}{p})$.

(iii) $\sup(\frac{x+y}{p}) \leq \sup(\frac{x}{p}) + \sup(\frac{y}{p})$.

(iv) $\inf(\frac{x+y}{p}) \geq \inf(\frac{x}{p}) + \inf(\frac{y}{p})$.

Proof. We first claim that $\inf(\frac{x}{p}) = \infty$ is not possible, for then, by definition $x \geq np$ for any positive integer n. That is to say, $\frac{x}{n} \geq p$, for all n. Thus $\frac{x}{n} - p \in K$. Since K is closed, $\frac{x}{n} - p \to -p \in K$. Since K is pointed, $p = 0$, which is a contradiction, and the claim is proved. Similarly, $\sup(\frac{x}{p})$ is finite, for otherwise, if $\sup(\frac{x}{p}) = -\infty$, then $x \le -np$ for all n positive. Thus $\frac{x}{n} + p \le 0$ for all n. Since K is closed, $p \le 0$, which is a contradiction. Lastly, let $\inf(\frac{x}{p})$ and $\sup(\frac{x}{p})$ be finite. Given $\epsilon > 0$, if $\sup(\frac{x}{p}) = \lambda_\circ$, $\inf(\frac{x}{p}) = \mu_\circ$, then $(\mu_\circ - \epsilon)p \le x \le (\lambda_\circ + \epsilon)p$. This shows that $(\lambda_\circ - \mu_\circ + 2\epsilon)p \ge 0$ for all $\epsilon > 0$. In the limit, $(\lambda_\circ - \mu_\circ)p \ge 0$. Since $-p \notin K$, $(\lambda_\circ - \mu_\circ) \ge 0$. This proves (i).

We prove the rest of the assertions assuming that $\sup(\frac{x}{p})$ and $\inf(\frac{x}{p})$ are finite. Using $\lambda_\circ$, $\mu_\circ$ as defined above, $(-\lambda_\circ - \epsilon)p \le -x \le (-\mu_\circ + \epsilon)p$. Since ϵ is arbitrary, $\sup(-\frac{x}{p}) = -\mu_\circ = -\inf(\frac{x}{p})$; $\inf(-\frac{x}{p}) = -\lambda_\circ = -\sup(\frac{x}{p})$. Let $\lambda_1 = \sup(\frac{x}{p})$, $\mu_1 = \inf(\frac{x}{p})$, $\lambda_2 = \sup(\frac{y}{p})$, and $\mu_2 = \inf(\frac{y}{p})$. Thus for $\epsilon > 0$ we have

$$(\mu_1 - \epsilon)p \le x \le (\lambda_1 + \epsilon)p$$

$$(\mu_2 - \epsilon)p \le y \le (\lambda_2 + \epsilon)p.$$

This shows that $[\mu_1 + \mu_2 - 2\epsilon]p \le x + y \le [\lambda_1 + \lambda_2 + 2\epsilon]p$. In particular,

$$\sup\left(\frac{x+y}{p}\right) \le \lambda_1 + \lambda_2 = \sup\left(\frac{x}{p}\right) + \sup\left(\frac{y}{p}\right)$$

and

$$\inf\left(\frac{x+y}{p}\right) \ge \mu_1 + \mu_2 = \inf\left(\frac{x}{p}\right) + \inf\left(\frac{y}{p}\right).$$

That completes the proof.   ∎

Let $p \in K \setminus \{0\}$. The oscillation between x and p for any $x \in R^n$ is defined by Osc $(\frac{x}{p}) = \sup(\frac{x}{p}) - \inf(\frac{x}{p})$. Since $\sup(\frac{x}{p}) < \infty$ and $\inf(\frac{x}{p}) > -\infty$, the oscillation is finite.

Lemma 6.3.2. *If $p \in K \setminus \{0\}$, then Osc $(\frac{x}{p}) \ge 0$. Furthermore, Osc $(\frac{x}{p}) = 0$ if and only if x is a scalar multiple of p.*

Proof. It follows from (i) of Lemma 6.3.1 that Osc $(\frac{x}{p}) \ge 0$. Let Osc $(\frac{x}{p}) = 0$. Then with $\mu_\circ$, $\lambda_\circ$ as defined in Lemma 6.3.1 we have $(\mu_\circ - \epsilon)p \le x \le (\lambda_\circ + \epsilon)p$ and $\mu_\circ = \lambda_\circ$. Thus in the limit as $\epsilon \to 0$, $\lambda_\circ p - x \ge 0$, $\lambda_\circ p - x \le 0$. Since K is pointed, $x = \lambda_\circ p$, which completes the proof.   ∎

Lemma 6.3.3. *Let $p \in K \setminus \{0\}$. For any $\beta > 0$ the following assertions hold:*

(i) $\beta \sup(\frac{x}{p}) = \sup(\frac{\beta x}{p})$.

(ii) $\sup(\frac{x}{\beta p}) = \frac{1}{\beta} \sup(\frac{x}{p})$.

(iii) $\sup(\frac{x+\alpha p}{p}) = \sup(\frac{x}{p}) + \alpha$ *for all real* α.

(iv) $\inf(\frac{x+\alpha p}{p}) = \inf(\frac{x}{p}) + \alpha$ *for all real* α.

(v) $Osc\,(\frac{x+\alpha p}{p}) = Osc\,(\frac{x}{p})$ *for all real* α.

(vi) *If* $p = x + y$, *then* $\sup(\frac{x}{p}) + \inf(\frac{x}{p}) = 1$.

(vii) *If* $x, y \in K \setminus \{0\}$, *then* $\sup(\frac{x}{y}) = [\inf(\frac{y}{x})]^{-1}$. *(We assume* $\frac{1}{0} = \infty$.)

Proof. (i). Let $\lambda_o = \sup(\frac{x}{p})$. For any $\epsilon > 0, x \leq (\lambda_o + \epsilon)p$. If $\beta > 0, \beta x \leq (\beta\lambda_o + \beta\epsilon)p$. Since ϵ is arbitrary, $\sup(\frac{\beta x}{p}) = \beta\lambda_o = \beta\sup(\frac{x}{p})$.

(ii). Note that $x \leq (\lambda_o + \epsilon)p \Rightarrow x \leq (\frac{\lambda_o}{\beta} + \frac{\epsilon}{\beta})\beta p$. Thus $\sup(\frac{x}{\beta p}) = \frac{\lambda_o}{\beta} = \frac{1}{\beta}\sup(\frac{x}{p})$.

(iii). Let $\sup(\frac{x+\alpha p}{p}) = \theta$. Then for $\epsilon > 0, x + \alpha p \leq (\theta + \epsilon)p$, i.e., $x \leq (\theta - \alpha + \epsilon)p$. Since ϵ is arbitrary, $\sup(\frac{x}{p}) = \theta - \alpha$. Thus $\theta = \sup(\frac{x}{p}) + \alpha$.

(iv). Let $\inf(\frac{x+\alpha p}{p}) = w$. For $\epsilon > 0, x + \alpha p \geq (w-\epsilon)p$. Therefore $x \geq ((w-\alpha) - \epsilon)p$. Since ϵ is arbitrary, $\inf(\frac{x}{p}) = w - \alpha$. Thus $\inf(\frac{x+\alpha p}{p}) = \alpha + \inf(\frac{x}{p})$.

(v). $Osc\,(\frac{x}{p}) = \sup(\frac{x}{p}) - \inf(\frac{x}{p}) = \sup(\frac{x+\alpha p}{p}) - \inf(\frac{x+\alpha p}{p}) = Osc\,(\frac{x+\alpha p}{p})$.

(vi). Let $p = x + y$. Then $\sup(\frac{x}{p}) = \sup(\frac{x+y-y}{p}) = \sup(\frac{-y+p}{p}) = \sup(\frac{-y}{p})+1$ by *(iv)*. Thus $\sup(\frac{x}{p}) - \sup(\frac{-y}{p}) = 1$. From *(ii)* of Lemma 6.3.1 we have $\sup(\frac{x}{p}) + \inf(\frac{y}{p}) = 1$.

(vii). Let $\inf(\frac{y}{x}) = \alpha$. Thus for $\epsilon > 0, y \geq (\alpha - \epsilon)x$. Since $y \in K \setminus \{0\}, \alpha \geq 0$. Further, $\inf(\frac{y}{x})$ is finite. Let $0 < \beta < \alpha$. Then $y \geq \beta x$. Also, by definition, $y \not\geq \beta x$ if $\beta > \alpha$. That is, $x \leq \frac{1}{\beta}y$ if $\frac{1}{\beta} > \frac{1}{\alpha}$ and $x \not\leq \frac{1}{\beta}y$ if $\frac{1}{\beta} < \frac{1}{\alpha}$. This is the same as saying $\sup(\frac{x}{y}) = \frac{1}{\alpha}$. Suppose $\alpha = 0$. Then $y \geq \beta x$ fails to hold for all $\beta > 0$. Thus $x \leq \frac{1}{\beta}y$ fails to hold for all $\beta > 0$ and therefore $\sup(\frac{x}{y}) = \infty$. ∎

Lemma 6.3.4. *Let* $p, q \in K \setminus \{0\}$. *If* $x \in K$, *then*

(i) $\sup(\frac{x}{q}) \leq \sup(\frac{x}{p})\sup(\frac{p}{q})$.

(ii) $\inf(\frac{x}{q}) \geq \inf(\frac{x}{p})\inf(\frac{p}{q})$.

Proof. Let $\sup(\frac{x}{p}) = \lambda, \sup(\frac{p}{q}) = \mu$. Thus for $\epsilon > 0, x \leq (\lambda + \epsilon)p, p \leq (\mu + \epsilon)q$. Thus $x \leq (\lambda + \epsilon)(\mu + \epsilon)q \leq (\lambda\mu + \epsilon(\lambda + \mu) + \epsilon^2)q$. Since ϵ arbitrary, the first inequality follows. The second inequality can be proved in a similar fashion. ∎

Lemma 6.3.5. *If* $p, q \in K \setminus \{0\}$, *then*

(i) $\sup(\frac{p}{p+q}) = \frac{1}{1+\inf(\frac{q}{p})} = \frac{\sup(\frac{p}{q})}{1+\sup(\frac{p}{q})} \leq 1$.

(ii) $\inf(\frac{p}{p+q}) = \frac{1}{1+\sup(\frac{q}{p})} = \frac{\inf(\frac{p}{q})}{1+\inf(\frac{p}{q})} \leq 1$.

Proof. We have

$$\sup\left(\frac{p}{p+q}\right) = \frac{1}{\inf\left(\frac{p+q}{p}\right)} \quad \text{by } (vii) \text{ of Lemma 6.3.3}$$

$$= \frac{1}{1 + \inf\left(\frac{q}{p}\right)}$$

$$= \frac{1}{1 + \frac{1}{\sup\left(\frac{p}{q}\right)}}$$

$$= \frac{\sup\left(\frac{p}{q}\right)}{1 + \sup\left(\frac{p}{q}\right)}$$

$$\leq 1.$$

Replacing sup by inf we get the second assertion, and the proof is complete. ∎

In the next result we introduce the definition of *Hilbert's projective metric* θ on $K \setminus \{0\}$.

Theorem 6.3.6. *Let* $x, y \in K \setminus \{0\}$*. Then* $\theta(x, y) = \log\{\sup(\frac{x}{y})\sup(\frac{y}{x})\}$ *is an extended real function satisfying*

(i) $\theta(x, y) = \theta(y, x)$.
(ii) $\theta(x, y) = 0 \Leftrightarrow x \sim y$, *that is,* $x = \alpha y$ *for some* $\alpha > 0$.
(iii) $\theta(x, y) + \theta(y, z) \geq \theta(x, z)$.

In essence, θ *is a pseudo-metric on* $K \setminus \{0\}$.

Proof. Assertion (i) follows from the definition of θ. Note that $\sup(\frac{x}{y}) = \frac{1}{\inf(\frac{x}{y})}$. Thus if

$$\theta(x, y) = \log \frac{\sup\left(\frac{x}{y}\right)}{\inf\left(\frac{x}{y}\right)} = 0,$$

then

$$\frac{\sup\left(\frac{x}{y}\right)}{\inf\left(\frac{x}{y}\right)} = 1.$$

That is, $\sup(\frac{x}{y}) = \inf(\frac{x}{y}) := \alpha$. Then for $\epsilon > 0$, $(\alpha - \epsilon)y \leq x \leq (\alpha + \epsilon)y$. Since ϵ is arbitrary, $x = \alpha y$. Hence we have assertion (ii).

For $x, y, z \in K \setminus \{0\}$, by Lemma 6.3.4,

$$\frac{\sup\left(\frac{x}{y}\right)}{\inf\left(\frac{x}{y}\right)} \frac{\sup\left(\frac{y}{z}\right)}{\inf\left(\frac{y}{z}\right)} \geq \frac{\sup\left(\frac{x}{z}\right)}{\inf\left(\frac{x}{z}\right)}.$$

By taking logarithms, *(iii)* follows. ∎

Lemma 6.3.7. *The projective metric* $\theta(x, y)$ *satisfies* $\theta(\alpha x, \beta y) = \theta(x, y)$ *for any* $\alpha, \beta > 0$.

Proof. By definition, $\theta(\alpha x, \beta y) = \log\{\sup(\frac{\alpha x}{\beta y})\sup(\frac{\beta y}{\alpha x})\}$. By *(ii)* of Lemma 6.3.3, this equals

$$\log\left[\left(\frac{\alpha}{\beta}\right)\sup\left(\frac{x}{y}\right)\left(\frac{\beta}{\alpha}\right)\sup\left(\frac{y}{x}\right)\right] = \log\left[\sup\left(\frac{x}{y}\right)\sup\left(\frac{y}{x}\right)\right]$$

$$= \theta(x, y),$$

and the proof is complete. ∎

From now on we restrict ourselves to the special cone $K_n = \{x : x \geq 0\} \in R^n$. The interior of K_n is the set $\{x : x > 0\}$. Let A be any $m \times n$ positive matrix. Then A maps all nonzero vectors of K_n into the interior of K_m. Our interest is in computing the change in the width between any two points of K_n, measured in Hilbert's projective metric, compared to the width between the corresponding points in the image of K_n under A. We could propose two different measures. The *diameter* of the image of K under A is defined by

$$\Delta(A) = \sup_{x, y \in K \setminus \{0\}} \theta(Ax, Ay).$$

We set $\Psi(A) = e^{\Delta(A)}$. The *contraction ratio* of Birkhoff is defined by

$$N(A) = \sup_{\substack{x, y \in K \setminus \{0\} \\ 0 < \theta(x, y) < \infty}} \left\{ \frac{\theta(Ax, Ay)}{\theta(x, y)} \right\}.$$

Another measure of variation is the *oscillation ratio* due to Hopf, defined by

$$\text{Osc }(A) = \sup_{x, y \in K \setminus \{0\}} \left\{ \frac{\text{Osc }(\frac{Ax}{Ay})}{\text{Osc }(\frac{x}{y})} \right\}.$$

We study some properties of $\Psi(A)$, Osc (A), $N(A)$, and Osc $(\frac{x}{y})$.

Lemma 6.3.8. *Let* $A = (a_{ij}) > 0$ *be an* $m \times n$ *matrix. Then*

$$\Delta(A) = \log \max_{i,j,k,l} \frac{a_{ik}a_{jl}}{a_{il}a_{jk}}.$$

Proof. For any $u, v > 0$, $\theta(u, v) = \max_{i,j} \frac{u_i v_j}{v_i u_j}$ and therefore,

$$\Psi(A) = \sup_{x, y \in K \setminus \{0\}} \left(\frac{\max\limits_{i,j} \sum\limits_{\alpha} a_{i\alpha}x_{\alpha} \sum\limits_{\beta} a_{j\beta}y_{\beta}}{\sum\limits_{\alpha} a_{j\alpha}x_{\alpha} \sum\limits_{\beta} a_{i\beta}y_{\beta}} \right).$$

Collecting the terms in the numerator and dividing and multiplying them by $a_{j\alpha}a_{i\beta}$, we get

$$\Psi(A) = \sup_{x,y\in K\setminus\{0\}} \max_{i,j} \sum_{\alpha}\sum_{\beta}\left(\frac{a_{i\alpha}a_{j\beta}}{a_{j\alpha}a_{i\beta}}\right)\cdot\left[\frac{a_{j\alpha}a_{i\beta}x_\alpha y_\beta}{\sum_{\alpha}\sum_{\beta}a_{j\alpha}a_{i\beta}x_\alpha y_\beta}\right].$$

Here, the second bracket consists of components of a probability vector. Let

$$\max_{\alpha,\beta}\frac{a_{i\alpha}a_{j\beta}}{a_{i\beta}a_{j\alpha}} = \frac{a_{ik}a_{jl}}{a_{il}a_{jk}}.$$

Thus by choosing x_k and y_l close to 1 we get the value of $\Psi(A)$. That completes the proof. ∎

Lemma 6.3.9. *Let X and Y be diagonal matrices with positive diagonal entries. If $B = XAY$, then $\Delta(B) = \Delta(A) = \Delta(A^T)$.*

Proof. We have

$$\begin{aligned}
\Delta(B) &= \sup_{p,q>0}\theta(XAYp, XAYq)\\
&= \sup_{u,v>0}\theta(XAu, XAv), \quad \text{where } u = Yp, \quad v = Yq\\
&= \sup_{u,v>0}\theta(A^T Xu, A^T Xv) \quad \text{(by Lemma 6.3.8)}\\
&= \sup_{z,w>0}\theta(A^T z, A^T w), \quad \text{where } z = Xu, \quad w = Xv\\
&= \sup_{z,w>0}\theta(Az, Aw)\\
&= \Delta(A).
\end{aligned}$$

That completes the proof. ∎

Lemma 6.3.10. *Let $x, y \in K\setminus\{0\}$. Then, for $A > 0$,*

$$0 \le \inf\left(\frac{x}{y}\right) \le \inf\left(\frac{Ax}{Ay}\right) \le \sup\left(\frac{Ax}{Ay}\right) \le \sup\left(\frac{x}{y}\right).$$

Proof. If $x = \alpha y$ for some α, the result is trivial. Indeed, $\mathrm{Osc}\left(\frac{Ax}{Ay}\right) = \mathrm{Osc}\left(\frac{x}{y}\right)$ in this case. Without loss of generality, we assume $\mathrm{Osc}\left(\frac{x}{y}\right) > 0$. Let $\lambda_\circ = \sup(\frac{x}{y})$, $\mu_\circ = \inf(\frac{x}{y})$, $\mu_1 = \inf(\frac{Ax}{Ay})$, and $\lambda_1 = \sup(\frac{Ax}{Ay})$. Observe that $\mu_\circ, \mu_1, \lambda_\circ$, and λ_1 are finite. Thus for any $\epsilon > 0$, we have $(\mu_\circ - \epsilon)y \le x \le (\lambda_\circ + \epsilon)y$ and consequently, since $A > 0$, $(\mu_\circ - \epsilon)Ay \le Ax \le (\lambda_\circ + \epsilon)Ay$. Thus

$$\inf\left(\frac{Ax}{Ay}\right) \ge \mu_\circ - \epsilon = \inf\left(\frac{x}{y}\right) - \epsilon,$$

$$\sup\left(\frac{Ax}{Ay}\right) \le \lambda_\circ + \epsilon = \sup\left(\frac{x}{y}\right) + \epsilon.$$

Since ϵ is arbitrary, the inequalities follow. ∎

Theorem 6.3.11. *Let $A > 0$ be an $n \times n$ matrix. Then the contraction ratio of Birkhoff never exceeds the oscillation ratio of Hopf on the cone of nonnegative vectors in R^n.*

Proof. Let $x, y \in K \setminus \{0\}$ with $0 < \theta(x, y) < \infty$. From Lemma 6.3.10 we get

$$\sup\left(\frac{Ay}{Ax}\right) - \inf\left(\frac{Ay}{Ax}\right) \leq \text{Osc}(A)\left\{\sup\left(\frac{y}{x}\right) - \inf\left(\frac{y}{x}\right)\right\}. \quad (6.3.1)$$

By *(vii)*, of Lemma 6.3.3, we have $\sup(\frac{x}{y}) \cdot \inf(\frac{y}{x}) = 1$. Thus the inequality (6.3.1) can also be written as

$$\left\{\left[\inf\left(\frac{Ax}{Ay}\right)\right]^{-1} - \left[\sup\left(\frac{Ax}{Ay}\right)\right]^{-1}\right\}$$

$$\leq \text{Osc}(A)\left\{\left[\inf\left(\frac{x}{y}\right)\right]^{-1} - \left[\sup\left(\frac{x}{y}\right)\right]^{-1}\right\}.$$

For any $\alpha > 0$, we substitute $\alpha x + y$ for x. Thus we get

$$\frac{\sup\left(\frac{Ax}{Ay}\right) - \inf\left(\frac{Ax}{Ay}\right)}{\left[\alpha \sup\left(\frac{Ax}{Ay}\right) + 1\right]\left[\alpha \inf\left(\frac{Ax}{Ay}\right) + 1\right]}$$

$$\leq \text{Osc}(A)\frac{\sup\left(\frac{x}{y}\right) - \inf\left(\frac{x}{y}\right)}{\left[\alpha \sup\left(\frac{x}{y}\right) + 1\right]\left[\alpha \inf\left(\frac{x}{y}\right) + 1\right]}.$$

Integrating with respect to α on $(0, \alpha)$, we get

$$\log\left(\frac{\alpha \sup\left(\frac{Ax}{Ay}\right) + 1}{\alpha \inf\left(\frac{Ax}{Ay}\right) + 1}\right) \leq \text{Osc}(A)\log\left\{\frac{\alpha \sup\left(\frac{x}{y}\right) + 1}{\alpha \inf\left(\frac{x}{y}\right) + 1}\right\}.$$

Let $\alpha \to \infty$. Then we see that $\theta(Ax, Ay) \leq \text{Osc}(A)\theta(x, y)$. Thus $N(A) \leq \text{Osc}(A)$, and the proof is complete. ∎

The following theorem is central to positive mappings on R^n equipped with Hilbert's projective metric.

Theorem 6.3.12. *Osc $(A) \leq \tanh \frac{\Delta(A)}{4}$.*

Proof. Consider $x, p \in K \setminus \{0\}$, such that x is not a multiple of p. Let $\mu = \sup(\frac{x}{p})$, $\lambda = \inf(\frac{x}{p})$, and $\delta = \mu - \lambda < \infty$. Let $\epsilon > 0$ be small such that $\bar{\lambda} = \lambda - \epsilon\delta > 0$, $\bar{\mu} = \mu + \epsilon\delta > 0$. Let $p_1 = x - \bar{\lambda}p$, $p_2 = \bar{\mu}p - x$. We have

$p_1 + p_2 = (\bar{\mu} - \bar{\lambda})p$. Let $\bar{\delta} = (\bar{\mu} - \bar{\lambda})$. By Lemma 6.3.3

$$\sup\left(\frac{Ax}{Ap}\right) = \sup\left(\frac{A(p_1 + \bar{\lambda}p)}{Ap}\right) = \sup\left(\frac{Ap_1}{Ap}\right) + \bar{\lambda},$$

$$\inf\left(\frac{Ax}{Ap}\right) = \inf\left(\frac{A(p_1 + \bar{\lambda}p)}{Ap}\right) = \inf\left(\frac{Ap_1}{Ap}\right) + \bar{\lambda}.$$

Further, $Ap = \frac{1}{\delta}(Ap_1 + Ap_2)$. Thus we have

$$\text{Osc}\left(\frac{Ax}{Ap} = \left[\sup\left(\frac{Ap_1}{Ap_1 + Ap_2}\right)\right] - \left[\inf\left(\frac{Ap_1}{Ap_1 + Ap_2}\right)\right]\bar{\delta}.$$

Again, by Lemma 6.3.3, the right-hand side of the above equation reduces to

$$\left(\frac{1}{\inf\left(\frac{Ap_2}{Ap_1}\right) + 1} - \frac{1}{\sup\left(\frac{Ap_2}{Ap_1}\right) + 1}\right)\bar{\delta}.$$

Let $u = \sup(\frac{Ap_1}{Ap_2})$, $v = \sup\left(\frac{Ap_2}{Ap_1}\right)$. The above expression is then

$$\left(\frac{1}{(1 + u^{-1})} - \frac{1}{(1 + v)}\right)\bar{\delta} = \left(\frac{vu - 1}{1 + vu + v + u}\right)\bar{\delta}.$$

Since $v, u > 0$, then $t = \sqrt{vu} \le \frac{v+u}{2}$ and

$$\text{Osc}\left(\frac{Ax}{Ap}\right) \le \frac{t^2 - 1}{1 + t^2 + 2t}\bar{\delta} = \frac{t - 1}{t + 1}\bar{\delta}.$$

Further, $\bar{\delta} \to \delta$ as $\epsilon \to 0$. Thus

$$\text{Osc}\left(\frac{Ax}{Ap}\right) \le \frac{t - 1}{t + 1}\text{Osc}\left(\frac{x}{p}\right).$$

Since $\frac{t-1}{t+1}$ is increasing in t,

$$t = \sqrt{\sup\left(\frac{Ap_1}{Ap_2}\right)\sup\left(\frac{Ap_2}{Ap_1}\right)} \le \exp\frac{\Delta(A)}{2}$$

satisfies

$$\text{Osc}\left(\frac{Ax}{Ap}\right) \le \frac{e^{\frac{\Delta(A)}{2}} - 1}{e^{\frac{\Delta(A)}{2}} + 1}\text{Osc}\left(\frac{x}{p}\right).$$

Dividing the numerator and the denominator by $\exp\frac{\Delta(A)}{4}$, we get

$$\frac{\text{Osc}\left(\frac{Ax}{Ap}\right)}{\text{Osc}\left(\frac{x}{p}\right)} \le \tanh\frac{\Delta(A)}{4},$$

and the proof is complete. ∎

As a consequence of Theorems 6.3.11 and 6.3.12, for $\beta = \tanh\frac{\Delta(A)}{4}$ we have $\theta(Ax, Ay) \le \beta\theta(x, y)$ for any $x, y \in K \setminus \{0\}$. Any positive $n \times n$ matrix A

is therefore a contraction on R^n_+ in the Hilbert metric θ and $Au \sim u$ for some $u \in K \setminus \{0\}$. Thus $Au = \lambda u$ for some $\lambda > 0, u > 0$; thus the basic assertion of Perron's Theorem follows from the Contraction Mapping Theorem.

We now apply Hilbert's projective metric to study the iterative scaling algorithm described in the beginning of this section. We first introduce some notation. Let $x, y \in R^n$. If x_i, y_i are nonzero for all coordinates i, let $z = \frac{x}{y}$, where $z_i = \frac{x_i}{y_i}, i = 1, 2, \ldots, n$. Let A be an $m \times n$ positive matrix and let $p \in R^m, q \in R^n$ be positive vectors with $\sum_{i=1}^m p_i = \sum_{j=1}^n q_j$. We multiply the columns of A by suitable positive constants (i.e., post-multiply A by a diagonal matrix with positive diagonal entries) to obtain the matrix A_0 with column sum vector q. We then row-normalize A_0 to obtain the matrix A_0' with row sum vector p. This process is continued and we obtain the sequence $A_0, A_0', A_1, A_1', A_2, A_2', \ldots$ such that $A_0, A_1, A_2, \ldots$ have column sum vector q and $A_0', A_1', A_2' \ldots$ have row sum vector p. The key idea is to show that the iterations A_k and A_k' are contractions in Hilbert's projective metric.

Let $r^{(k)}$ and $c^{(k)}$ denote the row sums and the column sums of A_k and A_k', respectively. Then $A_k' = S_k A_k$, where $S_k = \text{diag}(\frac{p_i}{r_i^{(k)}})$, and $A_{k+1} = A_k' T_k$, where $T_k = \text{diag}(\frac{q_j}{c_j^{(k)}})$. The next result uses the notation introduced so far.

Lemma 6.3.13. *Let* $\tau = N^2(A)$. *Then*

$$\theta(r^{(1)}, p) \leq \tau\theta(r^{(0)}, p), \qquad \theta(c^{(1)}, q) \leq \tau\theta(c^{(0)}, q).$$

Proof. We have $r^{(1)} = A_1 1 = A_0' T_0 1 = A_0'(\frac{q}{c^{(0)}})$. Also, $A_0' 1 = p$. Thus

$$\theta(r^{(1)}, p) = \theta\left(A_0'\left(\frac{q}{c^{(0)}}\right), A_0' 1\right)$$
$$\leq N(A)\theta\left(\frac{q}{c^{(0)}}, 1\right)$$
$$\leq N(A)\theta\left(q, c^{(0)}\right).$$

Further, $c^{(0)} = A_0'^T 1 = A^T S_0 1 = A^T(\frac{p}{r^{(0)}})$. Since $A^T 1 = q$, we have

$$\theta(q, c^{(0)}) = \theta\left(A^T 1, A^T\left(\frac{p}{r^{(0)}}\right)\right)$$
$$\leq N(A)\theta(r^{(0)}, p).$$

Thus $\theta(r^{(1)}, p) \leq \tau\theta(r^{(0)}, p)$. A similar argument shows that $\theta(c^{(1)}, q) \leq \tau\theta(c^{(0)}, q)$, and the proof is complete. ∎

A repeated application of Lemma 6.3.13 yields

$$\theta(r^{(k)}, p) \leq \tau^k\theta(r^{(0)}, p),$$
$$\theta(c^{(k)}, q) \leq \tau^k\theta(c^{(0)}, q).$$

With these preliminaries we are ready to prove the main result on iterative scaling. We continue to use the notation introduced above.

Theorem 6.3.14. *Let $A > 0$ be an $m \times n$ matrix. Let $\mathcal{X}$ denote the set of all $m \times n$ matrices $B > 0$ such that $B = XAY$ for some diagonal matrices X, Y with positive diagonal entries. For $B, C \in \mathcal{X}$, let a metric Λ be defined by $\Lambda(B, C) = \theta(u, \mathbf{1}) + \theta(v, \mathbf{1})$, where $C = UBV$, $U = \mathrm{diag}\{u_i\}$, $V = \mathrm{diag}\{v_j\}$. Let A_k, and A_k' denote the iterative scalings that achieve column sums q and the row sums p alternately. Then the sequence of matrices $A_k, A_k', A_{k+1}, A_{k+1}', \ldots \to \tilde{B}$ for some $\tilde{B}$. Further,*

$$\Lambda(A_k, \tilde{B}) \le \frac{1}{1-\tau}\{\theta(r^{(k)}, p) + \theta(c^{(k)}, q)\},$$

$$\Lambda(A_k', \tilde{B}) \le \frac{1}{1-\tau}\{\theta(r^{(k+1)}, p) + \theta(c^{(k)}, q)\},$$

where $\tau = N^2(A)$.

Proof. Let $S_k = \mathrm{diag}(s_i^{(k)})$, $s^{(k)} = \frac{p}{r^{(k)}}$ and $T_k = \mathrm{diag}(t_j^{(k)})$, $t^{(k)} = \frac{q}{c^{(k)}}$. Since $A_{k+1} = S_k A_k T_k$, we have

$$\begin{aligned}
\Lambda(A_k, A_{k+1}) &= \theta(s^{(k)}, \mathbf{1}) + \theta(t^{(k)}, \mathbf{1}) \\
&= \theta(r^{(k)}, p) + \theta(c^{(k)}, q) \\
&\le \tau^k\{\theta(r^{(0)}, p) + \theta(c^{(0)}, q)\}.
\end{aligned}$$

By the triangle inequality satisfied by θ,

$$\Lambda(A_0, A_{s+1}) \le (1 + \cdots + \tau^s)\{\theta(r^{(0)}, p) + \theta(c^{(0)}, q)\},$$

and we obtain for $s \to \infty$

$$\Lambda(A_0, \tilde{B}) \le \frac{1}{1-\tau}\{\theta(r^{(0)}, p) + \theta(c^{(0)}, q)\}.$$

This proves the first estimate for $k = 0$. The rest of the estimates follow along similar lines. ∎

Hilbert's projective metric was effectively used by Birkhoff (1957) to prove the Perron-Frobenius Theorem and its extension by Jentsch (1912) to positive integral operators in Banach spaces. Related references are Busemann and Kelly (1953) and Kohlberg and Pratt (1982). Oscillation between vectors was first introduced by Hopf (1963). Theorem 6.3.12 is due to Hopf (1963), and its proof is based on Bushell (1973), Bauer (1965), and Ostrowski (1964). Franklin and Lorenz (1989) were the first to view scaling as a contraction in projective metric. See Soules (1991) and Achilles (1993) for further results.

6.4. Algorithms for scaling

In this section we describe some heuristic improvements in scaling a nonnegative matrix with a doubly stochastic pattern to a doubly stochastic matrix. First we consider an example. Let

$$A = \begin{bmatrix} 5 & 8 & 7 \\ 6 & 4 & 15 \\ 1 & 3 & 7 \end{bmatrix}.$$

Normalizing the columns we get

$$A^{(0)} = \begin{bmatrix} \frac{5}{12} & \frac{8}{15} & \frac{7}{29} \\ \frac{6}{12} & \frac{4}{15} & \frac{15}{29} \\ \frac{1}{12} & \frac{3}{15} & \frac{7}{29} \end{bmatrix}.$$

Normalizing the rows we get

$$A^{(1)} = \begin{bmatrix} .3497 & .4477 & .2026 \\ .3894 & .2077 & .4029 \\ .1588 & .3812 & .4600 \end{bmatrix}.$$

We can repeat the above procedure and after a few iterations we get the approximate doubly stochastic matrix

$$B = \begin{bmatrix} .3535 & .4166 & .2300 \\ .4918 & .4157 & .2667 \\ .1547 & .3419 & .5034 \end{bmatrix}.$$

In general the algorithm could be quite slow. For example, it took 500 iterations before the matrix

$$A = \begin{bmatrix} 10^2 & 10^2 & 0 \\ 10^2 & 10^4 & 1 \\ 0 & 1 & 10^2 \end{bmatrix}$$

resulted in the approximate doubly stochastic matrix

$$B = \begin{bmatrix} .9090 & .0907 & .0000 \\ .0910 & .9080 & .0007 \\ .0000 & .0013 & .9993 \end{bmatrix}.$$

Indeed, the order of error in this case is much higher than in the previous example.

By appropriate choice of scale vectors, one can improve the speed of convergence. The following are some of the intuitive algorithms that have been suggested. The convergence of these algorithms can be rigorously proved but their rates of convergence are harder to compare except by simulation.

6.4.1. Deviation reduction algorithm

Given an $N \times N$ nonnegative matrix A with a doubly stochastic pattern we can proceed as follows:

Step 1. Compute $\gamma_i = \sum_j a_{ij}$, $c_j = \sum_i a_{ij}$; $i, j = 1, \ldots, N$. Compute $\mu = \frac{1}{N} \sum \gamma_i$, the average row sum.

Step 2. Let $\max_i |\gamma_i - \mu| = |\gamma_u - \mu|$, $\max_j |c_j - \mu| = |c_v - \mu|$ for some row u and column v. Let $\epsilon > 0$ be the tolerance. Test if $|\gamma_u - \mu| < \epsilon \mu$ and $|c_v - \mu| < \epsilon \mu$. If so, go to Step 5. If $|c_v - \mu| > |\gamma_u - \mu|$, then go to Step 4.

Step 3. Calculate the average of row sums for all but the u-th row. Let $\bar{\mu} = \frac{1}{N-1} \sum_{i \neq u} \gamma_i$ be this average. Scale row u to $\bar{\mu}$ by defining

$$a_{uj}^{(2)} = a_{uj} \cdot \frac{\bar{\mu}}{\gamma_u}, \quad j = 1, \ldots, N,$$

$$a_{ij}^{(2)} = a_{ij} \quad i \neq u, j = 1, \ldots, N.$$

Update row and column sums of $A^{(2)} = (a_{ij}^{(2)})$:

$$\sum_j a_{ij}^{(2)} = \gamma_i \quad i \neq u$$

$$a_{uj}^{(2)} = \bar{\mu}.$$

Let $c_j^{(2)} = c_j + (\frac{\bar{\mu}}{\gamma_u} - 1)a_{uj}$, $j = 1, \ldots, N$.

Go to Step 1 with the new matrix $A^{(2)}$.

Step 4. Calculate the average of column sums for all but the v-th column. Let $\bar{\mu} = \frac{1}{N-1} \sum_{j \neq v} c_j$ be this average. Scale column v to $\bar{\mu}$ by defining

$$a_{iv}^{(2)} = a_{iv} \cdot \frac{\bar{\mu}}{c_v}, \quad i = 1, \ldots, N.$$

Updating rows and columns of A, let

$$a_{ij}^{(2)} = a_{ij} \quad \text{if} \quad j \neq v$$

$$a_{iv}^{(2)} = a_{iv} \cdot \frac{\bar{\mu}}{c_v}.$$

Go to Step 1 with the new matrix $A = A^{(2)}$.

Step 5. Define $A^{(2)} = \frac{1}{\mu} \cdot A$, $i, j = 1, 2, \ldots, N$.
Stop.

6.4.2. Balancing algorithm

In a doubly stochastic matrix, the column sums and row sums are the same and they remain so under scalar multiplication. Taking this as the requirement one can look for the maximum deviation $|\gamma_u - c_v| = \max_{i,j} |\gamma_i - c_j|$ between the row sums and the column sums. If $|\gamma_u - c_v| < \epsilon(\frac{1}{N}\sum_i \gamma_i)$, then we can normalize A by $\frac{1}{\mu}A$ and stop; otherwise, we can balance row u and column v by scaling as follows. Let $f = (\frac{c_v - a_{uv}}{\gamma_u - a_{uv}})^{\frac{1}{2}}$. Multiply row u by f and column v by f. With the new matrix $A^{(2)}$ we can start all over again.

6.4.3. Deviation balancing algorithm

This third algorithm combines the above two algorithms. Start with the deviation reduction algorithm. Suppose that in the first iteration the algorithm went through Steps 1, 2, and 3 and in the second iteration it goes through the same steps. Instead of deviation reduction, go through the second algorithm of balancing the row sum r_u and column sum c_v, and again go to the deviation reduction algorithm. Similarly, if Step 4 is visited twice contiguously, in the deviation reduction algorithm, apply the second algorithm of balancing row sum and column sum for row r_u and column c_v and revert back to the deviation reduction. The following result can be proved for the above three algorithms [see Parlett and Landis (1982)].

Theorem 6.4.1. *Let $A \geq 0$ have doubly stochastic pattern. Let $A^{(n)} = (a_{ij}^{(n)})$, where $a_{ij}^{(n)} = a_{ij}x_{in}y_{jn}$. (Here x_{in}, y_{jn} are the scale factors induced by any of the above three algorithms.) Let $s_n = \prod_i x_{in}y_{in}$. Then*

(i) s_n is monotonically increasing.

(ii)

$$\lim_n \frac{s_n}{s_{n+1}} = 1,$$

$$\lim_n \alpha_{in} = \lim_n \sum_j x_{in}a_{ij}y_{jn} = 1,$$

$$\lim_n \beta_{jn} = \lim_n \sum_i x_{in}a_{ij}y_{jn} = 1.$$

(iii)

$$\mu_n = \frac{1}{N}\sum_i \alpha_{in} = 1, \quad n = 1, 2, \ldots.$$

The material in this section is based on Parlett and Landis (1982).

6.5. Maximum likelihood estimation

The outcome of a statistical survey is very often a final report in the form of a multidimensional matrix. Usually the entries represent frequency counts of certain qualitative or quantitative characteristics or their combinations observed in the sample. The aim of the survey is invariably to estimate certain parameters of the population based on the given data. Although the data can generally be reported in the form of a vector of counts $\{x_i; i \in I, \text{ card}(I) = q\}$, the overall survey can be better visualized when the data are reported in the form of a multidimensional matrix called a *contingency table*. An example will clarify what we have in mind.

Example 6.1. In a survey on car accidents, the responses from 4000 victims were summarized according to the type of accident and the severity of the accident as follows:

	not severe	moderately severe	severe
rollover	2012	765	406
not rollover	311	198	308

This form of reporting is called a 2×3 contingency table. The vector $(2012, 765, 406, 311, 198, 308)$ can yet well be considered yet another format for reporting the same data, once we know what each coordinate of this vector represents.

A problem of statistical interest is to estimate the expected number of accidents in each category or to estimate the chance for a random accident to belong to a particular category. In our particular example the survey might have reported the responses from 4000 victims out of an unknown number of victims. As a second possibility, the survey might be a report from 4000 predetermined victims from whom the results were extracted.

In the first case, an appropriate model would be to assume that accidents have a Poisson law and the type and severity of one accident has nothing to do with another accident. Thus, assuming statistical independence of cell counts, if X_{ij} is a random variable representing cell count at cell (i, j), the i-th row, j-th column entry in the data, our assumption that the X_{ij}s are independent Poisson random variates with $EX_{ij} = \lambda_{ij}$ gives the probability density

$$P\{X_{ij} = x_{ij}; \ i = 1, 2; \ j = 1, 2, 3\} = \left(\prod_{ij} e^{-\lambda_{ij}} \lambda_{ij}^{x_{ij}} \right) \Big/ \prod_{ij} x_{ij}!.$$

In the second model, if N is the total number of accidents observed, then

$$P\left\{ X_{ij} = x_{ij}; \ i, j \ \Big/ \ \sum_i \sum_j X_{ij} = N \right\} = \frac{\left(\prod_{ij} e^{-\lambda_{ij}} \lambda_{ij}^{x_{ij}} \right) \Big/ \prod_{ij} x_{ij}!}{e^{-\Sigma\Sigma\lambda_{ij}} \left(\Sigma\Sigma\lambda_{ij} \right)^N \Big/ N!}.$$

$$(6.5.1)$$

[Here, in evaluating the conditional probability, we have tacitly used the property that the sum of independent Poisson variates is a Poisson variate with the parameter as the sum of the component parameters (Ross (1985, p. 65).]

Equation (6.5.1) simplifies to the multinomial distribution

$$P\left\{ X_{ij} = x_{ij};\ i, j \middle/ \sum_i \sum_j X_{ij} = N \right\} = \frac{N!}{\prod_{ij} x_{ij}!} \prod_{ij} p_{ij}^{x_{ij}}. \qquad (6.5.2)$$

(Here $p_{ij} = \frac{\lambda_{ij}}{\Sigma\Sigma\lambda_{k\ell}}$.) One of the basic problems of statistical surveys is to propose estimates for the unknown parameters based on the data and to test the validity of the estimates. In the example above, one might like to estimate either the expected number λ_{ij} of accidents in a cell (i, j) or the chance p_{ij} for an accident to belong to the category (i, j), such as rollover, moderately severe, etc. A systematic approach to proposing estimates of parameters can be based on the principle of maximum likelihood. Among potential values for the parameters, the principle of maximum likelihood selects as estimates that set of allowable values that makes the given sample the most probable among all samples. Formally, if $f(x, \theta)$ is the joint density at x and we fix x (given the data), the function $L(\theta/x) = f(x; \theta)$ viewed as a function of θ is maximized by choosing a suitable value $\widehat{\theta}(x)$ for θ. When such a $\widehat{\theta}(x)$ exists, then the likelihood function $L(\theta/x) \leq L(\widehat{\theta}/x)$ for all allowable θ. In general, finding maximum likelihood estimates can be quite involved, especially when the parameter is restricted to certain subsets of Euclidean spaces called the parameter space. A certain notion of data reduction can be first implemented before searching for maximum likelihood estimates. This notion of data reduction, called the sufficiency principle, can be intuitively described as follows: Suppose the joint density function $f(x; \theta) = g(t_1(x), t_2(x), \dots t_k(x); \theta)h(x)$ for some real functions $t_1, t_2, \dots, t_k$. Then any information about θ that is contained in the original data x can also be recovered from the k functions $t_1, t_2, \dots, t_k$ instead of x. Notice that the second function h is independent of the true value of θ. We call $\{t_1, t_2, \dots, t_k\}$ *jointly sufficient* for θ.

We can illustrate these concepts with reference to our second accident model with an added assumption: The type of accident has nothing to do with the severity of the accident. Thus $p_{ij} = \alpha_i \beta_j$, where α_i is the probability for an accident to be of type i and β_j is the probability for an accident to have severity j; $i = 1, 2$; $j = 1, 2, 3$. Notice that

$$P\{X_{ij} = x_{ij};\ i, j\} = \frac{N!}{\prod_{i,j} x_{ij}!} \prod_{i,j} p_{ij}^{x_{ij}}$$

$$= \frac{N!}{\prod_{i,j} x_{ij}!} \prod_{i,j} (\alpha_i \beta_j)^{x_{ij}}$$

$$= \frac{N!}{\prod_{i,j} x_{ij}!} \prod_i \alpha_i^{x_{i.}} \prod_j \beta_j^{x_{.j}}.$$

(Here $x_{i.} = \sum_j x_{ij}$; $x_{.j} = \sum_i x_{ij}$.) Thus

$$P\{X_{ij} = x_{ij}/\alpha_i, \beta_j; i, j\} = h(x_{ij}; i, j)g(x_{i.}, x_{.j}, \alpha_i, \beta_j; i, j).$$

Hence $\{x_{i.}, x_{.j}; i, j\}$ are jointly sufficient for the unknown parameters $\{\alpha_i, \beta_j; i, j\}$.

By elementary calculus, the maximum likelihood estimates are functions of $x_{i.}$ and $x_{.j}$ and are given by $\widehat{\alpha_i} = \frac{x_{i.}}{N}$ and $\widehat{\beta_j} = \frac{x_{.j}}{N}$. Thus in our example the data propose $\widehat{\beta_2} = (765 + 198)/4000$. Incidentally, these are also the intuitive estimates.

In general, maximum likelihood estimates are difficult to determine in higher dimensional contingency tables. The following is a classic example by Bartlett (1935).

Example 6.2. An experiment was conducted to check on the propagation of root-stocks from root cuttings when plants were subjected to two type of treatments. Treatment 1 has to do with the time of planting. Treatment 2 has to do with the length of cutting. The following data summarize experimental observations:

	Planting Season			
	Rootstock (alive)		Rootstock (dead)	
	Length of cutting		Length of cutting	
	Long	Short	Long	Short
At once	156	107	84	133
Spring	84	31	156	209

The appropriate model for this $2 \times 2 \times 2$ table is to treat the status of root stock as a response variable with season and length of cutting as explanatory variables. We can identify the above data as the $2 \times 2 \times 2$ table

$$\begin{bmatrix} x_{111} & x_{112} \\ x_{121} & x_{122} \end{bmatrix} \quad \begin{bmatrix} x_{211} & x_{212} \\ x_{221} & x_{222} \end{bmatrix}.$$

The problem is to estimate the expected cell entries under the assumption that "there is no second-order interaction," which translates to

$$p_{ijk} = \alpha_{ij} \beta_{jk} \gamma_{ik} \quad \text{for all } i, j, k$$

for some parameters $\alpha_{ij}, \beta_{jk}, \gamma_{ik}$.

The multinomial model shows that

$$P\{X_{ijk} = x_{ijk};\ i, j, k\} = h(x_{ijk};\ i, j, k\} \prod_{ijk} \alpha_{ij}^{x_{ijk}} \prod_{ijk} \beta_{jk}^{x_{ijk}} \prod_{ijk} \gamma_{ik}^{x_{ijk}}$$

$$= h(x_{ijk};\ i, j, k) \prod_{ij} \alpha_{ij}^{x_{ij\cdot}} \prod_{jk} \beta_{jk}^{x_{\cdot jk}} \prod_{ik} \gamma_{ik}^{x_{i\cdot k}}$$

for some suitable function h of the data. (Here $x_{ij\cdot} = \sum_k x_{ijk}$, etc.) Unlike the previous example, here the maximum likelihood estimates are not easy to find even though we do know that $\{x_{ij\cdot}, x_{i\cdot k}, x_{\cdot jk};\ i, j, k\}$ are sufficient for $\{\alpha_{ij}, \beta_{jk}, \gamma_{ik};\ i, j, k\}$. Thus the problem is to devise a scheme for finding the maximum likelihood estimates. Even more important is knowing whether maximum likelihood estimates even exist or if they do, whether they are unique.

It is this problem of existence that is closely related to our matrix scaling problems. Before we take up this general issue we first consider a fundamental theorem due to Darroch and Ratcliff (1972).

Theorem 6.5.1. *Let a random vector of counts $\{x_j;\ j \in J\}$, $\mathrm{card}(J) = q$ have a probability distribution of the form*

$$p_j = \pi_j \mu \prod_{i=1}^{m} \mu_i^{a_{ij}}, \quad j \in J \tag{6.5.3}$$

where $\{\pi_j\}$, $\{a_{ij}\}$ are known and μ, $\{\mu_i\}$ are unknown parameters. Let $\{f_j : f_j > 0, j \in J\}$ be a random sample according to $p = \{p_j\}$, with $\sum_j f_j = N$. Let $\sum_i a_{ij} = 1$, $a_{ij} \geq 0$ for all i, j. Then the densities $\{p_j\}$ have unique maximum likelihood estimates $\{\widehat{p}_j\}$ of the type (6.5.3) satisfying the system of linear equations

$$\sum_j a_{ij} p_j = \sum_j a_{ij} \frac{f_j}{N} = h_i, \quad i = 1, 2, \ldots, m,$$

$$\sum_j p_j = 1, \quad \sum_j f_j = N. \tag{6.5.4}$$

The unique maximum likelihhood estimates $\{\widehat{p}_j\}$ are positive and are iteratively estimated by

$$p_j^{(0)} \equiv \frac{1}{N}; \quad h_i^{(0)} = \sum_j a_{ij} \left(\frac{f_j}{N}\right) = h_i$$

$$p_j^{(t+1)} = p_j^{(t)} \prod_{i=1}^{m} \left(\frac{h_i}{h_i^{(t)}}\right)^{a_{ij}}, \quad j \in J, \quad t = 0, 1, 2, \ldots.$$

Namely, $p_j^{(t)} \to \widehat{p}_j$, $j \in J$, the maximum likelihood estimates of $\{p_j\}$.

Proof. Let $\{p_j^{(t)}\}$ be defined as above. We claim that $\sum p_j^{(t)} \leq 1, t = 0, 1, 2, \ldots$
To see this, observe that by the arithmetic mean-geometric mean inequality
(Lemma 2.6.1),

$$
\begin{aligned}
\sum_j p_j^{(t+1)} &= \sum_j p_j^{(t)} \prod_i \left(\frac{h_i}{h_i^{(t)}} \right)^{a_{ij}} \\
&\leq \sum_j p_j^{(t)} \sum_i a_{ij} \left(\frac{h_i}{h_i^{(t)}} \right) \\
&\leq \sum_i \left(\frac{h_i}{h_i^{(t)}} \right) \sum_j a_{ij} p_j^{(t)} \\
&\leq \sum_i \frac{h_i}{h_i^{(t)}} h_i^{(t)} \\
&= 1.
\end{aligned}
$$

Since $\sum_j p_j^{(t)} \leq 1$ for all t, $\sum_i h_i^{(t)} \leq 1$. Further, for all t and j, $p_j^{(t)} > 0$ and
therefore $h_i^{(t)} > 0$. Thus by the Information Inequality (Lemma 2.6.2),

$$
\sum_{i=1}^m h_i \log \frac{h_i}{h_i^{(t)}} \geq 0.
$$

We claim that $\sum_{i=1}^m h_i \log \frac{h_i}{h_i^{(t)}} \to 0$ as $t \to \infty$. To see this let $q > 0$ be any
probability vector satisfying (6.5.4). Then

$$
\begin{aligned}
\sum_j q_j \log \frac{q_j}{p_j^{(t+1)}} &= \sum_j q_j \log \frac{q_j}{p_j^{(t)}} - \sum_j q_j \log \prod_{i=1}^m \left(\frac{h_i}{h_i^{(t)}} \right)^{a_{ij}} \\
&= \sum_j q_j \log \frac{q_j}{p_j^{(t)}} - \sum_i \log \frac{h_i}{h_i^{(t)}} \sum_j a_{ij} q_j \\
&= \sum_j q_j \log \frac{q_j}{p_j^{(t)}} - \sum_i h_i \log \frac{h_i}{h_i^{(t)}} \qquad (6.5.5) \\
&\leq \sum_j q_j \log \frac{q_j}{p_j^{(t)}} \qquad \text{(since the second term} \geq 0).
\end{aligned}
$$

Thus the sequence $\sum_j q_j \log \frac{q_j}{p_j^{(t)}}$, $t = 0, 1, 2, \ldots$ is a decreasing sequence,
bounded below by zero. From (6.5.5) it follows that $\sum h_i \log \frac{h_i}{h_i^{(t)}} \to 0$. We
conclude (see Exercise 6.5). that

$$
\sum_{i=1}^m h_i \log \frac{h_i}{h_i^{(t)}} \geq \frac{1}{2} \sum_{i=1}^m h_i (h_i - h_i^{(t)})^2 \to 0.
$$

Since $h_i > 0$ for all i, $h_i - h_i^{(t)} \to 0$ for all i.
   This shows that the bounded sequence $p^t = \{p_j^{(t)}\}$ has a limit point. Let, if
possible, p' and p'' be two limit points. Both p' and p'' are > 0, for otherwise,

$\sum q_j \log \frac{q_j}{p_j'} \to \infty$, contradicting the assertion that it converges. We claim that there is at most one solution satisfying (6.5.3) and (6.5.4). Suppose the claim is false. Then for any $q = \{q_j\} > 0$, $K(q, \pi) = \sum_j q_j \log \frac{p_j}{\pi_j}$ (where $0 \log 0 = 0$ by definition) satisfies

$$K(p', \pi) = \sum p_j' \left[\log \mu + \sum_i a_{ij} \log \mu_i \right]$$

$$= \log \mu \sum p_j'' + \sum \log \mu_i \sum a_{ij} p_j'$$

$$= \log \mu \sum p_j'' + \sum \log \mu_i \sum a_{ij} p_j''$$

$$= \sum p_j'' \log \frac{p_j'}{\pi_j}.$$

Hence

$$K(p'', \pi) - K(p'\pi) = K(p'', p') \geq 0. \tag{6.5.6}$$

Since $K(p, \pi) \geq 0$, equality holds in (6.5.6) only when $p'' = p'$.

It still remains to prove that the limits $\widehat{p}_k = \lim p_j^{(t)}$ are the maximum likelihood estimates of $\{p_j\}$. To see this consider the log-likelihood $\sum_i f_i \log p_i$ based on the data $\{f_j\}$. Let $\{\widehat{p}_j\}$ be the iterative limit of $\{p_j^{(t)}\}$. Consider

$$\sum f_j \log \widehat{p}_j - \sum f_j \log p_j = \sum_j f_j \cdot \log \frac{\widehat{p}_j}{p_j}$$

$$= N \sum_j \frac{f_j}{N} \cdot \left[\log \frac{\widehat{\mu}}{\mu} + \sum_i a_{ij} \log \frac{\widehat{\mu}_i}{\mu_i} \right]$$

$$= N \sum_j \widehat{p}_j \left[\log \frac{\widehat{\mu}}{\mu} + \sum a_{ij} \log \frac{\widehat{\mu}_j}{\mu_i} \right]$$

$$= N \sum \widehat{p}_j \log \frac{\widehat{p}_j}{p_j} \geq 0 \quad \text{(by Lemma 2.6.2).}$$

Equality holds only if $\widehat{p} = \{\widehat{p}_j\} = p = \{p_j\}$. Hence the $\widehat{p}$ obtained by iterative scaling maximizes the log-likelihood function and thus $\widehat{p}$ is the maximum likelihood estimate of p. ∎

It is not essential that all the equations in (6.5.4) are satisfied by the solutions $p^{(t)}$ in each iteration, only that this occurs in the limit. A sufficient condition is to demand that each equation in (6.5.4) be valid in finitely many iterations [see Darroch and Ratcliff (1972), Corollary 2] in a cyclic fashion. For example, the validity of (6.5.4) for $i = k$ may hold for $p^{(t)}$, where t is of the form $t \equiv (k - 1) \bmod m, k = 1, 2, \ldots, m$.

We now apply the above iterative scaling theorem to find the maximum likelihood estimates for the expected cell frequencies in Example 6.2. Under the assumption that there is no second-order interaction this translates to $\lambda_{ijk} = \alpha_{ij}\beta_{jk}\gamma_{ik}$ for some parameters α_{ij}, β_{jk}, γ_{ik}, where $\lambda_{ijk} = E(X_{ijk})$, $i, j, k = 1, 2$. Let us start with the initial estimates $\lambda_{ijk}^{(0)} \equiv 1$ for all i, j, k. Note that $\{x_{ij\cdot}, x_{i\cdot k}, x_{\cdot jk}; i, j, k\}$ are jointly sufficient for $\{\alpha_{ij}, \beta_{jk}, \gamma_{ik}; i, j, k\}$.

Before we initiate the algorithm we state the following result. The proof may be found in Theorem 3.1 of Haberman (1974) and is omitted.

Theorem 6.5.2. *Let an $m \times n \times \ell$ contingency table be derived from a multinomial trial of size N. Let the parameter space $\mu_{ijk} = \log p_{ijk}$ be of the type $\mu_{ijk} = \xi_{ij} + \eta_{jk} + \zeta_{ik}$ for all i, j, k. Given the data $X = x = (x_{ijk})$ the maximum likelihood estimate $\widehat{p} = (\widehat{p}_{ijk})$ must satisfy $P_M \widehat{p} = P_M x$, where P_M is the projection of $R^{mn\ell}$ onto the linear manifold generated by elements of the type $(\mu_{ijk}) = \xi_{ij} + \eta_{jk} + \zeta_{ik}$ for some $\xi_{ij}, \eta_{jk}, \zeta_{ik}; i, j, k$.*

Equivalently, in our $2 \times 2 \times 2$ example, if the parameter space is generated by parameters satisfying no second-order interactions, we have, for the maximum likelihood estimates,

$$\sum_k \widehat{p}_{ijk} = \frac{x_{ij\cdot}}{N}$$

$$\sum_i \widehat{p}_{ijk} = \frac{x_{\cdot jk}}{N}$$

$$\sum_j \widehat{p}_{ijk} = \frac{x_{i\cdot k}}{N}.$$

These are precisely the equations $P_M \widehat{p} = P_M x$. Incidentally, these equations correspond to (6.5.4) in Theorem 6.5.1.

Multiplying by N, we can write the iterative equations for the expected cell count estimates $\widehat{\lambda}_{ijk}$ as follows:

$$\lambda_{ijk}^{(3v)} = \frac{\widehat{\lambda}_{ijk}^{(3v-1)}}{\lambda_{\cdot jk}^{(3v-1)}} x_{\cdot jk}$$

$$\lambda_{ijk}^{(3v-1)} = \frac{\widehat{\lambda}_{ijk}^{(3v-2)}}{\lambda_{i\cdot k}^{(3v-2)}} x_{i\cdot k}$$

$$\lambda_{ijk}^{(3v-2)} = \frac{\widehat{\lambda}_{ijk}^{(3v-3)}}{\lambda_{ij\cdot}^{(3v-3)}} x_{ij\cdot}$$

$$\lambda_{ijk}^{(0)} = 1.$$

Theorem 6.5.1 can be modified to prove that the above cyclic iterative sequence also converges to the maximum likelihood estimates $\widehat{\lambda}_{ijk}$. Furthermore, since for subsequences the sums $\sum_i \lambda_{ijk}^{(3v)} = x_{.jk}$, then in the limit $\widehat{\lambda}_{.jk} = x_{.jk}$. Similarly, $\widehat{\lambda}_{i\cdot k} = x_{i\cdot k}$ and $\widehat{\lambda}_{ij\cdot} = x_{ij\cdot}$ are satisfied. Thus $\widehat{\lambda}$ is the unique maximum likelihood solution to which the above cyclic sequence converges. We now apply the iterative scheme to solve for maximum likelihood estimates for the expected cell entries in Bartlett's example:

cell =	(111)	(211)	(121)	(221)	(112)	(212)	(122)	(222)
$x_{ijk} =$	156	84	84	156	107	133	31	209
$\lambda_{ijk}^{(0)} =$	1	1	1	1	1	1	1	1

$$(x_{11\cdot}\ x_{12\cdot}\ x_{21\cdot}\ x_{22\cdot}) = (263,\ 217,\ 115,\ 365)$$

$$(\lambda_{11\cdot}^{(0)}\ \lambda_{12\cdot}^{(0)}\ \lambda_{21\cdot}^{(0)}\ \lambda_{22\cdot}^{(0)}) = (2,\ 2,\ 2,\ 2)$$

$$\lambda_{ijk}^{(1)} = (131.5\ 108.5\ 57.5\ 182.5\ 131.5\ 108.5\ 57.5\ 182.5)$$

$$(x_{1\cdot 1}\ x_{1\cdot 2}\ x_{2\cdot 1}\ x_{2\cdot 2}) = (240\ 138\ 240\ 342)$$

$$\lambda_{ijk}^{(2)} = (166.98\ 89.48\ 73.02\ 150.52\ 96.02\ 127.52\ 41.98\ 214.48)$$

$$(x_{\cdot 11}\ x_{\cdot 12}\ x_{\cdot 21}x_{\cdot 22}) = (240\ 240\ 240\ 240)$$

$$\lambda_{ijk}^{(3)} = (156.26\ 83.74\ 78.40\ 161.60\ 103.09\ 136.91\ 39.29\ 200.71).$$

Note that the maximum likelihood estimate differs from the cell counts in absolute value by the same unknown amount θ (if $\widehat{\lambda}_{111} = x_{111} + \theta$, then $\widehat{\lambda}_{111} + \widehat{\lambda}_{112} = x_{111} + x_{112} \Rightarrow \widehat{\lambda}_{112} = x_{112} - \theta$). Thus $|\widehat{\lambda}_{ijk} - x_{ijk}| \equiv \theta$, and the maximum likelihood estimate must satisfy this common difference θ for all cells i, j, k. Since we had variations in these differences, we repeat with another cycle of 3 iterations. In the 6-th cycle we find

$$\widehat{\lambda}_{111}^{(6)} = 160.18 \quad \widehat{\lambda}_{121}^{(6)} = 79.01 \quad \widehat{\lambda}_{112}^{(6)} = 101.94 \quad \widehat{\lambda}_{122}^{(6)} = 39.71$$

$$\widehat{\lambda}_{211}^{(6)} = 79.82 \quad \widehat{\lambda}_{221}^{(6)} = 160.99 \quad \widehat{\lambda}_{212}^{(6)} = 138.06 \quad \widehat{\lambda}_{222}^{(6)} = 200.29.$$

The variation $\max_{ijk} |\widehat{\lambda}_{ijk} - x_{ijk}|$ is reduced. Indeed, the estimates

$$\widehat{\lambda}_{111} = 161.096 \quad \widehat{\lambda}_{121} = 78.904 \quad \widehat{\lambda}_{112} = 101.904 \quad \widehat{\lambda}_{122} = 36.096$$

$$\widehat{\lambda}_{211} = 78.904 \quad \widehat{\lambda}_{221} = 161.096 \quad \widehat{\lambda}_{212} = 138.096 \quad \widehat{\lambda}_{222} = 203.904$$

are closer. The maximal difference $|x_{ijk} - \widehat{\lambda}_{ijj}|$ is 5.096 and is the same throughout. Thus the assumption of no second-order interaction gives a good fit for the maximum likelihood estimates.

One of the crucial steps in the algorithm is the assumption that all cells are occupied in the sample. It is quite conceivable that in certain problems the chance that a certain cell is occupied is nil (imagine sex as one variable and getting pregnant as another variable). Thus, it is important to know whether we could find density functions of the type (6.5.3) when certain cell entries are zero in the sample. It is not clear whether maximum likelihood estimates would exist in such circumstances.

Though historically, it was Karl Pearson's work on contingency tables and his celebrated χ^2-test that led to the subject of mathematical statistics, for more than half a century, there were no significant theoretical developments in this area. As early as 1935, Bartlett pointed out the difficulties in finding the maximum likelihood estimates for cell probabilities in a $2 \times 2 \times 2$ contingency table with the assumption of no second-order interactions. He showed that the maximum likelihood estimates have no closed form expressions in terms of sufficient statistics and reduced the problem to finding the root of a nonlinear equation in one variable. In fact he did not clarify as to which root to choose when there were more real roots present; Mitra (1954) showed in his Ph.D. thesis that the smallest real root for Bartlett's nonlinear equation is the relevant one. The theoretical foundations of log-linear models are rooted in the existence and uniqueness of maximum likelihood estimates for exponential families [see Barndorff–Neilson (1978)].

For a detailed treatment of categorical data analysis, see Haberman (1974), Bishop, Fienberg, and Holland (1975), Fienberg (1977), Gokhale and Kullback (1978), and Agresti (1990).

Exercises

1. Find the scale factors that reduce the matrix

$$\begin{bmatrix} 5 & 8 & 6 \\ 9 & 4 & 1 \\ 18 & 3 & 7 \end{bmatrix}$$

to a doubly stochastic matrix. Find the scale factors that reduce the same matrix to one with row sums 16, 12, and 31 and column sums 29, 13, and 17.

2. Let $A > 0$ be an $n \times n$ matrix. For a probability vector y of order n, let $f(y) = z$, where $z_j = \frac{u_j}{\sum_k u_k}$ and

$$\frac{1}{u_j} = \sum_i \frac{a_{ij}}{\sum_k a_{ik} y_k}.$$

Apply Brouwer's Fixed Point Theorem (Theorem 1.2.1) and show the existence of positive scale factors $x_i, y_i, i = 1, 2 \ldots, n$ such that $(a_{ij} x_i y_j)$ is doubly stochastic. [See Menon (1968)]. Use a similar technique to show that if A is an $n \times n$ symmetric, positive matrix then there exist positive numbers $z_1, \ldots, z_n$ such that $(a_{ij} z_i z_j)$ is doubly stochastic.

3. Let $A > 0$ be an $n \times n$ matrix. Consider the problem

$$\text{minimize} \sum_i \sum_j x_{ij} (\log x_{ij} - \log a_{ij})$$

subject to $X = (x_{ij})$ being doubly stochastic. Use the Kuhn-Tucker Theorem [see, for example, Bazarra and Shetty (1979)] to deduce the existence of positive scale factors x_i, y_i, $i = 1, 2 \ldots, n$ such that $(a_{ij} x_i y_j)$ is doubly stochastic. [See Brégman (1973).]

4. Let $A > 0$ be an $n \times n$ matrix. Consider the problem: Maximize $\prod_i x_i y_i$ subject to $x_i \geq 0$, $y_i \geq 0$, $i = 1, 2, \ldots, n$; $\sum_i \sum_j a_{ij} x_i y_j = 1$. Use the Kuhn-Tucker Theorem to deduce the existence of positive scale factors x_i, y_i, $i = 1, 2 \ldots, n$ such that $(a_{ij} x_i y_j)$ is doubly stochastic. [See Marshall and Olkin (1968).]

5. Let $A > 0$ be an $n \times n$ matrix. Consider the problem: Minimize $\sum_i \sum_j a_{ij} x_i y_j$ subject to $x_i \geq 0$, $y_i \geq 0$, $i = 1, 2, \ldots, n$; $\prod_i x_i y_i = 1$. Use the Kuhn-Tucker Theorem to deduce the existence of positive scale factors x_i, y_i, $i = 1, 2 \ldots, n$ such that $(a_{ij} x_i y_j)$ is doubly stochastic.

6. Let $\mathcal{K}$ and $\mathcal{L}$ be intervals on the real line and let $f : \mathcal{K} \times \mathcal{L} \to (0, \infty)$ be a continuous function such that $0 < c \leq f(x, y) \leq C$ for all $x \in \mathcal{K}, y \in \mathcal{L}$. Show that there exist $h : \mathcal{K} \to [0, 1]$, $g : \mathcal{L} \to [0, 1]$ such that

$$\int_\mathcal{K} f(x, y) h(x) g(y) dy = 1 \quad \text{for almost all } x$$

and

$$\int_\mathcal{L} f(x, y) h(x) g(y) dx = 1 \quad \text{for almost all } y.$$

[See Hobby and Pyke (1965).]

7. An individual is handed a deck of cards marked $1, 2$, and 3 and asked to shuffle it several times. The ordering of the cards is noted after each shuffle and it resuls in the following sequence:

$$123, 213, 231, 321, 132, 312, 132, 123, 231, 213, 321, 312, 123,$$

$$132, 231, 132, 213, 312, 321, 231, 213, 312, 321, 321, 132, 312.$$

Summarize the data in the form of a 6×6 matrix, indicating how many times the ordering (or permutation) τ was obtained, after shuffling the deck arranged according to the ordering σ. Estimate the transition matrix, keeping in view that in this example, it is natural to assume that the transition matrix is doubly stochastic.

8. For the matrix in Exercise 1, carry out three iterations of deviation reduction, balancing, and deviation balancing algorithms and compare the results.

9. Let p and q be positive probability vectors of order n. Show that

$$\sum_{i=1}^{n} p_i \log \frac{p_i}{q_i} \geq \sum_{i=1}^{n} p_i (q_i - p_i)^2.$$

[See Rao (1973), p. 357.]

10. In a survey following the outbreak of food poisoning at a marriage party in Karol Bagh, New Delhi, the following data were obtained: A total of 1576 persons responded. Out of 879 persons who ate both shahi paneer and kofta curry, 575 fell ill; out of 235 who ate only shahi paneer, 37 fell ill; out of 371 who ate only kofta curry, 265 fell ill; and out of 51 who ate neither of these two items, 4 fell ill. Express these data in the form of a $2 \times 2 \times 2$ contingency table. Estimate the cell frequencies assuming no second-order interactions. [See Bishop, Feinberg, and Holland (1975, p. 90) for a similar example.]

7

Special matrices in economic models

Mathematical precision in describing models of macroeconomies became prominent after the advent of Leontief's monumental work on input-output analysis [Leontief (1941)]. In a totally independent setting, von Neumann's model of an expanding economy gave a new impetus to the mathematical approach to economic models [von Neumann (1937)]. Economists of earlier centuries were often too ambitious in their tasks of incorporating many complex economic issues into their models. When it came to the analysis of their models they resorted to many heuristic arguments. In contrast, modern mathematical economists believe in the analytical rigor of their arguments with no ambiguity in the final conclusions. However, they too have to pay a price for the same. Often, their drastically simplified mathematical models tend to avoid the serious economic issues, such as production and capital accumulation over several periods.

In the study of an economy, many of the variables such as prices, costs of production, rates of return, intensity of operations, etc. are clearly nonnegative. With the introduction of Leontief's input-output analysis, a good linear approximation to the functioning of an economy controlled by a few firms or the state has been achieved with great empirical success [see Miller and Blair (1985)]. The theory of nonnegative matrices and M-Matrices play an important role in the study of these models. In the problem of Walrasian equilibrium in linear economic models, the Duality Theorem of linear programming plays a key role; see, for example, Lancaster (1968). In generalizing Leontief models, it is possible to give up the notion of constant returns to scale and still study the problems of wages and rates of return on capital goods via the Perron-Frobenius Theorem. The elegant linear model for the growth of an economy proposed by von Neumann in the late 1930s [von Neumann (1937)] remained dormant till an English translation of the same appeared in the late 1940s [von Neumann (1945)]. Such models can be analyzed by game theoretic methods. When certain goods become more expensive, buyers tend to look for other goods that would substitute for the more expensive ones. Hicks first initiated the study of gross substitutes [Hicks (1939)] to study this phenomenon, and the theory is intertwined with properties of P-matrices and M-matrices. When can the prices of factors of production be uniquely determined by knowing the prices of goods that are produced using these factors of production? This has been a key problem in the study of international trade [see Flam and Flanders (1991), Heckscher (1919), Ohlin (1933), Samuelson (1948), Samuelson (1953)]. Here one is once again led to important univalence theorems based on the nature of equilibrium costs and equilibrium prices.

7.1. Pure exchange economy

A *pure exchange economy* is a primitive economy where citizens bilaterally barter goods they own, in exchange for goods others own, to maximize their personal satisfaction (utility). Even in such economies with no production process, bartering of goods can be smoothly carried out by introducing prices for the goods owned. Consider the following example [Pasinetti (1977)]:

Example 7.1. Tom, Dick, and Harry are the only citizens of an island. Tom has 1 ounce of coffee powder. Dick has 1 pint of milk. Harry has 1 cup of sugar. Hot water is freely available. Tom likes .05 oz of coffee powder with .10 pint of milk and no sugar in every cup of coffee he drinks. Dick wants .05 oz of coffee powder and .15 cup of sugar and no milk in his cup of coffee. Harry wants .10 oz of coffee powder, .10 pint of milk, and .05 cup of sugar in his cup of coffee.

Problem: Find a set of prices that facilitates bilateral exchange.

Each one wishes to sell at the given price what one holds and buy the right mix of the three items they need with *all* the money earned. Money per se has no value for them. Observe that each person can drink 5 cups of coffee of the type they like with the given resources. Here we are actually looking for a *nonnegative* solution to the system of equations

$$p_1 = 5(.05p_1 + .10p_2),$$
$$p_2 = 5(.05p_1 + .15p_3), \qquad (7.1.1)$$
$$p_3 = 5(.10p_1 + .10p_2 + .05p_3).$$

For example, $p_1 = 6$, $p_2 = 9$, $p_3 = 10$ is a solution.

Table 7.1. *Flow of commodities for the production of the commodities*

	Farming	Dairy	Gardening
Rice	3600	600	300
Milk	1000	400	600
Vegetables	3000	1500	500

Table 7.2. *Flow of commodities, with labor included*

	Farming	Dairy	Gardening	
Rice	2400	400	100	1600
Milk	400	300	500	800
Vegetables	1800	1300	300	1600
Labor	600	100	100	0

From such a primitive barter society we now move to a simple agricultural society, where we encounter some serious complications in the analysis.

Example 7.2. Pulavanoor is a small village in India whose 800 citizens are perfectly homogeneous in their tastes and skills. Their basic needs consist of just rice, milk, and vegetables. Farming, Dairy, and Gardening are the three distinct cottage industries with rice, milk, and vegetables as their distinct outputs. Each citizen is involved in exactly one of these activities. In the village 600 citizens work in the Farming activity, 100 in Dairy activity, and 100 are involved in Gardening. It is a self-sufficient society. Each year the total output is 4500 units of rice, 2000 units of milk, and 5000 units of vegetables. These are consumed in the following quantities: Each citizen needs for his or her living, 2 units of rice, 1 unit of milk, and 2 units of vegetables. (Pretend that our units are perfectly divisible.) The total requirements are given in Table 7.1.

For example, Farming activity consumes 3600 units of rice, 1000 units of milk, and 3000 units of vegetables. Suppose land is available in plenty at no cost. Viewing Labor as an activity with no output, we can modify the above matrix to a 4 × 4 matrix given by Table 7.2.

(Here Labor with its limited supply is called a *primary good* or *factor of pro-*

duction. Primary goods are never the outputs of any activity.)

For example, the 600 workers in Farming will consume 1200 units of rice, and the leftover 2400 units are needed for this activity. The last column simply represents the total consumption of goods by the labor force as an activity with no output (zero for last row, last column entry). Thus the consumption is deducted out of the gross output. As long as there is no overproduction of goods or variation in their tastes and needs, and as long as they maintain their constancy of demands with their homogeneous tastes, the villagers can achieve their goals with a set of prices for the goods. Once there is any overproduction of goods at one of the activities, the insatiability and mutual jealousy of human beings can take their strange tolls in competing for the excess. The situation can be handled effectively by immediately rewarding the *entire* overproduction to the workers for consumption or to the state for reinvestment for more production. A different situation would arise if the workers were entrepreneurs receiving no wages from their workplace, but rather being rewarded for the value of goods they add to the system. We are thus led to the general problems of linear economies of the above extreme types.

In a stationary situation, when there are no technological changes in the production processes or in the consumption patterns, we could generalize our village economy as follows: Let the economy consist of n industries $I_1, I_2, \ldots, I_n$ producing, respectively, *distinct* goods $G_1, G_2, \ldots, G_n$ in positive quantities. Let $Q = (q_{ij})$ be the input matrix, where q_{ij} is the amount of good G_i directly consumed by industry I_j. In a steady state, where total production matches total consumption, one gets

$$\sum_j q_{ij} = s_i,$$

where s_i is the total supply of the i-th good G_i. We can rewrite these equations as

$$\sum_j \frac{q_{ij}}{s_j} s_j = s_i; \quad i = 1, 2, \ldots, n. \tag{7.1.2}$$

If we interpret $a_{ij} = \frac{q_{ij}}{s_j}$ as the average input of the i-th good needed in the production of one unit of the j-th good, then the nonnegative matrix $A = (a_{ij})$ has 1 as an eigenvalue with $s = (s_1, s_2, \ldots, s_n)^T$ as an eigenvector. The entries a_{ij} are called the *coefficients of production*. When the coefficients of production are independent of the q_{ij}s and s_is, the economy is said to have *constant returns to scale*. Clearly, by the Perron-Frobenius Theorem, $A^T p = \lambda_0 p$ for some λ_0, where $\lambda_0 \geq 0$, $p \geq 0$, $p \neq 0$. However,

$$\langle \lambda_0 p, s \rangle = \langle A^T p, s \rangle = \langle p, As \rangle = \langle p, s \rangle.$$

Since $s > 0$, then $\lambda_0 = 1$. Thus we have

$$\sum_{i=1}^{n} a_{ij} p_i = p_j, \quad j = 1, 2, \ldots, n$$

(7.1.3)

$$\sum_{i=1}^{n} q_{ij} p_i = s_j p_j, \quad j = 1, 2, \ldots, n.$$

If we take p_i as the unit price of good G_i, then the left-hand side of Equations (7.1.3) refer to the per unit and total cost of production for industry I_j. The right-hand side of Equations (7.1.3) refer to unit income and the gross income from industry I_j, selling, respectively, one unit and s_j units of its goods at production cost.

7.2. Linear slave economies

Generalizing the farm economies, where there is no excess production, we can think of a slave economy where a totalitarian state controls an unlimited supply of workers by paying a negligible wage rate for their identical skills. The state can aim at achieving some well-defined gross output for a distant future using all the excess output of the previous period as input for the next period.

The matrix A, introduced in the last part of the previous section, will change if there are changes in the technology or changes in the worker's consumption pattern or in the work force. Till now we have kept them exogenous to the system. When the workers are kept as slaves, it is possible to produce more and store the excess for the future. In our village model, labor is the only primary good and distinct goods are produced by distinct activities. Such a model is called a *simple Leontief model of production*. Consider, again, Example 7.2. If the workers work with unit intensity, then at the end of the year there will be an excess production of 1600 units of rice, 800 units of milk, and 1600 units of vegetables. (These are the entries in the last column of the 4×4 matrix above.) Suppose the workers worked with varying intensities (intensities can be any positive quantity under constant returns to scale!) x_1 in Farming, x_2 in Dairy, and x_3 in Gardening, then the net output of rice, milk, and vegetables under a negligible wage will be, respectively,

$$x_1 - \frac{2400}{4500} x_1 - \frac{400}{4500} x_2 - \frac{100}{4500} x_3,$$

$$x_2 - \frac{400}{2000} x_1 - \frac{300}{2000} x_2 - \frac{500}{2000} x_3,$$

$$x_3 - \frac{1800}{5000} x_1 - \frac{1300}{5000} x_2 - \frac{300}{5000} x_3.$$

We can reinterpret our earlier examples as follows. Under constant returns to scale, $(I - A)x$ will be the net yearly output when A is the production coefficient matrix and x is the intensity vector. We call an economy a *Slave economy* if the wages are negligible. We call an economy *productive* if for some intensity vector $x \geq 0$, $(I - A)x > 0$. Since $I - A$ is a Z-matrix, productivity is the same as condition (i) in Theorem 1.5.2. The equivalent condition (vi) says that for any vector $y \geq 0$, there exists an $x \geq 0$ such that $(I - A)x = y$, $y \geq 0$. Thus any target is achievable with an appropriate intensity x in such an economy.

We remark that the notion of arbitrary levels of intensity is quite unnatural as it amounts to producing any amount of goods with suitable intensities for productive economies. When primary factors like land and labor are in limited supply, this is never possible. Breakdown of the system is inevitable with wages unrelated to production and intensity of activities.

In our example of the village economy, the citizens may act totally differently if they are given the option of determining their own wages. With labor as the only primary factor of production for all the industries in a steady state, the workers can wipe out any excess production among themselves as additional wages, if the goods they produce are perishable, or else they might even think of storing the goods as savings of wages for future consumption.

Theorem 7.2.1. *Let there be n industries $I_1, I_2, \ldots, I_n$, producing perfectly divisible but distinct goods $G_1, G_2, \ldots, G_n$, respectively. Let A be the production coefficient matrix. Let G_0 be the limited supply of labor with input labor coefficient $a_{0j} > 0$ for all industries j. Let there be constant returns to scale. If the economy is productive, then there exists a unique price vector $p > 0$ such that the profit of operation for each industry is zero against a fixed wage rate.*

Proof. Suppose the wage rate is unity. Then the per unit cost of production for industry j at price p_j, $j = 1, 2 \ldots, n$, satisfies

$$p_j = \sum_{i=1}^{n} a_{ij} p_i + a_{0j}, \quad j = 1, 2, \ldots, n. \tag{7.2.1}$$

In matrix terms, if the net profit to industries is zero, this amounts to

$$(I - A)^T p = a_0,$$

where $a_0 = (a_{01}, a_{02}, \ldots, a_{0n})^T$. Our assumption that A is productive is equivalent to the assertion that $I - A$ as a Z-matrix satisfies condition (i) in Theorem 1.5.2. The equivalent condition (vi) in the theorem asserts that $p = (I - A^T)^{-1} a_0 > 0$ when $a_0 > 0$. Further, p is unique. Thus at such a price p, workers wipe out any profit of operation for the industries. ∎

7.3. Substitution Theorem

Theorem 7.3.1. *Let A be a productive Leontief system with just one primary factor of production. The system allows no activity to produce more than one good but allows more than one activity to produce the same good. Then given any feasible output by the system, it is possible to achieve the same level of output by an appropriate simple Leontief subsystem. The activities of this subsystem can be chosen to be the same for all feasible outputs.*

Proof. Let A be an $m \times n$ input matrix ($n \geq m$) of a general Leontief system that allows no joint production but permits more than one activity to produce the same good. Let labor be the only primary factor of production. Let the columns represent activities and let the rows represent goods. Let the total supply of labor be unity with labor input coefficients a_{0j}, $j = 1, 2, \ldots, n$. Let x_j be the intensity of activity j. Given a good G_i, let S_i be the set of activities (columns) producing good G_i. Since the economy is productive, for some outputs $\bar{u} > 0$, it is possible to attain these outputs with the given supply of labor. Let $H = (h_{ij})$ be an $m \times n$ matrix with $h_{ij} = 1$ if activity j produces good G_i and $h_{ij} = 0$ otherwise. Achieving the targeted output with the least effort amounts to

$$\min \sum_{j=1}^{n} a_{0j} x_j$$

such that

$$\sum_j (h_{ij} - a_{ij}) x_j = \bar{u}_i > 0, \quad i = 1, \ldots, m$$

$$x_j \geq 0, \quad j = 1, \ldots, n. \tag{7.3.1}$$

By assumption, the problem has an optimal solution with optimal value at most 1.

By choosing suitable columns $j_1, \ldots, j_m$ we have a basic optimal solution x^* to the above linear program. We will show that any potential output u of the general Leontief system is achievable by operating *just* the activities $j_1, \ldots, j_m$ and closing the rest. Let $\bar{A}$ be the $m \times m$ submatrix consisting of the columns $j_1, \ldots, j_m$ of A. Let $\bar{x}$ be the restriction of the basic optimal solution to the columns $j_1, \ldots, j_m$. Since there is no joint production, $(I - \bar{A})\bar{x} = \bar{u} > 0$. By Theorem 1.5.2, for the given $u \geq 0$, $(I - \bar{A})y = u$ for some $y \geq 0$. We will show that the activities $j_1, \ldots, j_m$ with intensity y will be possible with the given labor supply. Consider the dual to the above linear program (7.3.1) with dual variables $\pi = (\pi_1, \pi_2, \ldots, \pi_m)$. By complementary slackness we have for any optimal $\bar{\pi}$ for the dual,

$$\sum_{i=1}^{m}(h_{ij} - a_{ij})\bar{\pi}_i = a_{0j} \quad \text{for} \quad j = j_1, \ldots, j_m.$$

Also,

$$\sum_{i=1}^{m}(h_{ij} - a_{ij})\bar{\pi}_i \leq a_{0j} \quad \text{for} \quad j \notin \{j_1, \ldots, j_m\}.$$

Let y^* be the n-vector obtained from y by inserting zero coordinates for $j \notin \{j_1, \ldots, j_m\}$. Now y^* is feasible for the linear program

$$\min \sum_{j=1}^{n} a_{0j}x_j$$

such that

$$\sum_{j=1}^{n}(h_{ij} - a_{ij})x_j = u_i \quad i = 1, \ldots, m \tag{7.3.2}$$

$$x_j \geq 0 \quad j = 1, \ldots, n.$$

We claim that y^* is optimal to this problem. Consider the dual optimal $\bar{\pi}$ for (7.3.1). Since there is no joint production, $\bar{x}_j > 0$ for $j = j_1, \ldots, j_m$. Thus $\sum_i(h_{ij} - a_{ij})\bar{\pi}_i = a_{0j}$ for $j = j_1, \ldots, j_m$. If $\sum_i(h_{ij} - a_{ij})\bar{\pi}_i < a_{0j}$, then $j \notin \{j_1, \ldots, j_m\}$. Also, we have $y_j^* = 0$, $j \notin \{j_1, \ldots, j_m\}$. Thus the same $\bar{\pi}$ is feasible to the dual of (7.3.1) and satisfies complementary slackness with y^*. It follows that y^* is optimal to (7.3.2). Because u is a potential output of the general Leontief system, the optimal objective value is at most unity; thus the above intensity vector restricted to the simple Leontief system using only activities $j_1, \ldots, j_m$ achieves the same output and is even efficient. ∎

One of the serious limitations of the Leontief system is the assumption that the technical input coefficients are constants. In general the returns to scale are not constants. For many industries, decreasing returns or increasing returns to scale prevail. Often in such cases the coefficients are functions of production. A more serious limitation is often effected by changes in technologies that make the predictive aspects of the system meaningless. This section is based on Lancaster (1968).

7.4. Sraffa system

Sraffa (1963) proposed yet another system, called the *production of commodities by means of commodities*. The following is a brief description of this system:

(1) There are n industries producing n perfectly divisible distinct goods G_1, $G_2, \ldots, G_n$.

(2) Labor as a *primary good* is needed for the production of each good.
(3) The goods are completely used up by laborers each year. After a year the initial conditions of goods and services remain the same.
(4) The technology is constant and is defined by an input matrix $(A, a_0)^T$ of order $(n + 1) \times n$.
(5) The input matrix coefficients do not necessarily mean constant returns to scale. If q_{ij} is the input of the i-th commodity to the j-th industry producing s_j units, we simply define $a_{ij} = \frac{q_{ij}}{s_j}$, and that is all we say about them.

In Theorem 7.2.1 on simple linear models of production it was seen that with labor as the only primary factor of production, a suitable price wipes out any profit of operation, transferring anything and everything to the workers as wages. Sraffa's main thrust is to further decompose the net products of an economy into two distinct components called (i) wages and (ii) profits. Wages are directly related to the quantity of labor involved. Profits are distributed in proportion to the value of the means of production. The intuitive reasoning for the further refinement is the following.

Even though we could resolve the problem of associating prices to commodities including a wage for the labor where demand matches supply, as soon as there is an overproduction of some commodity, there will be a natural tendency to fight for some share of the excess. Thus there needs to be a distribution mechanism for distributing the surplus among the workers. An intuitive criterion is to assume that all the industries have the same rate of profit.

Otherwise, with reference to the example discussed in Section 7.2, if rice production in a year reaches 6000 units, then the workers in that industry, having supplied the regular requirement of 4500 units, would like to have a higher share of the excess, claiming that their efforts produced the surplus. This tempts workers in other industries, like Dairy, to compete with the farmers for the excesses. In a stationary state such a behavior cannot occur. If the rate of profit in each industry is the same, then the temptation to move from one industry to the other will not be there. Thus we need a fixed rate of profit α, wage rate p_0, and unit prices p_1, p_2, and p_3 for rice, milk, and vegetables. If the surplus is allotted before the prices are determined, this may not be consistent with the economic principle that the rate of profit is the same in all the industries. On the other hand, prices cannot be determined without knowing the rate of profit.

We begin by considering the following system of price equations assuming production and technology as stationary. Let

$$p_j = \sum_{i=1}^{n}(1 + \alpha)a_{ij} p_i + a_{0j}w, \quad j = 1, 2, \ldots, n. \tag{7.4.1}$$

Here α is the common rate of profit across activities and w is the wage rate. Citizens are simultaneously workers and entrepreneurs of the system. Thus we have a system of n equations in $n + 2$ unknowns $p_1, p_2, \ldots, p_n, \alpha$, and w. We have no unique solution for such a system. In the Leontief system we found a unique set of price ratios by imposing the condition that all industries make zero profit in monetary terms. Once we fix the wage rate as unity the price ratios are uniquely determined. This is one extreme. The other extreme is to demand that the rate of profit is a maximum with no restriction on wages. In such a system, wages, for all practical purpose, constitute an insignificant part of the system. There is a reward only for the management skill that maximizes the rate of profit. Workers are treated as entrepreneurs and the physical labor they put in has zero reward. This does not mean that workers are paid nothing. Entrepreneurs would like to take the full pie except for subsistence wages, and their aim will be to cluster around industries with the maximum rate of profit. At a stationary state, in such a system, all industries will have the same rate of profit. At this extreme entrepreneurial society with zero wages we will have

$$(1 + \alpha) \sum_{i=1}^{n} a_{ij} p_i = p_j, \quad j = 1, 2, \ldots, n. \tag{7.4.2}$$

Equivalently,

$$(I - (1 + \alpha)A)^T p = 0.$$

Let

$$\lambda = \frac{1}{1 + \alpha} > 0.$$

We obtain

$$(\lambda I - A)^T p = 0. \tag{7.4.3}$$

We are thus looking for a price system $p^* \geq 0$, $p^* \neq 0$ such that $A^T p^* = \lambda^* p^*$ for some $\lambda^* > 0$. The only natural object is the Perron eigenvector corresponding to the Perron eigenvalue λ^*. For example, when there is indirect dependence of inputs between any pair of industries, the matrix A is irreducible and then p^* is unique to within a scalar multiple. So is the maximum rate of profit α defined by $\alpha = \frac{1 - \lambda^*}{\lambda^*}$. Since

$$\sum_{j=1}^{n} \frac{q_{ij}}{s_j} < 1$$

for some i, the maximal eigenvalue of the nonnegative matrix A is less than one. Thus, for the rate of profit, the condition $0 \leq \lambda^* < 1$ is satisfied.

Whereas the earlier Leontief model proposes a pure labor theory of value, the price system p^* proposes a pure capital theory of value with workers treated on par with any other commodity. We could call the simple linear model of Leontief, the *Marxian system*, and the one by Sraffa maximizing the rate of return in all industries with zero wage, the *Ricardean system*. While the Marxian version demands $p = (I - A^T)^{-1}a_0$, the Ricardean version demands $(\lambda I - A)^T p = 0$. The two theories induce different sets of prices, which are rarely equal. When they are equal for a price $p = v$, then $v = (I - A^T)^{-1}a_0$ and $(I - (1 + \alpha)A^T)v = 0$. Since $v \geq 0$, $1 + \alpha > 0$, v is the Perron eigenvector of A^T and thus $a_0 = (I - A^T)v = v - \lambda^* v = \frac{\alpha}{1+\alpha}v$. Thus the ratio of capital to labor is uniform across industries.

Though the Leontief and the Sraffa systems look alike, their purposes and motivations are quite different. Sraffa is concerned with changes in the distributive shares and their resultant effects on prices; Leontief is concerned with the problem of achieving target outputs over a period of time, keeping wages low and fixed.

7.5. Dual Sraffa system on quantities

In our discussion so far we have considered a system of prices for goods with wages for workers satisfying certain linear equations. We get another new economic interpretation for the Sraffa system through the Perron-Frobenius Theorem when we consider a dual system on quantities of productive goods.

Let $s_1, s_2, \ldots, s_n$ be the gross output of goods, and let $y_1, y_2, \ldots, y_n$ be the excess after the consumption by the industries for productive purposes. Thus the system of equations

$$\sum_j a_{ij}s_j + y_i = s_i, \quad i = 1, 2, \ldots, n \qquad (7.5.1)$$

has two interpretations. In a Leontief system this can be taken as the excess production y_i that remains after consuming $s_i - \sum_j a_{ij}s_j$ units of the i-th good. A second interpretation of these equations is to assume that at a stationary state, the y_is constitute an allocation of goods in the system for total consumption by workers. Thus the surpluses are rewards to the workers. If

$$r_i = \frac{y_i}{s_i - y_i}, \quad i = 1, 2, \ldots, n,$$

then r_i measures the rate of surplus for industry i. Thus our system of equations (7.5.1) can be rewritten as

$$(1 + r_i) \sum_j a_{ij}s_j = s_i, \quad i = 1, 2, \ldots, n. \qquad (7.5.2)$$

In a Leontief system with slave labor we are looking for production rates that would result in specified excesses $y_1, \ldots, y_n$. In Sraffa's dual system of goods we are looking for specified rates of returns $r_1, \ldots, r_n$. Of course we want y_i, r_i nonnegative. Thus we get the system of equations

$$\sum_j a_{ij}s_j = \theta_i s_i, \quad i = 1, 2, \ldots, n. \tag{7.5.3}$$

(Here $\theta_i = \frac{1}{1+r_i}$. Our problem in this system is to look for s_is for given θ_is.)
The matrix

$$\begin{bmatrix} a_{11} - \theta_1 & a_{12} & \cdots & a_{1n} \\ a_{21} & a_{22} - \theta_2 & \cdots & a_{2n} \\ \vdots & \vdots & \ddots & \vdots \\ a_{n1} & a_{n2} & \cdots & a_{nn} - \theta_n \end{bmatrix}$$

has to be singular if we are looking for a nontrivial solution vector s to the system. Also, one has to guarantee a nonnegative solution s to the system for meaningful interpretations. Thus we may have to further restrict the freedom of choice among r_is in the hope of finding a nonnegative solution. Suppose we assume that all the r_is are equal to a fixed value r. Then we get $\theta_i = \theta = \frac{1}{1+r}$. We can now look for a nonnegative solution to the equations

$$\sum_j a_{ij}s_j = \theta s_i, \quad i = 1, 2, \ldots, n. \tag{7.5.4}$$

For example, if A is irreducible we have a unique θ, which is the Perron root corresponding to the essentially unique vector s. Clearly, θ has to be the same as the λ^* considered earlier with reference to the price system. Thus the Perron root has two different economic interpretations, one with reference to A and another with reference to A^T, namely, the maximum rate of profit equals the uniform rate of surplus. To determine the scale of operation we can now impose the labor constraint $\sum_j a_{0j}s_j = s_0$.

When A is irreducible we have the uniform rate of surplus $r = \frac{y_i}{s_i - y_i}$ with $y_i > 0, s_i > 0$. Also, $r(s_i - y_i) = y_i$ for all i. Thus the vectors $s^T = (s_1, s_2, \ldots, s_n)$, $y^T = (y_1, y_2, \ldots, y_n)$ are also Perron vectors. This has the following economic interpretation:

In a Sraffa system with fixed labor force, all the industries operate to give a maximum but identical rate of return. Thus at a steady state the relative rates of gross products, industrial consumption of the products, and workers' consumption of goods remain the same. Mathematically speaking, $\frac{s_i}{s_j} = \frac{y_i}{y_j} = \frac{s_i - y_i}{s_j - y_j}$ for any pair i, j. Such a system of outputs is called a *standard system*.

Sraffa's price system and physical quantity system are quite autonomous. Free labor does not necessarily determine a standard system nor a standard system of prices correspond to free labor. We would like to close this section with the following passage from Sraffa's preface to his monograph "Production of commodities by means of commodities" [Sraffa (1963)]. One may

"Suppose that the argument rests on a tacit assumption of constant returns to scale in all industries. If such a supposition is found helpful, there is no harm in adopting it as a temporary working hypothesis. However no such assumption is made. No changes in the proportions in which different means of production are used by an industry are considered, so that no question arises as to the variation or constancy of returns. The investigator is concerned exclusively with such properties of an economic system as do not depend on changes in the scale of production or in the proportions of factors".

7.6. A linear model of an expanding economy

In this section we consider the classical model of a competitive economy due to von Neumann (1937). The model is a dynamic model and contains some features of the economy not present in the more conventional Walrasian approach. In this model, prices and the rate of interest depend solely on the supply conditions and not on factors like tastes of consumers. The model can be intuitively described as follows:

Consider an economy with m activities and n goods. Each activity for its operation uses only the n goods as inputs and produces some of these n goods as outputs after a unit time of operation. The system is *closed* in the sense that there is no flow of goods to or from the model. Goods consumed at any later period must have been produced in an earlier period of time. There is no special role played by labor. By increasing the workload and by rewarding workers with consumer goods, labor is treated on par with any other good in the system. The net product at the end of each period is plowed back into the system to accumulate. Thus the growth rate of the system coincides with the rate of profit.

Let $I_1, I_2, \ldots, I_m$ be m industries (activities) producing goods $J_1, J_2, \ldots, J_n$, where joint production is allowed. We keep track of inputs and outputs in each period as two separate vectors for each industry. Let $(a_{i1}, a_{i2}, \ldots, a_{in})$ be the input of goods to the i-th industry, and let $(b_{i1}, b_{i2}, \ldots, b_{in})$ be the output of goods from the i-th industry when the activities are operating at unit intensity. We assume that constant returns to scale prevail so that if x is the intensity vector, for permissible intensities, $x^T A$ and $x^T B$ are the inputs and outputs of the goods. Assuming that no goods flow to or from the system, goods consumed are produced earlier. We plow back portions of savings for future production. We denote this model simply by (A, B).

We assume that the matrices A and B satisfy the following conditions:

(1) The matrix $B \geq 0$ has no column that is the null vector. (This means that each good is the output of some industry.)
(2) The matrix $A \geq 0$ has no row vector that is the null vector. (This means that at least one good is an input to each activity.)

We now describe two problems. Suppose the input activities are to be financed by borrowing at an interest rate $\beta - 1$, where \$1 borrowed results in \$$\beta$ payment in unit time. We would like to find a set of prices at which no industry makes any profit, even at the cheapest interest rate. Thus the problem is to

$$\text{minimize } \beta$$

such that

$$Bp \leq \beta Ap, \tag{7.6.1}$$

for some $p \geq 0$. The optimal β_0 is called the *economic expansion rate*; and the corresponding price vector is called the *optimal price vector*.

If an intensity vector x satisfies $x^T B \geq \alpha x^T A$, then we say that the model is *expanding* at a rate at least equal to α. Note that in this case the output of any good is at least α times its input. In the second problem we are in search of an intensity vector x such that the model is expanding at a rate α and α is the maximum possible. Thus the problem is to

$$\text{maximize } \alpha$$

such that

$$x^T (B - \alpha A) \geq 0, \tag{7.6.2}$$

for some $x \geq 0$. The optimal α_0 is called the *technological expansion rate*; the corresponding intensity x_0 is called an *optimal intensity vector*.

We remark that although only $\alpha_0, \beta_0 \geq 1$ are of economic interest (in which case the model is called an *expanding* model), we will not demand this restriction and thus the theory applies to *contracting* models as well.

Theorem 7.6.1. *For models (A, B) satisfying conditions 1 and 2, the technological expansion rate α_0 is finite and never exceeds the economic expansion rate β_0, which is also finite.*

Proof. Consider the matrix game with payoff $B - \alpha A$ with value $v(\alpha)$. For any completely mixed x for the maximizer (row player), $x^T B > 0$. Thus $v(0) > 0$.

For any completely mixed y for the minimizer (column player), $Ay > 0$. Thus for large α, $(B - \alpha A)y < 0$ and thus $v(\alpha) < 0$. As in Lemma 1.4.2, we have $v(\alpha_0) = 0$ for some $\alpha_0 > 0$. Let q be an optimal strategy for the minimizer. Then $(B - \alpha_0 A)q \le 0$. Since β_0 is the least such entry, we have $\beta_0 \le \alpha_0$. ∎

Theorem 7.6.2 (von Neumann). *If the model (A, B) satisfies conditions 1 and 2, then there exists $\theta > 0$ such that (a) for some intensity x^0, θ is the technological expansion rate and (b) for some price p^0, θ is the economic expansion rate that makes p^0 an optimal price vector. Furthermore, goods that are produced at higher than the technological expansion rate are priced zero and industries whose inputs at price p^0 cost more than the economic expansion rate θ are not operated at all.*

Proof. We continue to use the notation used in the proof of Theorem 7.6.1. By definition, $\alpha_0 = \sup\{\alpha : v(\alpha) = 0\}$. Choose $\theta = \alpha_0$. Thus $v(\theta) = 0$ and $(B - \theta A)p^0 \le 0$ for some optimal strategy p^0 of Player II. The required conditions are the usual complementary slackness conditions on optimal strategies. ∎

For the model (A, B) let $S \subset \{1, \ldots, n\}$ be a set of goods that can be produced using exclusively the goods in S. Then we call such a set of goods an *independent set*. In this case we will have a set of activities $T \subset \{1, \ldots, m\}$ such that $a_{ij} = 0$ if $i \in T$, $j \notin S$ and for all $j \in S$, $b_{ij} > 0$ for some $i \in T$. The model is called *irreducible* if T admits no proper independent subset.

Theorem 7.6.3. *If the model (A, B) is irreducible, then the technological expansion rate equals the economic expansion rate.*

Proof. In view of Theorem 7.6.1 it suffices to prove $\alpha_0 \le \beta_0$. Let x_0, p_0 be optimal for $B - \alpha_0 A$. We have $x_0^T B \ge \alpha_0 x_0^T A$ and $B p_0 \le \beta_0 A p_0 \le \alpha_0 A p_0$. Thus $\alpha_0 x_0^T A p_0 \le \beta_0 x_0^T A p_0$. It is enough to prove that $x_0^T A p_0 > 0$. If possible let $x_0^T A p_0 = 0$. Then $x_{0i} > 0 \Rightarrow (A p_0)_i = 0$. Let $U = \{i : x_{0i} > 0\}$, $V = \{j : ((x_0^T)B)_j > 0\}$. We have $x_0^T B_j = 0$ if $j \notin V$. Thus, by definition, $b_{ij} = 0$ if $i \in U$ and $j \notin V$. We claim that $a_{ij} = 0$ if $i \in U$, $j \notin V$; for otherwise $(x_0^T A)_j > 0$, contradicting our assumption that $(x_0^T B)_j \ge \alpha_0 (x_0^T A)_j$. Clearly, $b_{ij} > 0$ for some $i \in U$. Thus (A, B) is reducible, a contradiction. ∎

It is easy to construct a model with $\alpha_0 > \beta_0$. For example, we could consider two independent economies producing independent goods with independent technologies. Their technological expansion rate and economic expansion rate could be quite different. For the composite model the technological expansion rate will be the larger of the two rates and their economic expansion rate will be the smaller of the two. Thus the composite model will satisfy $\alpha_0 > \beta_0$.

The material in the present section is based on Gale (1960). Closely related to this problem is the problem of generalized eigenvalues as studied in Bapat, Olesky and van den Driessche (1995). We conclude the section by quoting the following *economic rules* satisfied by von Neumann's equilibrium for an irreducible model (A, B) [Champernowne (1945)]:

Profitability rule: Only those processes will be used that, with the actual prices and rate of interest, yield zero profits after payment of interest. These processes will be the most profitable ones available.

Free goods rule: In an equilibrium production system, those goods whose output exceeds their input more than the expansion rate α will be free goods.

System production rule: In equilibrium, the system of production will have the greatest possible rate of expansion of all possible productive systems.

Rate of interest rule: In equilibrium, the rate of interest equals the rate of expansion.

Price system rule: The price system in equilibrium will have a possible rate of interest smaller than or as small as that of any other price system.

7.7. Factor price equalization

Following Nikaido (1968), we first motivate a problem in international trade, initially formulated rigorously by Samuelson (1953) and which led to a well-known theorem of Gale and Nikaido (1965) on P-matrices and global univalence of mappings. We also point out other ramifications of the theorem under weaker conditions.

Suppose there are n factors of production, such as land, labor, etc., that are used in the production of n commodities, such as wheat, clothing, rice, etc. The factors of production may be different in different regions. A region may be rich in land resources while another may be rich in trained labor. Suppose the technology of production is the same in all the regions. For a given set of factor prices, such as prices for the unit of land and wage for labor, the cost of production is determined by the efficient use of the factors to produce the maximum output under the given technology. Under reasonable assumptions on the cost function this will be the same in all regions if the factor prices are the same in all the regions. Thus in a state of equilibrium, if it turns out that the factor prices are the same, then the commodity prices are also the same everywhere. However, the differing factor endowments and the differing demand for goods among the regions will, in general, only result in differing factor prices. Namely, two identical workers in different regions under the same technology may be rewarded differently. If there is free mobility of goods with no tariff restrictions and if the transportation costs are zero, then the prices in a state of equilibrium will be the same in all the regions. Goods move to regions fetching higher prices and indirectly affect the prices of the factors of production. In a perfectly competitive world market, under a single price system, the equilibrium price of a commodity must be equal to its cost of production. When the function defining commodity prices in terms of factor prices is *univalent* (we give the precise definition later in this section), then not only does one achieve uniform prices for the commodities, but also uniform prices for the factors of production. In a state of equilibrium, wage as the price for the factor labor will remain the same across regions. Higher wages can then no longer be an attraction for migration.

We now make our ideas precise by formulating an economic model of international trade among n countries with n goods and n factors of production.

Suppose there are n goods to be produced in amounts $x_1, \ldots, x_n$ using n factors of production. Let v_{ij} be the amount of factor j needed to produce an amount x_i of good i. Then the input per unit of output is $a_{ij} = \frac{v_{ij}}{x_i}$, $i, j = 1, \ldots, n$. Let the technology be the same in all the regions with common production functions

$$x_i = f_i(v_{i1}, v_{i2}, \ldots, v_{in}), \quad i = 1, 2, \ldots, n. \qquad (7.7.1)$$

We assume that f_i, $i = 1, \ldots, n$ are positively homogeneous of order 1, i.e.,

$$f_i(\lambda v_{i1}, \lambda v_{i2}, \ldots, \lambda v_{in}) = \lambda f_i(v_{i1}, \ldots, v_{in}), \quad i = 1, \ldots, n,$$

for all $\lambda \geq 0$. Further we assume that the f_is are concave with continuous first partial derivatives in the input space of nonnegative vectors. Let $p_1, \ldots, p_n$ denote the prices of goods. Given the factor prices $w_1, \ldots, w_n$, it costs $\sum_j a_{ij} w_j$ to produce one unit of good i. Perhaps the technology allows more than one way to utilize factors to produce goods, and thus the optimal costs are given by

$$c_i(w_1, \ldots, w_n) = \min \sum_j w_j a_{ij},$$

subject to

$$a_{ij} \geq 0, \quad j = 1, \ldots, n \quad \text{and} \quad f_i(a_{i1}, \ldots, a_{in}) = 1.$$

These optimal costs as functions of factor prices can also be shown [see, Lancaster (1968)] to be concave, continuous, and homogeneous of degree 1 given by

$$c_i = \sum_j w_j a_{ij}(w) = g_i(w_1, \ldots, w_n), \quad i = 1, \ldots, n. \qquad (7.7.2)$$

Given the common production functions f_i and positive prices p_i for the goods, at economic equilibrium it is true that $c_i = p_i$. Namely, the equilibrium price is the most economical cost of production. We have $p = g(w)$ for $p = (p_1, \ldots, p_n)$ and $g = (g_1, \ldots, g_n)$. The map $p = g(w)$ is defined on the nonnegative orthant excluding the origin. Let $J_g = (\frac{\partial g_i}{\partial w_j})$ be the Jacobian matrix. Now

$$\frac{\partial g_i}{\partial w_j} = a_{ij}(w) + \sum_k w_k \frac{\partial}{\partial w_j} a_{ik}(w).$$

Since the a_{ik}s are homogeneous of degree zero, by Euler's theorem on homogeneous functions, the second term vanishes and J_g is precisely the input-output matrix.

Let X and Y be subsets of Euclidean spaces. A map $\phi : X \to Y$ is called *globally univalent* if ϕ is a homeomorphism between X and Y. Thus ϕ is a one-to-one map of X to Y such that the two maps, ϕ and its inverse ϕ^{-1}, are both continuous. The map ϕ is *locally univalent* at $x^* \in X$ if there exists a neighborhood N of x^* and a neighborhood M of $\phi(x^*)$ such that the restriction $\phi : N \to M$ is globally univalent.

Now we formulate the following problem. Let $p_i = g_i(w_1, \ldots, w_n), i = 1, \ldots, n$ be n continuously differentiable functions of $w_1, \ldots, w_n$, defined in the interior of the positive orthant. When is the map $\phi : w \to g$ globally univalent?

If global univalence holds, the equilibrium prices for goods remain the same in all the regions and therefore the factors of production are priced the same everywhere. Thus without ever moving the workers from one place to another, by just moving the goods free of cost, one can achieve the same set of wages for labor. The main mathematical question is therefore to find sufficient conditions on the Jacobian and the domain that guarantee global univalence.

It must be remarked that although a univalence theorem, under certain assumptions on the region and the Jacobian, gives a one-to-one correspondence between commodity prices and factor prices, it may not reveal any kind of special relationship between commodity prices and factor prices. As an example, suppose Japan and Australia are involved in the production of electronic goods and wheat. Whereas electronic goods are labor intensive, wheat production is land intensive. Moreover, whereas Japan has more labor and less land, Australia has more land and less labor. Denoting w_1 as the wage for labor and w_2 as the rent for land, let p_1 and p_2 be the unit prices of electronic goods and wheat. Under a free market, the equilibrium prices $p_1 = g_1(w_1, w_2)$ and $p_2 = g_2(w_1, w_2)$ are the same in both countries when they have the same technology. Assume as before that g_1, g_2 satisfy concavity and the other requirements. Suppose the Jacobian $(\frac{\partial g_i}{\partial w_j})$ is nonsingular for all factor prices $w = (w_1, w_2) > 0$. By virtue of the continuity of $\frac{\partial g_i}{\partial w_j}$, the Jacobian is either positive everywhere or negative everywhere. If $u = \frac{w_2}{w_1}$, then the uniqueness of w_1, w_2, given p_1, p_2, is equivalent to

$$\frac{p_2}{p_1} = \frac{g_2(1, u)}{g_1(1, u)}.$$

The derivative of the right-hand side function $g(u) = g_2/g_1$ in the unknown u is given by $(g_1 g_2' - g_2 g_1')/g_1^2$. By assumption, it is of the same sign for all u. This shows that the function g is either increasing or decreasing for all u. Thus when the ratio of wheat to electronic good prices increase, there is a corresponding increase for the ratio of rent to wages. One tries to conjecture assertions of the following type for an n factor/n goods model. If an increase in commodity price p_i occurs, then one expects a more than proportionate increase in the corresponding factor price w_i, which is the most intensively used factor for the production of commodity i. Such assertions fail to hold in general for $n > 2$ without some strong conditions on the Jacobian.

7.8. P-matrices

As discussed in the previous section, we would like to find sufficient conditions on the Jacobian and the domain that guarantee global univalence. One condition that can be verified easily is that the Jacobian matrix is positive definite and the domain is an open convex set. The sufficiency of this condition is proved in the next result.

Theorem 7.8.1. *Let X be an open convex set in $\mathbf{R}^n$. Let $\mathbf{F} : X \to \mathbf{R}^n$ be a map with a positive definite Jacobian at all points in X. Then $\mathbf{F}$ is globally univalent.*

Proof. Let $a, b \in X$ and let $\mathbf{F}(a) = \mathbf{F}(b)$. For any $t, 0 \leq t \leq 1$, define $x(t) = a + t(b - a) \in X$. Let, if possible, $a \neq b$. Consider

$$\phi(t) = \sum_i (b_i - a_i)(f_i(x(t)) - f_i(a)).$$

Then

$$\phi'(t) = \sum_i \sum_j (b_i - a_i)(b_j - a_j) f_{ij}(x(t)). \quad \left(\text{Here } f_{ij} = \frac{\partial f_i}{\partial x_j}. \right)$$

Since the Jacobian (f_{ij}) is positive definite at all $x \in X$, $\phi'(t) > 0$ if $b \neq a$. However, $\phi(0) = 0$ and $\phi(1) = 0$, and it follows by Rolle's Theorem that $\phi'(\xi) = 0$ for some $0 < \xi < 1$. This is a contradiction, and the proof is complete. ∎

In trying to extend the previous theorem to more general cases, one is led to some further generalization of positive definiteness. In particular, a positive definite matrix has positive principal minors. This motivates the following definition.

A real $n \times n$ matrix A is called a *P-matrix* if each principal minor of A is positive. It is called a P_0-*matrix* if each principal minor is nonnegative.

We remark that not every P-matrix gives rise to a positive definite quadratic form. For example, the matrix

$$\begin{bmatrix} 1 & 5 \\ 0 & 2 \end{bmatrix}$$

is a P-matrix, although the induced quadratic form $x^2 + 5xy + 2y^2$ is negative at $(x, y) = (1, -1)$.

Theorem 7.8.2. *Let A be a P-matrix of order n. Then for any $n \times n$ diagonal matrix D with nonnegative diagonal entries, the matrix $P + D$ is a P-matrix.*

Proof. We use induction on the order of the matrix. Assume the result for P-matrices of order $k < n$ and consider the case $k = n$. By expanding the determinant along the first column we get

$$|A + D| = \begin{vmatrix} a_{11} & a_{12} & \cdots & a_{1n} \\ a_{21} & a_{22} + d_2 & \cdots & a_{2n} \\ \vdots & \vdots & \ddots & \vdots \\ a_{n1} & a_{n2} & \cdots & a_{nn} + d_n \end{vmatrix}$$

$$+ d_1 \begin{vmatrix} a_{22} + d_2 & \cdots & a_{2n} \\ \vdots & \ddots & \vdots \\ a_{n2} & \cdots & a_{nn} + d_n \end{vmatrix}.$$

By the induction assumption, the second term in the above expression is nonnegative. Expanding the first term along the second column we get

$$\begin{vmatrix} a_{11} & a_{12} & a_{13} & \cdots & a_{1n} \\ a_{21} & a_{22} & a_{23} & \cdots & a_{2n} \\ a_{31} & a_{32} & a_{33} + d_3 & \cdots & a_{3n} \\ \vdots & \vdots & \vdots & \ddots & \vdots \\ a_{n1} & a_{n2} & a_{n3} & \cdots & a_{nn} + d_n \end{vmatrix} + d_3 \begin{vmatrix} a_{33} + d_3 & \cdots & a_{3n} \\ \vdots & \ddots & \vdots \\ a_{n3} & \cdots & a_{nn} + d_n \end{vmatrix}.$$

Again the second term is nonnegative and we need to prove that the first term is positive. Each time a d_i disappears from the first term, after n steps we are left with $|A|$, which is positive. ∎

When A is a positive definite matrix, $\langle x, Ax \rangle > 0$ for any $x \neq 0$. Thus $x_i(Ax)_i > 0$ for some coordinate i. Interestingly, this weaker property characterizes P-matrices, as shown in the next result.

Theorem 7.8.3. *Let A be an $n \times n$ matrix. Then the following conditions are equivalent:*

(i) A is a P-matrix.
(ii) For any n-vector $x \neq 0$, $x_i(Ax)_i > 0$ for some coordinate i.
(iii) The real eigenvalues of A and its principal minors are positive.

Proof. $i \Rightarrow (ii)$. Let, if possible, $x_i(Ax)_i \leq 0$ for some $x \neq 0$ and for all $i = 1, \ldots, n$. We can choose $\alpha_i > 0$ satisfying $(Ax)_i = -\alpha_i x_i$ for those i for which $x_i(Ax)_i < 0$. We assume, without loss of generality, that $x_i(Ax)_i = 0$ precisely for $i = k + 1, \ldots, n$. Set $\alpha_i = 0, i = k + 1, \ldots, n$. Thus the leading principal submatrix B of order k of A satisfies $(B + D)u = 0$, where $D = \text{diag}\{\alpha_1, \ldots, \alpha_k\}$ and $u = (x_1, \ldots, x_k)^T$. This contradicts Theorem 7.8.2 and hence (ii) must hold.

$(ii) \Rightarrow (iii)$. It is enough to prove this assertion for leading principal submatrices. Let B be the leading $k \times k$ submatix of A. Let $By = \alpha y$ for an eigenvalue α. Let $x = (y^T, 0)^T$. Since $x_i = 0$ for $i > k$, by assumption for some $i \leq k$ we have $x_i(Ax)_i = y_i(By)_i > 0$. That is, $\alpha y_i^2 > 0$ for some $i \leq k$ and it follows that $\alpha > 0$.

$(iii) \Rightarrow (i)$. For any principal submatrix M of A, $|M|$ is the product of all its eigenvalues. Since complex eigenvalues occur with their conjugates with the same multiplicity, our assumption implies that $|M| > 0$. ∎

Theorem 7.8.4. *The value of a P-matrix payoff is positive.*

Proof. We use induction on the order of the matrix. Let A be a P-matrix of order n. As usual, we denote the value of A by $v(A)$. For any $\alpha \geq 0$, we know from Theorem 7.8.2 that $A + \alpha I$ is a P-matrix. In case $v(A) < 0$ we can find $\alpha \geq 0$ such that $v(A + \alpha I) = 0$. Since $|A + \alpha I| > 0$, we know by Theorem 1.3.6 that the game $A + \alpha I$ is not completely mixed. Let v_{ij} be the value of the matrix game obtained by deleting row i and column j in $A + \alpha I$. By Theorem 1.3.8

$$\min_i \max_j v_{ij} = \max_j \min_i v_{ij} = 0.$$

In particular, $\max_j v_{kj} = 0$ for some k and thus $v_{kk} \leq 0$. This contradicts our induction assumption for P-matrices of order less than n and completes the proof. ∎

Corollary 7.8.5. *Let A be a P-matrix. Then $Ax > 0$, $A^T y > 0$ for some $x > 0$, $y > 0$.*

Proof. By Theorem 7.8.4 the value of the game with payoff A is positive. Let u be optimal for player I. Define $x = u + \epsilon e > 0$. Observe that for sufficiently small ϵ, $Ax > 0$. A similar argument applies for A^T, which is also a P-matrix. That completes the proof. ∎

Theorem 7.8.6. *Let $A = (a_{ij})$ be an $n \times n$ matrix with all diagonal entries positive, and let $B = (b_{ij})$ be given by*

$$b_{ij} = \begin{cases} -|a_{ij}| & \text{if } i \neq j \\ |a_{ii}| & \text{if } i = j. \end{cases}$$

If B is a nonsingular M-matrix, then A is a P-matrix.

Proof. We first prove that A is nonsingular. Suppose A is singular. Then there exists $u \neq 0$ such that

$$\sum_{i=1}^{n} a_{ij} u_j = 0, \quad i = 1, \ldots, n.$$

By Theorem 1.5.2 the value of B is positive, and for some positive p_is we have

$$a_{ii} p_i > \sum_{j \neq i} |a_{ij}| p_j, \quad i = 1, \ldots, n.$$

Let $\theta = \max \frac{|u_j|}{p_j}$ and suppose the maximum is attained at $\frac{|u_k|}{p_k}$. Then $|u_k| > 0$. Since

$$a_{kk} u_k = -\sum_{j \neq k} a_{kj} u_j,$$

we have

$$a_{kk}|u_k| \leq \sum_{j \neq k} |a_{kj}| |u_j| \leq \theta \sum_{j \neq k} |a_{kj}| p_j < a_{kk} p_k = a_{kk}|u_k|,$$

a contradiction. Thus A is nonsingular. Since the matrix $\lambda I + A$ also satisfies the hypotheses of the theorem for $\lambda \geq 0$, then $|\lambda I + A| \neq 0$ for $\lambda \geq 0$. Since $|\lambda I + A| \to \infty$ as $\lambda \to \infty$ it follows that $|A| > 0$. A similar argument shows that each principal minor of A must be positive. Thus A is a P-matrix. ∎

We remark that matrices A satisfying the hypotheses of Theorem 7.8.6 are called *diagonally dominant matrices*.

The next result contains some characterizations of P_0-matrices. Condition *(iv)* of the result is due to Arrow (1974).

Theorem 7.8.7. *Let A be an $n \times n$ matrix. Then the following conditions are equivalent:*

(i) *A is a P_0-matrix.*

(ii) *For any $n \times n$ diagonal matrix D with positive diagonal entries, the matrix $A + D$ is a P-matrix.*

(iii) *For any n-vector $x \neq 0$, there exists some coordinate i such that $x_i \neq 0$ and $x_i(Ax)_i \geq 0$.*

(iv) *For any $n \times n$ diagonal matrix D with positive diagonal entries, every real eigenvalue of DA is nonnegative.*

Proof. The proof of $(i) \Rightarrow (ii)$ is similar to that of Theorem 7.8.2. Conversely, if (ii) holds, then for any diagonal matrix D with positive diagonal entries, the matrix $A + D$ is a P-matrix and therefore all principal minors of $A + D$ are positive. It follows by a continuity argument that the principal minors of A are all nonnegative and thus A is a P_0-matrix, establishing (i).

$(ii) \Rightarrow (iii)$. Let x be an n-vector, $x \neq 0$. We assume, without loss of generality, that $x_i \neq 0$, $i = 1, \ldots, k$ and $x_i = 0$, $i = k+1, \ldots, n$. Let B be the $k \times k$ leading principal submatrix of A. Since we have shown that (i) and (ii) are equivalent, A is a P_0-matrix and hence B is a P_0-matrix. Let, if possible, $x_i(Ax)_i < 0$, $i = 1, \ldots, k$ and choose $\alpha_i > 0$ satisfying $(Ax)_i = -\alpha_i x_i$, $i = 1, \ldots, k$. Let $D = \mathrm{diag}\{\alpha_1, \ldots, \alpha_k\}$ and let $u = (x_1, \ldots, x_k)$. Then $(B + D)u = 0$ and therefore $B + D$ is singular. This is not possible since, in view of (ii), $B + D$ must be a P-matrix.

$(iii) \Rightarrow (iv)$. Let D be a diagonal matrix with positive diagonal entries, and let λ be a real eigenvalue of DA with x as a real eigenvector. Thus $DAx = \lambda x$. By (iii), there exists $x_i \neq 0$ such that $x_i(Ax)_i = \frac{\lambda x_i^2}{d_{ii}} \geq 0$. Since $\frac{x_i^2}{d_{ii}} > 0$, it follows that $\lambda \geq 0$.

$(iv) \Rightarrow (i)$. First observe that if A satisfies (iv) then so does PAP^T for any permutation matrix P. Thus it will be sufficient to prove that any leading principal minor of A is nonnegative. Let B be the $k \times k$ leading principal submatrix of A. If $|B| = 0$, then there is nothing to prove, so suppose $|B| \neq 0$. For each $t > 0$, define the diagonal matrix

$$D(t) = \begin{bmatrix} I_k & 0 \\ 0 & tI_{n-k} \end{bmatrix}.$$

Then the nonzero eigenvalues of $D(0)A$ are precisely the eigenvalues of B, with the same multiplicities. Let $\epsilon > 0$ be less than the least absolute value of an eigenvalue of B. As $t \to 0$, some of the eigenvalues of $D(t)A$ approach those of B whereas the remaining approach 0. For t sufficiently small, those approaching the eigenvalues of B will be above ϵ in absolute value; the remainder are below.

Call the first set the *large* eigenvalues. Clearly, the product of the complex large eigenvalues is positive. By (*iv*), the real eigenvalues of $D(t)A$ are nonnegative for $t > 0$. Thus the product of the large eigenvalues is nonnegative for $t > 0$. It follows that the product of the eigenvalues of B is nonnegative and hence $|B| \geq 0$. That completes the proof.   ∎

Let A be an $n \times n$ matrix and suppose the following condition, stronger than the one in part (*iv*) of Theorem 7.8.7, holds: For any $n \times n$ diagonal matrix D with positive diagonal entries, every real eigenvalue of DA is positive. Then it is not necessarily true that A is a P-matrix as shown by the following example [Arrow (1974)]: Let

$$A = \begin{bmatrix} 1 & 1 \\ -1 & 0 \end{bmatrix}.$$

Then the eigenvalues of DA are given by

$$\frac{d_{11} \pm \sqrt{d_{11}(d_{11} - 4d_{22})}}{2}.$$

Thus if $d_{11} < 4d_{22}$, then DA has no real eigenvalues, whereas if $d_{11} \geq 4d_{22}$, then both the eigenvalues are positive. However, A is clearly not a P-matrix.

7.9. *N*-matrices

Yet another class of matrices that are closely related to P-matrices are the so-called N-matrices.

A real $n \times n$ matrix A is called an *N-matrix* if all the principal minors of A are negative. An N-matrix is of the *first category* if it has at least one positive entry; otherwise it is an N-matrix of the *second category*

Theorem 7.9.1. *Let A be an $n \times n$ N-matrix of the first category. Then the value of A is positive.*

Proof. Since A is an $n \times n$ matrix of the first category, $n \geq 2$. It suffices to prove that $Ax \leq 0, x \geq 0 \Rightarrow x = 0$. (Otherwise, normalizing x, we would get a mixed strategy ξ for player II, and then the value of the game A would not be positive.) By hypothesis, some entry of A, say a_{jk}, must be greater than 0. Clearly, $j \neq k$. For $x = (x_1, \ldots, x_n)^T$ and for $\theta \geq 0$, let $y = y(\theta) = (x_1, x_2, \ldots, x_k + \theta, \ldots, x_n)^T$. We have $(Ay)_i = (Ax)_i + \theta a_{ik}, i = 1, \ldots, n$. Clearly, $y \geq x$, and for some suitable $\theta \geq 0$, we have $Ay \leq 0$ and one of the coordinates of Ay say, $(Ay)_1 = 0$.

Since $a_{11} < 0$, we could use this as the pivot and reduce the rest of the entries in the pivotal first column to zero by row operations using the pivotal first row. Equivalently, we can write

$$
\begin{bmatrix}
1 & 0 & 0 & \cdots & 0 \\
-\frac{a_{21}}{a_{11}} & 1 & 0 & \cdots & 0 \\
-\frac{a_{31}}{a_{11}} & 0 & 1 & \cdots & 0 \\
\vdots & \vdots & \vdots & \ddots & \vdots \\
-\frac{a_{n1}}{a_{11}} & 0 & 0 & \cdots & 1
\end{bmatrix}
\begin{bmatrix}
a_{11} & a_{12} & \cdots & a_{1n} \\
a_{21} & a_{22} & \cdots & a_{2n} \\
\vdots & \vdots & \ddots & \vdots \\
a_{n1} & a_{n2} & \cdots & a_{nn}
\end{bmatrix}
=
\begin{bmatrix}
a_{11} & a_{12} & \cdots & a_{1n} \\
0 & a_{22}^* & \cdots & a_{2n}^* \\
\vdots & \vdots & \ddots & \vdots \\
0 & a_{n2}^* & \cdots & a_{nn}^*
\end{bmatrix}.
$$

Further, any principal submatrix containing row index 1 of the right-hand side matrix can be got by multiplying the corresponding principal submatrices of the left-hand side. Since $a_{11} < 0$ this means that for any set of indices $\{\alpha, \beta, \ldots, \tau\} \subset \{2, \ldots, n\}$,

$$
0 >
\begin{vmatrix}
a_{\alpha\alpha} & a_{\alpha\beta} & \cdots & a_{\alpha\tau} \\
a_{\beta\alpha} & a_{\beta\beta} & \cdots & a_{\beta\tau} \\
\vdots & \vdots & \ddots & \vdots \\
a_{\tau\alpha} & a_{\tau\beta} & \cdots & a_{\tau\tau}
\end{vmatrix}
= a_{11}
\begin{vmatrix}
a_{\alpha\alpha}^* & a_{\alpha\beta}^* & \cdots & a_{\alpha\tau}^* \\
a_{\beta\alpha}^* & a_{\beta\beta}^* & \cdots & a_{\beta\tau}^* \\
\vdots & \vdots & \ddots & \vdots \\
a_{\tau\alpha}^* & a_{\tau\beta}^* & \cdots & a_{\tau\tau}^*
\end{vmatrix}.
$$

Thus the principal minors of the right-hand side matrix that skips row index 1 are positive and hence is a P-matrix. Since

$$a_{11}y_1 + a_{12}y_2 + \cdots + a_{1n}y_n = 0, \tag{7.9.1}$$

the elimination of y_1 from the other inequalities still preserves the inequalities and we have

$$
\begin{bmatrix}
a_{22}^* & \cdots & a_{2n}^* \\
\vdots & \ddots & \vdots \\
a_{n2}^* & \cdots & a_{nn}^*
\end{bmatrix}
\begin{bmatrix}
y_2 \\
\vdots \\
y_n
\end{bmatrix}
\leq 0.
$$

By Theorem 7.8.3, $y_i = 0$, $i = 2, \ldots, n$. From (7.9.1) we conclude that $y = 0$. It follows that $x = 0$, and the proof is complete. ∎

A diagonal matrix is called a *signature matrix* if all the diagonal entries are ± 1.

Theorem 7.9.2. *Let $A < 0$ be a square matrix of order n. Then the following conditions are equivalent:*

(i) A is an N-matrix.

(ii) For any signature matrix S other than I or $-I$, the value of the payoff SAS is positive.

(iii) A reverses the sign of only nonnegative or nonpositive vectors, i.e., for any $x \neq 0$, if $x_i(Ax)_i \leq 0$ for all coordinates i, then $x \geq 0$ or $x \leq 0$.

Proof. $(i) \Rightarrow (ii)$. For any signature matrix S, the matrix SAS is again an N-matrix. It is of the first category if $S \neq I$, $S \neq -I$. By Theorem 7.9.1, the value of SAS is positive.

$(ii) \Rightarrow (iii)$. Suppose $x_i(Ax)_i \leq 0$ for all coordinates i, for some $x \neq 0$. There is nothing to prove if $x \geq 0$ or $x \leq 0$. Suppose on the contrary we have a signature matrix S other than I or $-I$, such that $Sx \geq 0$. By assumption, $SAx \leq 0$. Thus the vector $u = Sx$ satisfies $SASu = SASSx = SAx \leq 0$. By assumption, $x \neq 0$. Thus the game SAS has nonpositive value, which is a contradiction.

$(iii) \Rightarrow (i)$. Consider the matrix

$$C(\lambda) = \begin{bmatrix} a_{11} & a_{12} & \cdots & \cdots & a_{1n} \\ a_{21} & a_{22} + \lambda & \cdots & \cdots & a_{2n} \\ a_{31} & a_{32} & a_{33} + \lambda & \cdots & a_{3n} \\ \vdots & \vdots & \vdots & \ddots & \vdots \\ a_{n1} & a_{n2} & a_{n3} & \cdots & a_{nn} + \lambda \end{bmatrix}.$$

Let M_{1j} be the cofactor of the $(1, j)$-entry of the above matrix $C(\lambda)$. Whereas M_{11} is a polynomial of degree $n-1$ in λ, the rest of the cofactors are polynomials of degree $\leq n-2$. Thus for sufficiently large λ, the sign of $|C(\lambda)|$ is determined by the sign of M_{11}. The principal submatrix obtained by deleting the $(1, 1)$-entry of $C(\lambda)$ is a nonsingular M-matrix for sufficiently large λ. Thus $M_{11} > 0$ and therefore $|C(\lambda)| = a_{11}M_{11} < 0$. We claim that $|C(0)| = |A| < 0$. Otherwise, we can find $\mu > 0$ such that $|C(\mu)| = 0$, where $0 < \mu < \lambda$. Let $C(\mu)x = 0$ for some $x \neq 0$. We have, $x_i(C(\mu)x)_i = x_i(Ax)_i + \mu x_i^2 = 0$ for $i \geq 2$ and $x_1(Ax)_1 \leq 0$. Thus by assumption, $x \geq 0$ or $x \leq 0$. Without loss of generality we assume $x \geq 0$. This is impossible with the first row of $C(\mu)$ being negative. Thus $|A| < 0$. A similar argument applies for all principal submatrices of A, and we conclude that A is an N-matrix.   ■

Theorem 7.9.3. *Let A be an N-matrix of the first category. Then every proper principal minor of A^{-1} is a P-matrix.*

Proof. Consider the principal minor formed by the last k rows and k columns of A^{-1}. Partitioning both A and A^{-1} conformally, let us write

$$A = \begin{bmatrix} A_{11} & A_{12} \\ A_{21} & A_{22} \end{bmatrix} \quad \text{and} \quad A^{-1} = \begin{bmatrix} L_{11} & L_{12} \\ L_{21} & L_{22} \end{bmatrix}.$$

We have

$$|A||L_{22}| = \begin{vmatrix} A_{11} & A_{12} \\ A_{21} & A_{22} \end{vmatrix} \begin{vmatrix} I & L_{12} \\ O & L_{22} \end{vmatrix} = |A_{11}|.$$

Thus $|L_{22}| > 0$. A similar argument shows that every proper principal minor of A^{-1} is positive, and the proof is complete. ∎

Theorem 7.9.4. *Let A be an N-matrix of the first category. Then there exists a permutation matrix P such that*

$$PAP^T = \begin{bmatrix} A_{JJ} & A_{JJ'} \\ A_{J'J} & A_{J'J'} \end{bmatrix},$$

where $A_{JJ} < 0$, $A_{J'J'} < 0$ are square matrices and $A_{J'J} > 0$, $A_{JJ'} > 0$.

Proof. Since $n \geq 2$, and since the case $n = 2$ is trivial, we assume $n \geq 3$. No row or column of an N-matrix A of the first category can be negative, for otherwise the value $v(A)$ or $v(A^T) < 0$, contradicting Theorem 7.9.1. Also, no entry can be zero and all the diagonal entries are negative. Let us assume $J = \{j : a_{1j} < 0\}$ and let $J' = \{j : a_{1j} > 0\}$. We claim that the decomposition of A claimed in the theorem holds for the choice J, J'. To see this, let $\alpha, \beta \in J$ and consider the 3×3 principal submatrix using rows and columns $1, \alpha, \beta$. Since $a_{11} < 0$, $a_{1\alpha} < 0$, and $a_{1\beta} < 0$, the first row is negative. As before, the principal submatrix has to be an N-matrix of the second category and hence $a_{\alpha\beta} < 0$. Since α and β are arbitrary elements in J, we have $A_{JJ} < 0$. Now let $\alpha \in J$, $\beta \in J'$, so that $a_{1\alpha} < 0$, $a_{1\beta} > 0$. Since all the 2×2 principal minors are negative, the sign pattern of the 3×3 principal submatrix

$$\begin{bmatrix} a_{11} & a_{1\alpha} & a_{1\beta} \\ a_{\alpha 1} & a_{\alpha\alpha} & a_{\alpha\beta} \\ a_{\beta 1} & a_{\beta\alpha} & a_{\beta\beta} \end{bmatrix}$$

must be

$$\begin{bmatrix} - & - & + \\ - & - & * \\ + & * & - \end{bmatrix},$$

where the two starred entries $a_{\alpha\beta}$ and $a_{\beta\alpha}$ are either both positive or both negative. If $a_{\alpha\beta} < 0$, then the second row is negative, which contradicts Theorem 7.9.1 for the 3×3 N-matrix of the first category. Thus the entries $a_{\alpha\beta} > 0$, and $a_{\beta\alpha} > 0$, and thus the matrices $A_{JJ'} > 0$ and $A_{J'J} > 0$. Finally, let $\alpha \in J'$, $\beta \in J'$. Then $a_{1\alpha} > 0$, $a_{1\beta} > 0$, and the sign pattern of the above 3×3 matrix is

$$\begin{bmatrix} - & + & + \\ + & - & * \\ + & * & - \end{bmatrix}.$$

As before, the two starred entries $a_{\alpha\beta}$ and $a_{\beta\alpha}$ are either both positive or both negative. Suppose they are both positive. Then the determinant of the 3×3 matrix must be positive as can be seen by expanding along row 1 and taking account of the fact that all the principal 2×2 minors are negative $((-) \times (-) - (+) \times (-) + (+) \times (+) > 0)$. This is a contradiction. Thus both the starred entries $a_{\alpha\beta}$ and $a_{\beta\alpha}$ are negative and hence $A_{J'J'} < 0$. That completes the proof.   ∎

7.10. Global univalence

Consider the system of equations

$$p_i = g_i(x_1, x_2, \ldots, x_n), \quad i = 1, \ldots, n,$$

where the g_is are continuously differentiable functions in an open domain $X \subset \mathbf{R}^n$. When the Jacobian matrix

$$J = \begin{bmatrix} \frac{\partial g_1}{\partial x_1} & \frac{\partial g_1}{\partial x_2} & \cdots & \frac{\partial g_1}{\partial x_n} \\ \vdots & \vdots & \ddots & \vdots \\ \frac{\partial g_n}{\partial x_1} & \frac{\partial g_n}{\partial x_2} & \cdots & \frac{\partial g_n}{\partial x_n} \end{bmatrix}$$

is nonsingular at a point $x^\circ = (x_1^\circ, x_2^\circ, \ldots, x^\circ) \in X$, then by the Local Univalence Theorem [see Apostol (1957, p. 144] there exists a neighborhood N of x° in X such that the map $g = (g_1, g_2, \ldots, g_n) : N \to g(N)$ is univalent.

If the Jacobian is nonsingular at all points of X it does not necessarily guarantee global univalence over X. The following is a simple example to illustrate this.

Example 7.3. Let X be the complement of the coordinate axes in $\mathbf{R}^2$, i.e., $X = \{(x, y) : x \neq 0, y \neq 0\}$. Let

$$p_1 = g_1(x, y) = x^2 - y^2$$

$$p_2 = g_2(x, y) = x^2 + y^2.$$

The Jacobian matrix

$$\begin{bmatrix} 2x & -2y \\ 2x & 2y \end{bmatrix}$$

is nonsingular for all (x, y) in X. However, the two points $(4, 3)$ and $(-4, 3)$ are mapped to the same point $(7, 25)$. Indeed, the functions g_1 and g_2 are polynomials.

It is a deep and unresolved conjecture, known as the *Jacobian Conjecture*, that when the domain X is the entire $\mathbf{R}^n$, and the functions g_is are polynomials in n variables, the nonsingularity of the Jacobian over X implies global univalence

[Keller (1939)]; also see van den Essen (1990)]. Even for $n = 2$ the problem remains open.

In case the functions are not necessarily polynomials, then one can construct examples where the Jacobian is positive even over the entire plane $\mathbf{R}^2$, but the map is not globally univalent. The following is a simple example.

Example 7.4. Let $X = \mathbf{R}^2$. The map $g : (x, y) \to (g_1, g_2)$ given by

$$p_1 = g_1(x, y) = e^{2x} - y^2$$
$$p_2 = g_2(x, y) = 4ye^{2x} - y^3$$

has a positive Jacobian everywhere, though $(0, 2)$, $(0, -2)$ are both mapped to $(-3, 0)$. Thus, besides nonsingularity, we might have to impose some further conditions on the Jacobian matrix and the region X to assert global univalence. We now proceed to develop a result of this type. We first prove a preliminary result.

Theorem 7.10.1. *Let X be an open rectangle in $\mathbf{R}^n$. Let*

$$\mathbf{F} = (f_1, f_2, \ldots, f_n) : X \to \mathbf{R}^n$$

be a continuously differentiable map. Let the Jacobian $J(x)$ at each $x \in X$ be a P-matrix. Then the inequalities

$$\mathbf{F}(x) \leq \mathbf{F}(a), \quad x \geq a$$

have only the trivial solution $x = a$.

Proof. For $n = 1$ the theorem is trivial by elementary calculus. We proceed by induction on n. By a translation we can retain the conditions of the theorem and also assume $a = 0$. By defining $\mathbf{G}(x) = \mathbf{F}(x) - \mathbf{F}(0)$ we might as well assume $\mathbf{F}(0) = 0$. We prove that $\mathbf{F}(x) \leq 0$, $x \geq 0$ has no nontrivial solution. Let $S = \{x : \mathbf{F}(x) \leq 0, x \geq 0\}$. We use the l_1-norm $\|x\| = \sum |x_i|$. For each i, $f_i(x)$, the i-th coordinate function is differentiable and

$$\frac{f_i(x) - f_i(0) - (J(0)x)_i}{\|x\|} \to 0 \quad \text{as} \quad x \to 0. \tag{7.10.1}$$

When $x \geq 0$, $x \neq 0$, the vector $\frac{x}{\|x\|}$ is a probability vector. Applying Theorem 7.8.4 to the transpose of $J(0)$, we get

$$\frac{(J(0)x)_i}{\|x\|} \geq \delta > 0$$

for some i. Here δ can be chosen independently of the particular choice of $x \geq 0, x \neq 0$. Thus for $x \geq 0$ and sufficiently near 0, from Equation (7.10.1), it follows that

$$\max_i \frac{f_i(x)}{\|x\|} \geq \frac{\delta}{2}.$$

This shows that 0 is not a limit point of the set S. Since S is closed, 0 is an isolated point of S and $S \setminus \{0\}$ is closed in $\mathbf{R}^n$. We claim that $S = \{0\}$. Let, if possible $c \in S \setminus \{0\}$. Consider the set $Q = \{x : 0 \leq x \leq c, \mathbf{F}(x) \leq 0\}$. Since X is rectangular, $Q \subset X$. The set $Q \setminus \{0\}$ is closed in Q and hence compact in $\mathbf{R}^n$. Let $u \in Q \setminus \{0\}$ have the least norm guaranteed by the compactness of $Q \setminus \{0\}$.

Case (i): $u > 0$. By Theorem 7.8.4 there exists $w < 0$ with $J(u)w < 0$. For t sufficiently small, $x(t) = u + tw$ satisfies $u > x(t) > 0$ and $x(t) \in X$. Moreover, $\mathbf{F}(x(t)) = \mathbf{F}(u) + tJ(u)w + O(t\|w\|) \leq 0$ and $x(t) \in Q \setminus \{0\}$. However, $\|u\| > \|x(t)\|$, a contradiction.

Case (ii): $u_i = 0$ for some i. Without loss of generality, let $u_1 = 0$. Let X_1 be the restriction of the domain X to those points with the first coordinate zero. This is again a rectangular domain, and the new functions $\overline{f}_i(x_1, x_2, \ldots, x_n) = f_i(0, x_2, \ldots, x_n), i = 2, 3, \ldots, n$, whose Jacobian is again a P-matrix, are continuously differentiable. By induction, $\overline{f}_i = \overline{f}_i(x_2, \ldots, x_n) = f_i(0, x_2, \ldots, x_n) \leq 0, x_i \geq 0, i = 2, \ldots, n$ has only the trivial solution. This shows that $u = 0$, which is a contradiction. Thus the set $S = \{0\}$, and the theorem is proved. ∎

We are now ready to prove the fundamental theorem of Gale and Nikaido (1965).

Theorem 7.10.2. *Let X be an open rectangle in R^n, and let $\mathbf{F} : X \to R^n$ be a continuously differentiable map. Further, for each $x \in X$ let the Jacobian $J(x)$ be a P-matrix. Then $\mathbf{F}$ is globally univalent.*

Proof. Let $u, w \in X$ with $\mathbf{F}(u) = \mathbf{F}(w)$, and we show that $u = w$. Let us assume, without loss of generality, that precisely the first k coordinates of u are greater than the first k coordinates of w. Consider the matrix D obatined from the identity matrix I by changing the sign of the first k columns. Let $Y = D^{-1}X, \mathbf{G}(y) = D^{-1}\mathbf{F}(Dy)$. The Jacobian of $\mathbf{G}$ is $H = D^{-1}J(Dy)D$. All the principal minors of H are positive and Y is rectangular. With $u^\circ = D^{-1}u, w^\circ = D^{-1}w$, we find $\mathbf{G}(u^\circ) = \mathbf{G}(w^\circ), u^\circ \leq w^\circ$. By Theorem 7.10.1 $u^\circ = w^\circ$ and therefore $u = w$. Now for any $z = \mathbf{F}(u)$, we know that $u = \mathbf{F}^{-1}(z)$. Also, by the local univalence theorem, $\mathbf{F}^{-1}$ is continuous at z. Thus $\mathbf{F}^{-1}$ is continuous in a small neighborhood of z. Because continuity is a local property we are done. ∎

For Jacobians that are M-matrices we can say something stronger.

Theorem 7.10.3. *Let X be an open rectangle in R^n, and let $\mathbf{F} : X \to R^n$ be a continuously differentiable map. Let the Jacobian of $\mathbf{F}$ be a nonsingular M-matrix. Then the inverse map $\mathbf{F}^{-1}$ is nondecreasing, i.e., $\mathbf{F}(x) \le \mathbf{F}(y) \Rightarrow x \le y$.*

Proof. We first remark that by Theorem 1.5.2, any nonsingular M-matrix is a P-matrix. The result is trivial for $n = 1$. We proceed by induction on n. Let $\mathbf{F}(x) \le \mathbf{F}(y)$. By Theorem 7.10.2, $x_k \le y_k$ for some coordinate k, say $k = 1$. Since $\frac{\partial f_i}{\partial x_1} \le 0, i = 2, \ldots, n$ we know that for $x = (x_1, x_2, \ldots, x_n)$, $y = (y_1, y_2, \ldots, y_n)$,

$$f_i(y_1, x_2, \ldots, x_n) \le f_i(x_1, x_2, \ldots, x_n)$$

$$\le f_i(y_1, y_2, \ldots, y_n), \quad i = 2, \ldots, n.$$

Let $\mathbf{G} : \overline{X} \to \mathbf{R}^{n-1}$, where $\overline{X}$ is the image of X under the projection map $(x_1, x_2, \ldots, x_n) \to (x_2, \ldots, x_n)$ and

$$G(x_2, \ldots, x_n) = ((f_2(y_1, x_2, \ldots, x_n), \ldots, f_n(y_1, x_2, \ldots, x_n)).$$

The Jacobian of $\mathbf{G}$ is again a nonsingular M-matrix, and we know that $G(x_2, \ldots, x_n) \le G(y_2, \ldots, y_n)$. By the induction assumption, $x_i \le y_i, i = 2, \ldots, n$. Since $x_1 = y_1$ the proof is complete. ∎

Let p be the equilibrium price of goods, and let $c = \mathbf{F}(w)$ be the vector of optimal costs of production with factor prices w. Since at equilibrium $c = p$, when the map F is globally univalent, w is uniquely fixed by p. If $\mathbf{F}$ has a nonsingular Jacobian that is also an M-matrix, then global univalence is guaranteed. In addition, by Theorem 7.10.3, $\mathbf{F}^{-1}$ is nondecreasing. Thus a rise in the price of the i-th good results in a rise in the price of the i-th factor with a lowering of prices of other factors. In this section we have followed the treatment in Nikaido (1968).

7.11. Stability and market prices

Given non-numeraire goods (i.e., goods other than money) $i = 1, \ldots, n$, let $z_i(p)$ denote the excess of demand over supply at the price vector $p \ge 0$. The price vector p^* is a *Walrasian equilibrium* if

- $z_i(p^*) \le 0 \quad (i = 1, \ldots, n)$
- $p_i^* = 0$ if $z_i(p^*) < 0$.

Thus if good i fetches a positive equilibrium price, then the excess demand is zero and that demand equals supply. The second condition says that goods whose demands fall short of supply become free goods at the equilbrium price. Individual consumers with a fixed income I look for the best commodity mix $\phi(p, I)$ for each price vector p. Given a bundle of goods a and a price q let $\sigma(q, a)$ denote the bundle that is as good as a for any given consumer and costs the least. Thus $\sigma(p, \phi(p, I)) = \phi(p, I)$. Any price change affects the consumer in two ways. It encourages the consumer to look for substitutes that are cheaper and it causes a change in real income in terms of purchasing power. These two effects are captured by the so-called *Slutsky equation* [Slutsky (1915), Nikaido (1968)]:

$$\frac{\partial \phi_i}{\partial p_j} = -\phi_j \frac{\partial \phi_i}{\partial I} + \sigma_{ij}, \tag{7.11.1}$$

where

$$\sigma_{ij} = \lim_{\Delta p_j \to 0} \frac{\sigma_i(p + \Delta_j p, \phi(p, I)) - \sigma_i(p, \phi(p, I))}{\Delta p_j}. \tag{7.11.2}$$

[Here $\Delta_j p = (0, \ldots, \Delta p_j, \ldots, 0)$.] In (7.11.1) the first term on the right-hand side measures the *income effect*; and the second term σ_{ij} measures the *substitution effect*.

When $\frac{\partial \phi_i}{\partial p_j} > 0$, the i-th good's demand increases acting as a gross substitute for the j-th good when the j-th good's price increases. One could define this effect also for the market at large by considering the excess demand functions $z_i(p)$ ($i = 1, \ldots, n$).

The notion of Walrasian equilibrium and changes in prices due to changes in taste, costs, and methods of production can also be described by modeling the prices of goods in multiple markets via a dynamical system. The aim of such a system will be to derive properties of the equilibrium from the stability of the dynamical system. For a single commodity market it is somewhat easy. Displacement from equilibrium price could be self-correcting if any rise above equilibrium price creates excess supply and any fall from equilibrium price generates excess demand, which in turn sets economic forces into action to restore equilibrium. In trying to extend this notion of stability to multiple markets one faces many formidable problems. Two key problems facing multiple markets are the interdependence of the markets and differential speeds of adjustments between markets.

In order to extend the stability concept for multiple markets, Hicks (1939), in his classic monograph, defined the notions of imperfect and perfect stability. The market according to Hicks is *imperfectly stable* if a fall in the price of commodity i creates an excess demand for that commodity after the rest of the market has adjusted to zero excess demand.

Consider any arbitrary subset of goods and the market for any arbitrary good, say, good i. Suppose the equilibrium price for good i falls. The market is called *perfectly stable*, if the excess demand for good i shows an increase after allowing for the market prices for other goods in the subset to vary to reach zero excess demand. To make this more precise, given continuously differentiable excess demand functions $z_i(p)$, $i = 1, \ldots, n$, consider the differential form

$$dz_i = \sum_j \frac{\partial z_i}{\partial p_j} dp_j, \quad (i = 1, \ldots, n). \tag{7.11.3}$$

The conditions of imperfect stability stipulate $dz_j = 0$ and $dp_j = 0$ for all $j \neq i$. Thus by Cramer's rule we have

$$\frac{dz_i}{dp_i} = \frac{\Delta}{\Delta_{ii}}, \quad (i = 1, \ldots, n), \tag{7.11.4}$$

where Δ is the determinant of the Jacobian ($\frac{\partial z_i}{\partial p_j}$ and Δ_{ii} is its i-th principal minor). Imperfect stability stipulates $\frac{dz_i}{dp_i} < 0$ and thus Δ and Δ_{ii} are of opposite signs. The stronger perfect stability means the same thing applied to submarkets, and this implies that any principal minor of order m must satisfy the property that it has sign opposite to any of its $(m-1)$-rowed minors. Thus perfect stability according to Hicks means that the sign of any r-th order principal minor of the Jacobian ($\frac{\partial z_i}{\partial p_j}$) is $(-1)^r$. We say that the $n \times n$ matrix A is *Hicksian* (or *Hicksian stable*) if every i-th order principal minor of A has sign $(-1)^i$, $i = 1, 2, \ldots, n$. In particular, Hicksian stability implies that the negative of the Jacobian matrix is a P-matrix. Unfortunately, the Hicksian condition comes more out of a definition than as a consequence of a true dynamic model for multiple markets. Any dynamic model on the price behavior must take into account not only the excess demand functions but also the varying speeds of adjustments between individual markets. For example, let

$$\frac{dp_i}{dt} = H(z_i(p)), \quad (i = 1, \ldots, n)$$

be a dynamical system representing the prices in multiple markets $i = 1, \ldots, n$ for respective commodities. (We assume that all goods are non-numeraire goods.) Here as usual z_i denotes the excess demand function and H is an increasing function with $H(0) = 0$. Local stability of the dynamical system requires that any solution to the above system whose starting point lies in some neighborhood of the equilibrium price p° defined by $z(p^\circ) = 0$ converge to p°. Thus, apart from some degenerate cases, when the equilibrium price is stable then all the eigenvalues for the Jacobian of the map $p \to z(p)$ have negative real parts.

Perfect stability is neither necessary nor sufficient for dynamic stability. Although perfect stability imposes the condition that the negative of the Jacobian is a P-matrix, which in turn guarantees that all real roots are positive, it could still happen that the Jacobian may possess complex roots with negative real parts. The following is a counter example due to Samuelson (1944):

Example 7.5. Let

$$
A = \begin{bmatrix} \epsilon & 1 & 0 & 0 \\ 0 & \epsilon & 1 & 0 \\ 0 & 0 & \epsilon & 1 \\ -1 & 1 & -1 & 1+\epsilon \end{bmatrix}.
$$

For $0 < \epsilon < \cos\frac{\pi}{5}$, the matrix A, though a P-matrix, has two eigenvalues whose real parts are negative.

The following theorem rehabilitates Hicksian stability within the framework of dynamic stability.

Theorem 7.11.1. *Let a multiple market for non-numeraire goods be dynamically stable for all possible sets of speeds of adjustments k_i with the law of motion given by*

$$
\frac{dp_i}{dt} = k_i z_i (p - p^\circ), \quad (i = 1, \ldots, n), \tag{7.11.5}
$$

where p° is the equilibrium price and z_i the excess demand functions. Then the system must be perfectly stable in the Hicksian sense.

Proof. The characteristic equation of the varying speed system (7.11.5) is given by $|(k_i \frac{\partial z_i}{\partial p_j} - \delta_{ij}\lambda)| = 0$. Let $a_{ij} = \frac{\partial z_i}{\partial p_j}$. If the system (7.11.5) is dynamically stable for all speeds k_i, then for large λ the sign of $|A - \lambda I|$ is $(-1)^n$. Also, the equation $|A - \lambda I| = 0$ has no roots with nonnegative real parts. Thus the sign of $|A|$ is $(-1)^n$. However, A is the same as the matrix in the Hicks static equations except for the positive speed product $\Pi_i k_i$. If the market system has to be stable for all possible speeds, by letting some of the k_i go to zero and keeping others fixed, we get the sign of the determinant of any principal submatrix as $(-1)^r$, where r is the order of the minor. Thus Hicks's conditions on perfect stability are satisfied. ∎

Theorem 7.11.2. *If all commodities are gross substitutes in the multiple market, then the dynamic stability of the system*

$$
\frac{dp_i}{dt} = k_i \sum_j \frac{\partial z_i}{\partial p_j} (p - p_j^\circ), \quad (i = 1, \ldots, n) \tag{7.11.6}
$$

is equivalent to Hicksian perfect stability.

Proof. If all the commodities are gross substitutes, it is equivalent to assuming $a_{ij} > 0$ for all $i \neq j$ and $a_{ii} < 0$ for all i, where $a_{ij} = \frac{\partial z_i}{\partial p_j}$ for all i, j. Consider the dynamical system (7.11.6). Here k_i are different speeds of adjustments. Without loss of generality, $|k_i a_{ii}| < 1$. Consider the difference equations

$$y_i(t) = \sum_j (k_i a_{ij} + \delta_{ij}) y_i(t-1), \quad (i = 1, \ldots, n). \tag{7.11.7}$$

The proof is accomplished in two stages. In the first stage we show that the differential equations (7.11.6) are stable if and only if the difference equation (7.11.7) is stable. In the second stage we show that perfect Hicksian stability is equivalent to the stability of the difference equation. The characteristic equation for the difference equations is

$$\begin{vmatrix} k_1 a_{11} + 1 - \rho & k_1 a_{12} & \cdots & k_1 a_{1n} \\ k_2 a_{21} & k_2 a_{22} + 1 - \rho & \cdots & k_2 a_{2n} \\ \vdots & \vdots & \ddots & \vdots \\ k_n a_{n1} & \cdots & \cdots & k_n a_{nn} + 1 - \rho \end{vmatrix} = 0. \tag{7.11.8}$$

If $\lambda = \rho - 1$, we obtain $|(k_i a_{ij} - \lambda \delta_{ij})| = 0$, the characteristic equation of the dynamical system. The roots of (7.11.8) are one more in number than the roots of the characteristic equation (7.11.6). When the difference equation (7.11.7) is stable then the roots of (7.11.8) are less than unity. Therefore, the real parts of the characteristic roots associated with (7.11.6) are negative. Hence the stability of the difference equation guarantees the stability of the differential equation.

We now show the converse. By assumption, matrix A has $a_{ij} > 0$ for all $i \neq j$ and $a_{ii} < 0$ for all i. Mimicking the proof of the Perron-Frobenius Theorem, one can conclude that the spectral radius is an eigenvalue of A. Thus in case a root of (7.11.8) is greater than unity, then there must be a positive root greater than unity. This would violate stability of the difference equations. Thus the differential equation (7.11.6) is unstable.

The proof will be completed by showing the equivalence of Hicks conditions and the conditions for the stability of difference quations. We can, after dividing the rows $i = 1, \ldots, n$ by $k_1, \ldots, k_n$, reduce Equation (7.11.8) to $|(b_{ij} - \alpha_i))| = 0$, where $b_{ij} > 0$ if $i \neq j$ and $b_{ii} < 0$, $\alpha_i > 0$ for all i. We know that the difference equation is stable if B has no nonnegative eigenvalue. This is equivalent to $-B$ being a P-matrix. In turn, this is equivalent to Hicks' conditions on the positivity of all principal minors of $-B$, or equivalently, of $-A$. Since the stability of the difference equations is inherited by its subsystems, Hicks' conditions are also necessary. ∎

7.12. Historical notes

The origins of developing a detailed accounting of interindustrial activity date back to the French school of economists and notably to the works of Francois Quesnay. In 1758 Quesnay published [see Kuczynski and Meek (1972) and Pressman (1994)] a monograph on " Tableau Économique," where he described how a land owner who receives a sum of money as rent spends portions on agricultural products and the rest on products of artisans. In turn, artisans buy food, raw materials, and other items. Leontief (1941) introduced his own empirical work as an attempt to construct a similar table for the United States. economy at large. This monumental contribution won him the Nobel price in economics in 1973. For Leontief's model, the substitution theorem was conceived by Samuelson (1951) and the elegant proof using the duality and complementary slackness in linear programming was given by Gale (1960).

Although the Leontief system presupposes constant input coefficients, attempts to weaken this condition have been ongoing at both theoretical and empirical levels. Viewing a productive economy as run by entrepreneurs who are rewarded more for their productive skills than for their pure labor, Sraffa (1963) proposed another linear economic model. The elegant application of the Perron-Frobenius Theorem to this model was greatly expanded by Pasinetti (1977). Adjusting the input coefficients based on past coefficients was the main thrust of Bacharach (1970), who develops the theory of scaling of nonnegative matrices. The matrix scaling problem was discussed in Chapter 6 in greater detail.

The fundamental paper on balanced growth by von Neumann remained dormant in an obscure journal (*Ergebnisse eines Mathematischen Kolloqueiums*, No. 8, 1935–36, Franz-Deuticke, Leipzig, Wien 1937) until the first English translation of the same work in *Review of Economic Studies* 13 (1945–46) triggered a great deal of excitement among economists. The economic rules governing von Neumann's expanding economy given at the end of Section 7.6 were first stated by Champernowne (1945). Many socialist governments were keen on using his model for planning purposes. The model presented here is a refinement based on the work of Gale (1960). The proof technique using matrix games is due to Kemeny, Morgenstern, and Thompson (1956). Kaneko (1989) has proposed a noncooperative game theoretic formulation involving a continuum of producers who live from the infinite past to the infinite future with a linear production technology. Heckscher (1919) first pointed out how differences in geographical factor endowments affect income distribution and determine regional patterns and how the resulting trade in movable goods will serve as a partial substitute for the mobility of factors in reducing factor price differentials. The precise formulation had to wait at least four decades and could not have been possible without some serious discussion on global univalence theory. Expanding his Ph.D. thesis, Bertil Ohlin (1933), a student of Heckscher, formulated the essentials of interregional and international trade. It was the pioneering article of Heckscher (1919) and Ohlin's thesis on international trade that triggered Samuelson (1948) to take a closer look at their work on international trade. Neither Heckscher nor Ohlin tried to descend from their full generalities to some special cases, which was necessary to appreciate the technical issues involved in the factor price equalization assertion. While Heckscher's generality kept him doubting his own assertions, Ohlin won the Nobel price in economics in 1979 for his pioneering work on factor price equalization in international trade.

It was the work of Gale and Nikaido (1965) that set the pace for correcting Samuelson's conjectures and proving global univalence for maps whose Jacobians are P-matrices over an open box in an Euclidean space. The theorems per se have no additional insight to offer for various economic models. Perhaps Theorem 7.10.3 is one that offers such an insight. For related theorems, see Stolper and Samuelson (1941).

The class of P-matrices were first studied by Fiedler and Ptak (1966). In Theorem (7.8.7) the last assertion is due to Arrow (1974). Arrow came to this problem while studying global stability of systems with varying speeds. The idea was to generalize an earlier work of Metzler (1945) on Hicksian stability. The connection between nonsingular M-matrices and P-matrices to completely mixed matrix games were exploited by Raghavan (1978) to give alternate proofs of well-known theorems of Fiedler and Ptak (1966), Fan (1958), and others. It was McKenzie (1960) who exploited diagonally dominant matrices to study the stability of dynamical systems. Besides P-matrices yet another important class of matrices that share many properties common to P matrices is that of N-matrices due to Inada (1971), where it is shown that N-matrices of the first kind have positive value. Whereas for P-matrices one is able to apply game theoretic proofs, this theorem is somewhat elusive to game theoretic treatment. The sign reversal property (*ii*) in Theorem 7.8.3 for P-matrices was extended to N-matrices by Parthasarathy and Ravindran (1990) and is given in Theorem (7.9.2). The paper by Olech, Parthasarathy, and Ravindran (1991) contains some related results. Theorem 7.9.3 is due to Kojima and Saigal (1979) and is a further generalization of a theorem of Chipman [See Theorem 21.3, Nikaido (1968)]. Theorem 7.9.4 is contained in Mohan and Sridhar (1992).

Global univalence theorems have been approached entirely differently by algebraic geometers and economists. While algebraic geometers are trying to resolve the yet unresolved Jacobian Conjecture even in two-dimensional complex space, economists have been extending the Gale-Nikaido Theorem by weakening the nature of the Jacobian and by considering restricted subsets of the regions where the P-matrix or N-matrix property is valid for the Jacobian. For a survey on the Jacobian Conjecture see Bass, Connell, and Write

(1982). The conjecture can be stated simply as follows: Let $f : C^n \to C^n$ be a polynomial whose Jacobian is a nonzero constant. Then f has a polynomial inverse. An extension of the Gale-Nikaido Theorem to N-matrices of the first kind was proved by Inada (1971). For other refinements and extensions see Nikaido (1968). For more information about the Jacobian Conjecture see the surveys by van den Essen (1990, 1992).

The notion of Walrasian equilibrium price, though elegant and central to the study of competitive market economy, says very little about how one arrives at them. An algorithmic or iterative method for arriving at an equilibrium is both qualitatively and computationally useful. The process of price adjustments and variations in excess demand due to price fluctuations is what is called a tatonnement process. Hicks (1939) introduced the notion of imperfect and perfect stability. His stability notion was not properly grounded in dynamical systems or in a tatonnement process. It was more of a definition for multiple markets based on the clear analysis he could carry out in a dynamic version for a single market. The notions of dynamic stability looked unrelated to Hicksian stability. In a fundamental paper, Metzler (1945) showed the interrelationship between dynamic stability and Hicksian stability. The main results in Section 7.11 are due to Metzler (1945). Hicks (1939) introduced the notion of gross substitutes, in which an increased price of one good encourages customers to look for substitute goods. Metzler (1945) observed that for a multiple market where all goods are gross substitutes, dynamic stability with arbitrary speeds of market adjustments is equivalent to Hicksian stability. The eigenvalues of P-matrices and M-matrices play a key role in these discussions. Metzler's results were extended in full generality to global stability for gross substitute multiple markets by Arrow and Hurwicz (1958). In the case of Arrow, establishing the formal existence of competitive equilibrium and the above global stability theorems were cited as fundamental contributions to economics by the Nobel prize committee. These global stability theorems can also be obtained via the theory of diagonally dominant matrices; see McKenzie (1960). Also, see Morishima (1964), who extends the comparative static theorems and Hicksian laws to global laws. The volume in honor of Metzler, edited by Horwich and Samuelson (1974), contains many articles showing the power of special classes of matrices in resolving conceptual issues in economics.

Exercises

1. Show that the production matrix

$$\begin{bmatrix} a & b \\ c & d \end{bmatrix}$$

is productive if and only if $(1 - a)(1 - d) - bc > 0$.

2. Is the following matrix productive?

$$\begin{bmatrix} 0 & .5 & 1 \\ 1.3 & 0 & .4 \\ .3 & .2 & 0 \end{bmatrix}$$

3. An economy can produce three goods. The production technology for the first two goods is unique with input vectors

$$\begin{bmatrix} 0 \\ \frac{1}{6} \\ \frac{1}{4} \end{bmatrix}, \begin{bmatrix} \frac{1}{4} \\ 0 \\ \frac{1}{5} \end{bmatrix}.$$

For the third good there are two possible technologies with input vectors

$$\begin{bmatrix} \frac{1}{2} \\ \frac{1}{6} \\ 0 \end{bmatrix}, \begin{bmatrix} \frac{1}{5} \\ \frac{1}{5} \\ 0 \end{bmatrix}.$$

The labor input requirements are 1 and 2 for the first and the second goods and 1 for both technologies of the third good. Use linear programming to find the optimal level of activity for the third industry. Given an arbitrary vector of final demands, check that the labor input with the optimal system is less.

4. Consider an economy organized into three industries: lumber and wood products, paper and allied products, and machinery and transportation equipment. A consulting firm estimates that last year the lumber industry had an output valued at $50 (assume units in $100,000), 5% of which it consumed itself, 70% was consumed by final demand, 20% by the allied paper and product industry, and 5% by the equipment industry. The equipment industry consumed 15% of its own products out of a total of $100, 25% went to final demand, 30% to the lumber industry, and 30% to the paper and allied industry. Finally, the paper and allied products industry produced $50 of which it consumed 10%, 80% went to final demand, 5% went to the lumber industry, and 5% to the equipment industry.

(a) Construct the input-output matrix for this economy on the basis of these estimates from last year's data.

(b) Find the corresponding matrix of technical coefficients, and show that the Hawkins-Simon conditions are satisfied, namely, that all principal minors of $I - A$ are positive for the input-output matrix A.

5.   For a von Neumann model (A, B) given by

$$A = \begin{bmatrix} 0 & 1 & 0 & 0 \\ 1 & 0 & 0 & 1 \\ 0 & 0 & 1 & 0 \end{bmatrix}, \qquad B = \begin{bmatrix} 1 & 0 & 0 & 0 \\ 0 & 0 & 2 & 0 \\ 0 & 1 & 0 & 1 \end{bmatrix},$$

show that the optimal intensity vectors are

$$x^* = \left(2^{-\frac{1}{3}}, 2^{-\frac{2}{3}}, 1\right) : p^*\left(1, 2^{-\frac{2}{3}}, 0\right).$$

Find the rates α^* and β^*.

6.   Construct a von Neumann model (A, B) with $\alpha^* < \beta^*$.

7.   Consider the example on the village economy of Pulavanoor producing rice, vegetables, and farm equipment, discussed in Section 7.1. (See Tables 7.1 and 7.2 in p. 277). Suppose in a year there is an overproduction of 1000 units of rice, 500 units of milk, and 400 units of vegetables. What is the maximum rate of profit that would keep the workers in their present industry with no incentive to move to other industries?

8.   If A is a symmetric matrix, prove that A is stable (i.e., all eigenvalues of A have negative real part) if and only if A is Hicksian. [See Morishima (1964).]

9.   If A is Hicksian, prove that there exists a diagonal matrix D with positive diagonal entries such that DA is stable. (see Fisher and Fuller (1958), where the following more general result is proved: If A has all leading principal minors nonzero (or, more generally, if PAP^T has all leading principal minors nonzero for some permutation matrix P), then there exists a diagonal matrix D such that DA is stable. Conversely, if DA is stable for some diagonal matrix D, then A has at least one nonzero principal minor of each order.

10.   Consider the dynamical system for two goods:

$$\frac{dp_1}{dt} = -k_1(p_1 - p_1^*) - k_1(p_2 - p_2^*)$$

$$\frac{dp_2}{dt} = 2k_2(p_1 - p_1^*) + k_2(p_2 - p_2^*).$$

Show that this system is dynamically stable if $k_1 > k_2$ and dynamically unstable if $k_1 < k_2$. Show that the system neither possesses perfect stability nor imperfect stability in the sense of Hicks.

REFERENCES

Achilles, E. (1993): Implications of convergence rates in Sinkhorn balancing, *Linear Algebra Appl.*, 187, 109–12.

Afriat, S. N. (1974): On sum-symmetric matrices, *Linear Algebra Appl.*, 8, 129–40.

Agresti, A. (1990): *Categorical Data Analysis*, Wiley, New York.

Alexandroff, A. D. (1938): On the theory of mixed volumes of convex bodies IV, *Mat. Sb. (N. S.)* 3(45), 227–51 (in Russian).

Alexandroff, P. and Hopf, H. (1935): *Topologie*, Springer-Verlag, Berlin.

Anderson, W. N. Jr. (1971): Shorted operators, *J. Math. Anal. Appl.*, 20, 520–25.

Anderson, W. N. Jr. and Duffin, R. J. (1963): Series and parallel addition of matrices, *J. Math. Anal. Appl.*, 11, 576–94.

Anderson, W. N. Jr., Morley, T. D., and Trapp, G. E. (1984): Symmetric function means of positive operators, *Linear Algebra Appl.*, 60, 129–43.

Anderson, W. N. Jr. and Trapp, G. E. (1975): Shorted operators II, *J. Math. Anal. Appl.*, 28, 60–71.

Ando, T. (1983): An inequality between symmetric function means of positive operators, *Acta Sci. Math.*, 45, 19–22.

Ando, T. (1989): Majorization, doubly stochastic matrices, and comparison of eigenvalues, *Linear Algebra Appl.*, 118, 163–248.

Ando, T. (1995): Majorization relations for Hadamard products, *Linear Algebra Appl.*, 223/224, 57–64.

Ando, T. and Kubo, F. (1989): Some matrix inequalities in multiport network connections, *Operator Theory: Adv. Appl.*, 40, 111–31.

Apostol, T. M. (1957): *Mathematical Analysis: A Modern Approach to Advanced Calculus*, Addison Wesley, New York.

Arrow, K. J. (1974): Stability independent of adjustment speed. In *Trade, Stability and Macroeconomics*, G. Horwich and P. A. Samuelson, eds., pp. 181–202, Academic, New York.

Arrow, K. J. and Hurwicz, L. (1958): On the stability of competitive equilibrium, *Econometrica*, 26, 522–52.

Baccelli, F. L., Cohen, G., Olsder, G. J., and Quadrat J.-P. (1992): *Synchronization and Linearity: An Algebra for Discrete Event Systems*, Wiley, Chichester.

Bacharach, M. (1965): Estimating nonnegative matrices from marginal data, *Internat. Econom. Rev.*, 6, 294–310.

Bacharach, M. (1970): *Biproportional Matrices and Input Output Change*, Cambridge University Press, Monograph 16, Dept. Appl. Economics, Cambridge University, England.

315

Bapat, R. B. (1981): *On Permanents and Diagonal Products of Doubly Stochastic Matrices*, Ph.D. Thesis, Univ. Illinois at Chicago.

Bapat, R. B. (1982): $D_1 A D_2$ theorems for multidimensional matrices, *Linear Algebra Appl.*, 48, 437–42.

Bapat, R. B. (1986a): Applications of an inequality in information theory to matrices, *Linear Algebra Appl.*, 78, 107–17.

Bapat, R. B. (1986b): Multinomial probabilities, permanents and a conjecture of Karlin and Rinott, *Proc. Amer. Math. Soc.*, 102(3), 467–72.

Bapat, R. B. (1987a): Two inequalities for the Perron root, *Linear Algebra Appl.*, 85, 241–48.

Bapat, R. B. (1987b): A refinement of Oppenheim's inequality. In *Current Trends in Matrix Theory*, B. Grone, F. Uhlig, eds., pp. 29–32, North Holland, New York.

Bapat, R. B. (1987c): Discrete multivariate distributions and generalized log-concavity, *Sankhya Ser. A* 50, 98–110.

Bapat, R. B. (1989a): Comparing the spectral radii of two nonnegative matrices, *Amer. Math. Monthly*, 96(2), 137–39.

Bapat, R. B. (1989b): Mixed discriminants of positive semidefinite matrices, *Linear Algebra Appl.*, 126, 107–24.

Bapat, R. B. (1990): Permanents in probability and statistics, *Linear Algebra Appl.*, 127, 3–25.

Bapat, R. B. (1992a): Interpolating the determinantal and the permanental Hadamard inequality (research problem), *Linear and Multilinear Algebra*, 32, 335–37.

Bapat, R. B. (1992b): Mixed discriminants and spanning trees, *Sankhya, Special volume in memory of Professor R. C. Bose*, 54, 49–55.

Bapat, R. B. (1993a): Symmetric function means and permanents, *Linear Algebra Appl.*, 182, 101–8.

Bapat, R. B. (1993b): *Linear Algebra and Linear Models*, Hindustan Book Agency, Delhi.

Bapat, R. B. (1994): König's Theorem and bimatroids, *Linear Algebra Appl.*, 212/213, 353–65.

Bapat, R. B. (1995): Permanents, max algebra and optimal assignment, *Linear Algebra Appl.*, 226–228; 73–86.

Bapat, R. B. and Ben-Israel, Adi (1995): Singular values and maximum rank minors of generalized inverses, *Linear Multilinear Algebra*, 40, (2), 153–62.

Bapat, R. B. and Constantine, G. M. (1992): An enumerating function for spanning forests with color restrictions, *Linear Algebra Appl.*, 173, 231–37.

Bapat, R. B., Jain, S. K., and Pati, S. (1995): Weighted Moore-Penrose inverse of Boolean matrices, *Linear Algebra Appl.*, to appear.

Bapat, R. B. and Kwong, M. K. (1987): A generalization of $A \circ A^{-1} \geq I$, *Linear Algebra Appl.*, 93, 107–12.

Bapat, R. B. and Lal, A. K. (1991): Path-positive graphs, *Linear Algebra Appl.*, 149, 125–49.

Bapat, R. B. and Lal, A. K. (1993a): Path-positivity and infinite Coxeter groups, *Linear Algebra Appl.*, 196, 19-36.

Bapat, R. B. and Lal, A. K. (1993b): Inequalities for the q-permanent, *Linear Algebra Appl.*, 197/198, 397–410.

Bapat, R. B., Olesky, D. D., and P. van den Driessche (1995): Perron-Frobenius theory for generalized eigenvalue problem, *Linear and Multilinear Algebra*, 40, 141–52.

Bapat, R. B. and Raghavan, T. E. S. (1980): On diagonal products of doubly stochastic matrices, *Linear Algebra Appl.*, 31, 71–75.

Bapat, R. B. and Raghavan, T. E. S. (1989): An extension of a theorem of Darroch and Ratcliff in loglinear models and its applications to scaling multidimensional matrices, *Linear Algebra Appl.*, 114/115, 705–15.

Bapat, R. B., Stanford, D., and Van den Driessche, P. (1993): Pattern properties and spectral inequalities in max algebra, *SIAM J. Matrix Anal. Appl.*, to appear.

Bapat, R. B. and Sunder, V. S. (1985): On majorization and Schur products, *Linear Algebra Appl.*, 72, 107–17.

Bapat, R. B. and Sunder, V. S. (1986): An extremal property of the permanent and the determinant, *Linear Algebra Appl.*, 76, 153–63.

Bapat, R. B. and Sunder, V. S. (1991): On hypergroups of matrices, *Linear and Multilinear Algebra*, 29, 125–40.

Bapat, R. B. and Tijs, S. (1995): Incidence matrix games, preprint.

Barndorff-Neilson, O. E. (1978): *Information and Exponential Families in Statistical Theory*, Wiley, New York.

Bartlett, M. S. (1935): Contingency table interactions, *Suppl. J. R. Stat. Soc.*, 2, 248–52.

Bass, H., Connell, E. H., and Wright, D. (1982): The Jacobian conjecture: Reduction of degree and formal expansion of the inverse, *Bull. Amer. Math. Soc.*, 7(2), 287–330.

Bauer, F. L. (1965): An elementary proof of the Hopf inequality for positive operators, *Numer. Math.*, 7, 331–37.

Baum, L. E. and Eagon, J. A. (1967): An inequality with applications to statistical estimation for probabilistic functions of Markov processes and to a model for ecology, *Bull. Amer. Math. Soc.*, 73, 360–63.

Baxter, B. J. C. (1991): Conditionally positive functions and p-norm distance matrices, *Constr. Approx.*, 7, 427–40.

Bazarra, M. S. and Shetty, C. M. (1979): *Nonlinear Programming*, Wiley, New York.

Bellman, R. (1970): *Introduction to Matrix Analysis*, McGraw-Hill, New York.

Berman, A. and Plemmons, R. J. (1994): *Nonnegative Matrices in the Mathematical Sciences*, SIAM, Philadelphia.

Bhatia, R. (1987): *Perturbation Bounds for Matrix Eigenvalues*, Longman, Scientific and Technical; Wiley, New York.

Bhatia, R. and Bhattacharyya, T. (1993): A generalization of the Hoffman-Wielandt Theorem, *Linear Algebra Appl.*, 179, 11–17.

Biggs, N. L. (1974): *Algebraic Graph Theory*, Cambridge Tracts in Mathematics, vol. 67, Cambridge University Press, Cambridge.

Birkhoff, G. (1946): Tres observaciones sobre el algebra lineal, *University Nac. Tucuman Rev. Ser.* A5, 147–50.

Birkhoff, G. (1957): Extensions of Jentzch's theorem. *Trans. Amer. Math. Soc.*, 85, 219–27.

Birkhoff, G. and Varga, R. S. (1958): Reactor criticality and nonnegative matrices, *J. Soc. Ind. Appl. Math.*, 6, 354–77.

Bishop, Y. M. M., Feinberg, S. E., and Holland, P. W. (1975): *Discrete Multivariate Analysis: Theory and Practice*, MIT Press, Cambridge, Mass.

Blackwell, D. (1961): Minimax and irreducible matrices, *J. Math. Anal. Appl.*, 3, 37–39.

Blumenthal, L. M. (1970): *Theory and Application of Distance Geometry*, Chelsea, New York.

Bondareva, O. N. (1962): The core of an n-person game, *Vestnik Leningrad University*, 17, 141–2.

Bondy, J. A. and Murty, U. S. R. (1976): *Graph Theory with Applications*, Macmillan, London.

Bożejko, M. and Speicher, R. (1991): An example of a generalized Brownian motion. *Comm. Math. Phys.*, 137, 519–31.

Braker, J. G. and Olsder, G. J. (1993): The power algorithm in max algebra, *Linear Algebra Appl.*, 182, 67–89.

Brauer, A. (1961): On the characteristic roots of power-positive matrices, *Duke Math. J.*, 28, 439–45.

Brégman, L. M. (1973): Certain properties of nonnegative matrices and their permanents, *Soviet Math. Dokl.*, 14, 945–9.

Brown, D. T. (1959): A note on approximations to discrete probability distributions, *Information and Control*, 2, 386–92.

Brualdi, R. A. (1974): The DAD theorem for arbitray row sums, *Proc. Amer. Math. Soc.*, 45, 189–94.

Brualdi, R. A. (1982): Notes on the Birkhoff algorithm for doubly stochastic matrices, *Can. Math. Bull.*, 25, 191–9.

Brualdi, R. A. (1988): Some applications of doubly stochastic matrices, *Linear Algebra Appl.*, 107, 77–89.

Brualdi, R. A. and Csima, J. (1992): Butterfly embedding proof of a theorem of König, *Amer. Math. Monthly*, 99(3), 228–30.

Brualdi, R. A., Parter, S. V., and Schneider, H. (1966): The diagonal equivalence of a nonnegative matrix to a stochastic matrix, *J. Math. Anal. Appl.*, 31, 31–50.

Brualdi, R. and Ryser, H. (1991): *Combinatorial Matrix Theory*, Cambridge University Press, New York.

Burago, Yu. D. and Zalgaller, V. A. (1988): *Geometric Inequalities*, Springer-Verlag, Berlin.

Busemann, H. and Kelly, P. J. (1953): *Projective Geometry and Projective Metrics*, Academic, New York.

Bushell, P. J. (1973): On the projective contraction ratio of positive linear mappings, *J. London Math. Soc. (2)*, 6, 256–58.

Cao, Z.-Q., Kim, K. H., and Roush, F. W. (1984): *Incline Algebra and Applications*, Ellis Horwood, Chichester.

Carré, B. (1979): *Graphs and Networks*, Clarendon, Oxford.

Chaiken, S. (1982): A combinatorial proof of the all minors matrix tree theorem, *SIAM J. Algebra Discrete Meth.*, 3, 319–29.

Chaiken, S. and Kleitman, D. J. (1978): Metrix-tree theorems, *J. Combin. Theory*, 24, 319–81.

Champernowne, D. G. (1945): A note on J. von Neumann's article, *Rev. Economic Studies*, 12, 10–18.

Chvátal, V. (1983): *Linear Programming*. Freeman, New York.

Cohen, J. E. (1979): Random evolutions and the spectral radius of a non-negative matrix, *Math. Proc. Cambridge Philos. Soc.*, 86, 345–50.

Cohen, J. E. (1981): Convexity of the dominant eigenvalue of an essentially nonnegative matrix, *Proc. Amer. Math. Soc.*, 81, 657–8.

Cohen, G., Dubois, D., Quadrat, J.-P., and Viot, M. (1983): Analyse du comportment périodique des systèmes de production par la théorie des dioïdes, *INRIA Rep. 191*, Le Chesnay, France.

Cohen, G., Dubois, D., Quadrat, J.-P., and Viot, M. (1985): A linear-system-theoretic view of discrete-event processes and its use for performance evaluation in manufacturing, *IEEE Trans. Automat. Control*, 30, 210–20.

Comtet, L. (1974): *Advanced Combinatorics*, Reidel, Boston.

Constantine, G. M. (1987): *Combinatorial Theory and Statistical Design*, Wiley, New York.

Cottle, R. W. and Ferland, J. A. (1972): Matrix-theoretic criteria for the quasi-convexity and pseudo-convexity of quadratic functions. *Linear Algebra Appl.* 5, 123–36.

Cottle, R. W. and Veinott, A. F. (1972): Polyhedral sets having a least element. *Math. Programming.*, 3, 238–49.

Coxeter, H. S. M. and Moser, W. O. J. (1980): *Generators and Reflections for Discrete Groups*, Springer-Verlag, Berlin.

Cunninghame-Green, R. A. (1979): *Minimax Algebra*, Lecture Notes in Economics and Math. Systems, 166, Springer-Verlag, Berlin.

Curiel, I., Potters, J., Rajendra Prasad, V., Tijs, S., and Veltman, B. (1994): Sequencing and cooperation, *Oper. Res.* 42, 366–68.

Cvetković, D. M., Doob, M., and Sachs, H. (1979): *Spectra of Graphs*, Academic, New York.

Darroch, J. N. and Ratcliff, D. (1972): Generalized iterative scaling for loglinear models, *Ann. Math. Statist.*, 43, 1470–80.

Deming, W. E. and Stephan, F. F. (1940): On a least squares adjustment of a sampled frequency table when the expected marginal totals are known, *Ann. Math. Statist.*, 11, 427–44.

Deodhar, V. V. (1982): On the root system of a Coxeter group, *Comm. Algebra*, 10(6), 611–30.

Deutsch, E. and Neumann, M. (1984): Derivatives of the Perron-root of an essentially non-negative matrix and the group inverse of an M-matrix, *J. Math. Anal. Appl.*, 102, 1–29.

Dey, A., Hande, S. N., and Tiku, M. L. (1994): Statistical proofs of some matrix results, *Linear Multilinear Algebra*, 38, 109–16.

Dharmadhikari, S. and Joag-dev, K. (1988): *Unimodality, Convexity, and Applications*, Academic, New York.

Dudnikov, P. I. and Samborski, S. N. (1992): Endomorphisms of semimodules over semirings with an idempotent operation, *Math. USSR Izvestiya*, 38(1), 91–106.

Duff, I. S., Erisman, A. M., and Reid, J. K. (1986): *Direct Methods for Sparse Matrices*, Oxford Science Publ., Oxford.

Dwyer, P. S. (1941): The Doolittle Technique, *Ann. Math. Statist.*, 12, 449–58.

Eaves, B. C., Hoffman, A. J., Rothblum, U. G., and Schneider, H. (1985): Line-sum-symmetric scalings of square nonnegative matrices, *Math. Programming Stud.*, 25, 124–41.

Egorychev, G. P. (1981): A solution of van der Waerden's permanent problem, *Soviet Math. Dokl.* 23, 619–22.

Egorychev, G. P. (1990): Mixed discriminant and parallel addition, *Dokl. Akad. Nauk SSSR*, 312(3), 528–31 (in Russian); English translation in *Soviet Math. Dokl.*, 41(3)1990, 451–55, (1991).

Elsner, L. (1984): On convexity properties of the spectral radius of nonnegative matrices, *Linear Algebra Appl.*, 61, 31–35.

Elsner, L. (1993): A note on the Hoffman-Wielandt theorem, *Linear Algebra Appl.*, 182, 235–7.

Elsner, L. and Johnson, C. R. (1989): Nonnegative matrices, zero patterns, and spectral inequalities, *Linear Algebra Appl.*, 120, 225–36.

Falikman, D. I. (1981): A proof of van der Waerden's conjecture on the permanent of a doubly stochastic matrix, *Math. Notes*, 29, 475–9.

Fan, K. (1958): Topological proofs for certain theorems on matrices with nonnegative elements. *Monatsh. Math.*, 62, 219–37.

Feller, W. (1968): *An Introduction to Probability Theory and Its Applications*, Vol. 1, Wiley, New York.

Ferland, J. A. (1972): Maximal domains of quasi-convexity and pseudo-convexity for quadratic functions, *Math. Programming*, 3, 178–92.

Ferland, J. A. (1981): Matrix-theoretic criteria for the quasiconvexity of twice continuously differentiable functions, *Linear Algebra Appl.*, 38, 51–63.

Fiedler, M. (1983): A note on the Hadamard product of matrices, *Linear Algebra Appl.*, 49, 233–5.

Fiedler, M., Johnson, C. R., Markham, T. L., and Neumann, M. (1985): A trace inequality for M-matrices and the symmetrizability of a real matrix by a positive diagonal matrix, *Linear Algebra Appl.*, 71, 81–94.

Fiedler, M. and Ptak, V. (1962): On matrices with non-positive off diagonal elements and positive principal minors. *Czechoslovak Math. J.*, 12(87), 382–400.

Fiedler, M. and Ptak, V. (1966): Some generalizations of positive definiteness and monotonicity, *Numerich. Math.*, 9, 163–72.

Fienberg, S. E. (1977): *The Analysis of Cross-Classified Categorical Data*, MIT Press, Cambridge, Mass.

Fisher, M. E. and Fuller, A. T. (1958): On the stabilization of matrices and the convergence of linear iterative processes, *Proc. Cambridge Philos. Soc.*, 54, 417–25.

Flam, H. and Flanders, M. J. (1991): *Heckscher-Ohlin Trade Theory: Translated, Edited and Introduced*, MIT Press, Boston, Mass.

Franklin, J. and Lorenz, J. (1989): On the scaling of multidimensional matrices, *Linear Algebra Appl.*, 114/115, 717–35.

Friedland, S. (1981): Convex spectral functions, *Linear and Multilinear Algebra*, 9, 299–316.

Friedland, S. (1986): Limit eigenvalues of nonnegative matrices, *Linear Algebra Appl.*, 74, 173–78.

Friedlander, D. (1961): A technique for estimating a contingency table, given the marginal totals and some supplementary data, *J.R. Stat. Soc., Ser. A*, 124, 412–20.

Frobenius, G. (1908): Über Matrizen aus Positiven Elementen, *S.-B. Preuss. Akad. Wiss.* (Berlin), 471–6.

Frobenius, G. (1912): Über Matrizen aus nichtnegativen Elementen, *S.-B. Preuss. Akad. Wiss.* (Berlin), 456–77.

Gale, D. (1960): *Theory of Linear Economic Models*, McGraw-Hill, London.

Gale, D. and Nikaido, H. (1965): The Jacobian matrix and global univalence of mappings, *Math. Annalen*, 159, 81–93.

Gantmacher, F. R. (1959): *The Theory of Matrices*. vols. I and II, Chelsea, New York.

Gaubert, S. (1992): *Théorie Linéarie des Systèmes dans des Dioïdes*, Thèse, École des Mines de Paris, Paris.

Girko, V. L. (1990): *Theory of Random Determinants*, Kluwer Acdemic Publishers, Dordrecht.

Girko, V. L. (1995): The method of random determinants for estimating the permanent, *Random Operators and Stoch. Equations*, 3, 181–92.

Gokhale, D. V. and Kullback, S. (1978): *The information in Contingency Tables*, Marcel Dekker, New York.

Gondran, M. and Minoux, M. (1984): *Graphs and Algorithms*, Wiley, New York.

Gonzales, T. (1979): A note on open shop preemptive schedule, *IEEE Trans. Comput.*, C-28, 782–6.

Gonzales, T. and Sahni, S. (1976): Open shop scheduling to minimize finish time, *J. Assoc. Comput. Mach.*, 23, 665–79.

Gregory, D. A., Kirkland, S., and Pullman, N. J. (1993): Power convergent boolean matrices, *Linear Algebra Appl.*, 179, 105–17.

Grove, L. C. and Benson, C. T. (1985): *Finite Reflection Groups*, Springer-Verlag, Berlin.

Haberman, S. J. (1974): *The Analysis of Frequency Data*, University Chicago Press, Chicago.

Hadamard, J. (1893): Résolution d' une question relative aux déterminants, *Bull. Sci. Math.*, 2, 240–8.

Hall, P. (1935): On representatives of subsets, *J. London Math. Soc.*, 10, 26–30.

Hardy, G. H., Littlewood J. E., and Pólya, G. (1952): *Inequalities*, 2nd ed., Cambridge University Press, Cambridge.

Hartfiel, D. J. and Loewy, R. (1984): A determinantal version of the Frobenius-König Theorem, *Linear and Multilinear Algebra*, 16, 155–65.

Heckscher, E. (1919): The effect of foreign trade on the distribution of income, *Economisk Tidscrift*, XXI(2), 1–32.

Heilman, O. J. and Lieb, E. H. (1972): Theory of monomer-dimer systems, *Comm. Math. Phys.*, 25, 190–232.

Herstein, I. N. (1964): *Topics in Algebra*, Lexington, Massachuetts.

Hicks, J. R. (1939): *Value and Capital: An Enquiry into some Fundamental Principles of Economic Theory*, Oxford University Press, London.

Hobby, C. and Pyke, R. (1965): Doubly stochastic operators obtained from positive operators, *Pacific J. Math.*, 15(1), 153–7.

Hoffman, A. J. and Wielandt, H. W. (1953): The variation of the spectrum of a normal matrix, *Duke Math. J.*, 20, 37–39.

Holladay, J. C. and Varga, R. S. (1958): On powers of nonnegative matrices. *Proc. Amer. Math. Soc.*, 9, 631–4.

Hopf, E. (1963): An inequality for positive integral operators, *J. Math. Mech.*, 12, 683–92.

Horn, R. and Johnson, C. R. (1985): *Matrix Analysis*, Cambridge University Press, Cambridge.

Horn, R. and Johnson, C. R. (1991): *Topics in Matrix Analysis*, Cambridge University Press, Cambridge.

Horwich, G. and Samuelson, P., A., eds. (1974): *Trade, Stability and Macroeconomics*, Academic, New York.

Hotelling, H. (1943): Some new methods in matrix calculation, *Ann. Math. Statist.*, 14, 1–34.

Householder, A. S. (1964): *The Theory of Matrices in Numerical Analysis*, Dover, New York.

Howlett, R. B. (1982): Coxeter groups and *M*-Matrices, *Bull. London Math. Soc.* 14, 137–41.

Humphreys, J. E. (1990): *Reflection Groups and Coxeter Groups*, Cambridge Studies in Advanced Mathematics 29, Cambridge University Press, Cambridge.

Inada, K. (1971): The production coefficient matrix and Stolper-Samuelson condition, *Econometrica*, 39, 219–39.

Jentzsch, R. (1912): Ueber integralgeichungen nit positivem kern., *J. fur Math.*, 141, 235–44.

Johnson, C. R. (1987): The permanent-on-top conjecture: A status report. In *Current Trends in Matrix Theory*, B. Grone, F. Uhlig eds., pp. 167–74, North Holland, New York.

Johnson, C. R. and Bapat, R. B. (1988): A weak multiplicative majorization conjecture for Hadamard products (problem proposed at the 1987 Utah State University matrix theory conference), *Linear Algebra Appl.*, 104, 246–7.

Johnson, D. M., Dulmage A. L., and Mendelsohn, N. S. (1960): On an algorithm of G. Birkhoff concerning doubly stochastic matrices, *Can. Math. Bull.*, 3, 237–42.

Kaneko, M. (1989): *A game theoretical description of the von Neumann growth economy*, Technical Report, Hitotsubashi University, Japan.

Kaplansky, I. (1945): A contribution to von Neumann's theory of games, *Ann. Math.*, 46, 474–9.

Karlin, S. (1959): Positive operators, *J. Math. Mech.*, 8, 907–37.

Karlin, S. (1984): Mathematical models, problems and controversies of evolutionary theory, *Bull. Amer. Math. Soc.*, (10)2, 221–73.

Karlin, S. and Ost, F. (1985): Some monotonicity properties of Schur powers of matrices and related inequalities, *Linear Algebra Appl.*, 68, 47–65.

Karlin, S. and Rinott, Y. (1981): Entropy inequalities for classes of probability distributions II. The multivariate case. *Adv. Appl. Probab.*, 13, 325–51.

Keller, O. (1939): Ganze Cremona-Transformationen, *Monatsch. Math. Phys.*, 47, 299–306.

Kemeny, J. G. and Snell, J. L. (1960): *Finite Markov Chains*, van Nostrand, Princeton, NJ.

Kemeny, J. G., Morgenstern, O., and Thompson, G. L. (1956): A generalization of the von Neumann model of an expanding economy, *Econometrica*,.24, 115–35.

Kim, K. H. (1982): *Boolean Matrix Theory and Applications*, Marcel Dekker, New York.

Kingman, J. F. C. (1961): A convexity property of positive matrices, *Quart. J. Math. Oxford Ser. (2)*, 12, 283–4.

Kirkland, S. and Pullman, N. J. (1992): Boolean spectral theory, *Linear Algebra Appl.*, 175, 177–90.

Kohlberg, E. and Pratt, J. W. (1982): The contraction mapping theory to Perron Frobenius theory: Why Hilbert's metric? *Math. Oper. Res.*, 7, 198–210.

Kojima, M. and Saigal, R. (1979): On the number of solutions to a class of linear complementarity problems, *Math. Programming*, 17, 136–9.

Kolmogorov, A. N. (1935): Zur theorie der Markoffschen Ketten., *Math. Ann.*, 10, 155–60.

König, D. (1936): *Theorie der Endlichen und Unendlich/en Graphen*, Akademische, Verlags Gesellschaft, Leipzig.

Koyak, R. (1987): On measuring internal dependence of a set of random variables, *Ann. Statist.*, 15, 1215–8.

Krein, M. G. (1950): On an application of the fixed-point principle in the theory of linear transformations of spaces with an indefinite metric. *Uspehi. Mat. Nauk*, *N.S.* 5, No. 2 (36), 180–90 (in Russian). (AMS translation Ser. 2, Vol. 1, (1955) 27–35.)

Krein, M. G. and Rutman, M. A. (1948): Linear operators leaving invariant a cone in a Banach space. *Usp. Mat. Nauk. (N.S.)* , 3, 3–95 (in Russian). (AMS Translation Ser. 1, Vol. 10, (1962) 199–325.)

Kuczynski, M. and Meek, R. L. (1972): *Quesnay's Tableau Économique* (Translations and notes) MacMillan, New York.

Kuhn, H. W. (1955): The Hungarian method for the assignment problem, Naval Research Logistic Quarterly, 2, 83–97.

Kuhn, H. W. (1967): On games of fair division. In *Essays in Mathematical Economics*, ed. M. Shubik, pp. 29–37, Princeton University Press, Princeton, NJ.

Lal, A. K. (1993): Inequalities for the q-permanent, II, *Linear Algebra Appl.*, to appear

Lancaster, K. (1968): *Mathematical Economics*, Macmillan, New York.

Leontief, W. (1941): *The Structure of American Economy: 1919–1929*, Oxford University Press, New York.

Levinger, B. W. (1970): *Notices Amer. Math. Soc.*, 17(119), 260.

Lewin, M. (1971): On nonnegative matrices, *Pacific J. Math.*, 36, 753–9.

Liu, J. (1992): Spectral radius, Kronecker products and stationarity, *J. Time Ser. Anal.*, 13(4), 319–25.

London, D. (1971): Some notes on the van der Waerden conjecture, *Linear Algebra Appl.* 4, 155–60.

Lovász, L. (1979): *Combinatorial Problems and Exercises*, North-Holland, Amsterdam.

Lovász, L. and Plummer, M. D. (1986): *Matching Theory*, Elsevier, New York.

Marcus, M. and Lopes, L. (1957): Inequalities for symmetric functions and Hermitian matrices, *Can. J. Math.*, 9, 305–12.

Marcus, M. and Minc, H. (1964): *A Survey of Matrix Theory and Matrix Inequalities*, Allyn & Bacon, Boston.

Marcus, M. and Newman, M. (1959): On the minimum of the permanent of a doubly stochastic matrix, *Duke. Math. J.*, 26, 61–72.

Marek, I. (1984): Perron roots of a convex combination of a cone preserving map and its adjoint, *Linear Algebra Appl.*, 58, 185–200.

Marshall, A. W. and Olkin, I. (1968): Scaling of matrices to achieve specified row and column sums, *Numer. Math.*, 12, 83–90.

Marshall, A. W. and Olkin, I. (1979): *Inequalities: Theory of Majorization and its Applications*, Academic, New York.

Martos, B. (1969): Subdefinite matrices and quadratic forms, *SIAM J. Appl. Math.*, 17, 1215–23.

Martos, B. (1971): Quadratic programming with a quasi-convex objective function, *Oper. Res.*, 19, 87–97.

Maschler, M., Peleg, B., and Shapley, L. S. (1979). Geometric properties of the kernel, nucleolus and related solution concepts, *Math. Oper. Res.*, 4, 303–38.

Mason, J. H. (1972): Matroids: unimodal conjectures and Motzkin' Theorem. In *Combinatorics*, D. J. A. Welsh and D. R. Woodall, eds., pp. 207–21, Institute of Math. and Appl., Southend-on-Sea. 207–20.

McKenzie, L. W. (1960): Matrices with dominant diagonals and economic theory. In *Mathematical Methods in the Social Sciences, 1959*, K. J. Arrow, S. Karlin, and P. Suppes, eds., pp. 47–62, Stanford University Press, Stanford, CA.

Mehta, M. L. (1989): *Matrix Theory: Selected Topics and Useful Results*, enlarged re-edition, Hindustan Publ. Co., Delhi.

Menon, M. V. (1968): Matrix links, an extremaziation problem and the reduction of a nonnegative matrix to one with prescribed row and column sums, *Can. J. Math.*, 20, 225–32.

Merris, R. (1987): The permanental dominance conjecture. In *Current Trends in Matrix Theory*, B. Grone, F. Uhlig eds., pp. 213–23, Elsevier.

Merris, R. (1993): Laplacian matrices of graphs: A survey, *Linear Algebra Appl.*, 197/198, 143–76.

Metzler, L. A. (1945): Stability of multiple markets: the Hicks conditions, *Econometrica*, 13, 277–92.

Meyer, C. D., Jr. (1975): The role of the group generalized inverse in the theory of finite Markov chains, *SIAM Rev.* 17, 443–64.

Micchelli, C. A. (1986): Interpolation of scattered data: distance matrices and conditionally positive definite matrices, *Constr. Approx.*, 2, 11–22.

Miller, R. E. and Blair, P. D. (1985): *Input-Output Analysis: Foundations and Extensions*, Prentice-Hall, Englewood Cliffs, NJ.

Minc, H. (1963): Upper bounds for permanents of zero one matrices, *Bull. Amer. Math. Soc.*, 69, 789–91.

Minc, H. (1988): *Nonnegative Matrices*, Wiley, New York.

Mirsky, L. (1971): *Transversal Theory*, Academic, New York.

Mitra, S. K. (1954): Ph.D. Thesis, Dept. Statistics, University North Carolina, Chapel Hill.

Mitra, S. K. and Odell, P. L. (1986): On parallel sumability of matrices, *Linear Algebra Appl.*, 74, 239–55.

Mitra, S. K. and Puri, M. L. (1982): Shorted matrices—An extended concept and some applications, *Linear Algebra Appl.*, 42, 57–79.

Mohan, S. R. and Shridhar, R. (1992): On characterizing N-matrices using linear complementarity, *Linear Algebra Appl.*, 160, 231–45.

Morishima, M. (1964): *Equilibrium, Stability and Growth: A Multi-Sectoral Analysis*, Oxford Clarendon,

Morley, T. D. (1989): Linear programming and operator means, *Adv. Appl. Math.*, 10, 497–506.

Mulmuley, K., Vazirani, U., and Vazirani, V. (1987): Matching is as easy as matrix inversion, *Combintorica*, 7, 105–14.

Murota, K. (1987): *Systems Analysis by Graphs and Matroids*, Springer-Verlag, Berlin.

Murota, K. (1993): Mixed matrices: irreducibility and decomposition. In *Combinatorial and Graph-Theoretical Problems in Linear Algebra*, R. A. Brualdi, S. Friedland, and V. Klee eds., The IMA Volumes in Mathematics and Its Applications, Vol. 50, pp. 39–71, Springer-Verlag, Berlin.

Nijenhuis, A. (1976): On permanents and the zeros of rook polynomials, *J. Combin. Theory Ser. A*, 21, 240–4.

Nikaido, H. (1968): *Convex Structures and Economic Theory*, Academic, New York.

Ohlin, B. (1933): *Interregional and International Trade*, Harvard University Press, Cambridge, Mass.

Olech, C., Parthasarathy, T., and Ravindran, G. (1991): Almost N-matrices and linear complementarity, *Linear Algebra Appl.*, 145, 107–25.

Olsder, G. J. and Roos, C. (1988): Cramer and Cayley-Hamilton in max algebra, *Linear Algebra Appl.*, 101, 87–108.

Oppenheim, A. (1930): Inequalities connected with definite hermitian forms, *J. London Math. Soc.*, 5, 114–9.

Ortega, J. M. and Rheinboldt, W. C. (1970): *Iterative Solution of Nonlinear Equations in Several Variables*, Academic, New York.

Ostrowski, A. M. (1937–1938): Über die Determinenten mit überwiegender Hauptdiagonale. *Comment Math. Helv.*, 10, 175–210.

Ostrowski, A. M. (1963): On positive matrices, *Math. Ann.*, 150, 276–84.

Ostrowski, A. M. (1964): Positive matrices and Functional analysis. *In Recent Advances in Matrix Theory*, Ed. H. Schneider, University Wisconsin Press, Madison, Wisconsin.

Panov, A. A. (1985a): On mixed discriminants connected with positive semidefinite quadratic forms, *Dokl. Akad. Nauk. SSSR*, 282, 273–6 (in Russian); English translation in *Soviet Math. Dokl.*, 31 (1985).

Panov, A. A. (1985b): Some properties of mixed discriminants, *Mat. Sb. (N.S.)*, 128(170), no. 3, 291–305 (in Russian); English translation in *Math USSR - Sb.*, 56(2), 279–93(1985).

Parlett, B. N. and Landis, T. L. (1982): Methods for scaling to doubly stochastic matrices, *Linear Algebra Appl.*, 48, 53–79.

Parlett, B. N. and Reinsch, C. (1969): Balancing a matrix for calculation of eigenvalues and eigenvectors, *Numer. Math.*, 13, 293–304.

Parthasarathy, K. R. and Schmidt, K. (1972): *Positive Definite Kernels, Continuous Tensor Products and Central Limit Theorems of Probability Theory*, Springer-Verlag, Berlin.

Parthasarathy, T. and Raghavan, T. E. S. (1971): *Some Topics in Two Person Games*, Amer. Elsevier, New York.

Parthasarathy, T. and Ravindran, G. (1990): N-matrices, *Linear Algebra Appl.*, 139, 89–102.

Pasinetti, L. L. (1977): *Essays on the Theory of Production*, Columbia University Press, New York.

Perkins, P. (1961): A note on regular matrices, *Pacific J. Math.*, 11, 1529–34.

Perron, O. (1907): Zur Theorie der Über Matrizen, *Math. Ann.* , 64, 248–63.

Phelps, R. R. (1966): *Lectures on Choquet's theorem*, van Nostrand, Princeton, NJ.

Plemmons, R. J. (1971): Generalized inverses of boolean matrices, *SIAM J. Appl. Math.*, 20, 426–33.

Polya, G. and Szegö, S. (1976): *Problems and Theorems in Analysis*, vol. II, Springer-Verlag, Berlin, Heidelberg, New York.

Prasada Rao, P. S. S. N. V. and Bhaskara Rao, K. P. S. (1975): On generalized inverses of boolean matrices, *Linear Algebra Appl.*, 11, 135–53.

Pressman, S. (1994): *Quesnay's Tableau Économique: A Critique and Reassessment*, Kelley, West Caldwell, NJ.

Putnam, C. R. (1958): On bounded matrices with non-negative elements, *Can. J. Math.*, 10, 587–91.

Quandt, R. E. and Baumol, W. J. (1966): The demand for abstract transport modes: theory and measurement, *J. Regional Science*, 6(2), 13–26.

Raghavan, T. E. S. (1978): Completely mixed games and M-matrices, *Linear Algebra Appl.*, 21, 35–45.

Raghavan, T. E. S. (1979): Some remarks on matrix games and nonnegative matrices, *SIAM J. Appl. Math.*, 36, 83–5.

Raghavan, T. E. S. (1984a): On a pair of multidimensional matrices, *Linear Algebra Appl.*, 62, 263–8.

Raghavan, T. E. S. (1984b): On pairs of multidimensional matrices and their applications, *Contemp. Math.*, 46, 339–54.

Rao, C. R. (1973): *Linear Statistical Inference and Its Applications*, Wiley, New York.

Rao, C. R. and Mitra, S. K. (1971): *Generalized Inverse of Matrices and Its Applications*, Wiley, New York.

Rathore, R. K. S. and Chetty, C. S. K. (1981): Some angularity and inertia theorems related to normal matrices, *Linear Algebra Appl.*, 40, 69–77.

Recski, A. (1989): *Matroid Theory and Its Applications in Electric Network Theory and in Statics*, Springer-Verlag, Berlin.

Reijnierse, H. (1995): *Games, Graphs and Algorithms*, Ph.D. Thesis, Univ. Nimegen, The Netherlands.

Romanovskii, V. I. (1970): *Discrete Markov Chaions*, Wolters-Noordhoff, Groningen, The Netherlands.

Ross, S. (1985): *Stochastic Processes*, Wiley, New York.

Rota, G. C. (1961): On the eigenvalues of positive operators, *Bull. Amer. Math. Soc.*, 67, 556–8.

Rothaus, O. S. (1974): Study of the permanent conjecture and some of its generalizations *Israel J. Math.*, 18, 75–96.

Rothblum, U. G. (1989): Generalized scalings satisfying linear equations, *Linear Algebra Appl.*, 114/115, 765–83.

Rothblum, U. G. and Schneider, H. (1989): Scaling of matrices which have prescribed row sums and column sums, *Linear Algebra Appl.*, 114/115, 737–64.

Sahi, S., (1993): Logarithmic convexity of Perron-Frobenius eigenvectors of positive matrices, *Proc. Amer. Math. Soc.*, 118(4), 1035–6.

Samuelson, P. A. (1944): The relation between Hicksian stability and true dynamic stability, *Econometrica*, 12, 256–7.

Samuelson, P. A. (1948): International trade and the equalization of factor prices, *Economic J.*, 58, 163–84.

Samuelson, P. A. (1951): Abstract of a theorem concerning substitutability in open Leontief models. In *Activity Analysis of Production and Allocation*, T. C. Koopmans, ed., pp. 142–6, Wiley, New York.

Samuelson, P. A. (1953): Prices of factors and goods in general equilibrium, *Rev. Economic Studies*, 21, 1.

Schmeidler, D. (1969): The nucleolus of a characteristic function game, *SIAM J. Appl. Math.*, 17, 1163–70.

Schneider, H. (1977): The concepts of irreducibility and full indecomposability of a matrix in the works of Frobenius, König and Markov, *Linear Algebra Appl.*, 18, 139–62.

Schneider, H. (1986): The influence of the marked reduced graph of a nonnegative matrix on the Jordan form and on related properties: a survey, *Linear Algebra Appl.*, 84, 161–89.

Schneider, M. H. (1989): Matrix scaling, entropy minimization and conjugate duality 1. Existence conditions, *Linear Algebra Appl.*, 114/115, 785–813.

Schneider, R. (1966): On A. D. Aleksandrov's inequalities for mixed discriminants, *J. Math. Mech.*, 15(2), 285–90.

Schneider, R. (1993): *Convex Bodies: The Brunn-Minkowski Theory*, Cambridge University Press, Cambridge.

Schrijver, A. (1978a): A short proof of Minc's conjecture, *J. Combin. Theory Ser. A*, 25, 80–81.

Schrijver, A. (1978b): *Matroids and Linking Systems*, Mathematical Centre Tracts, 88, Amsterdam.

Schur, I. (1918): Über endliche Gruppen und Hermitesche Formen, *Math. Z.*, 1, 184–207.

Schwarz, B. (1964): Rearrangement of square matrices with non-negative elements, *Duke Math. J.*, 31, 45–62.

Schwartz, J. T. (1980): Fast probabilistic algorithms for verification of polynomial identities, *J. Assoc. Comput. Mach.*, 27(4):701–17.

Seidel, J. J. (1989): Graphs and their spectra. In *Combinatorics and Graph Theory*, Banach Center Publications, vol. 25, 147–162., PWN Polish Scientific Publishers, Warsaw.

Seidel, J. J. (1992): A note on path-zero graphs, *Discrete Math.*, 106/107:435–8.

Seneta, E. (1973): *Non-Negative Matrices*, Wiley, New York.

Shapley, L. S. (1965): On balanced sets and cores, *Rand Corporation Memorandum* RM4601-PR.

Shapley, L. S. and Scarf, H. E. (1974): On cores and indivisibility, *J. Math. Econom.*, 1, 23–38.

Shapley, L. S. and Shubik, M. (1972): The assignment game, I: The core, *International J. Game Theory*, 2, 111–30.

Sinkhorn, R. (1964): A relationship between arbitrary positive matrices and doubly stochastic matrices, *Ann. Math. Statist.*, 35, 876–9.

Sinkhorn, R. and Knopp, P. (1967): Concerning nonnegative matrices and doubly stochastic matrices, *Pacific J. Math.*, 21, 343–8.

Skeel, R. D. (1981): Effect of equilibration on residual size for partial pivoting, *SIAM J. Numer. Math.*, 18, 449–54.

Slutsky, E. (1915): Sulla teoria del bilancio del consumetore, *Georn. Econ.*, 51.

Solymosi, T. and Raghavan, T. E. S. (1994): An algorithm for finding the nucleolus of assignment games, *International J. Game Theory*, 23, 119–43.

Soules, G. W. (1979): Constructing symmetric nonnegative matrices, *Linear Multilinear Algebra*, 7, 343–57.

Soules, G. W. (1991): The rate of convergence of Sinkhorn balancing, *Linear Algebra Appl.*, 150, 3–40.

Soules, G. W. (1994): An approach to the permanental-dominance conjecture, *Linear Algebra Appl.*, 201, 211–29.

Sraffa, P. (1963): *Production of Commodities by Means of Commodities: Prelude to a critic of Economic Theory*, Cambridge University Press, Cambridge.

Stanciu, S. (1992): The energy operator for infinite statistics, *Comm. Math. Phys.*, 147, 211–16.

Stanley, R. P. (1981): Two combinatorial applications of the Alexandrov-Fenchel inequalities, *J. Combin. Theory Ser. A*, 31, 56–65.

Stillwell, J. (1980): *Classical Topology and Combinatorial Group Theory*, Springer Verlag, New York.

Stopher, P. R. and Meyburg, A. H. (1975): *Urban Transportation Modeling and Planning*, Lexington, Lexington, Mass.

Stolper, W. A. and Samuelson, P. A. (1941): Protection and real wages, *Rev. Economic Studies*, 9, 1.

Strang, G. (1980): *Linear Algebra and Its Applications*, Harcourt Brace Jovanovich, San Diego.

Tigelaar, H. H. (1991): On monotone linear operators and the spectral radius of their representing matrices, *SIAM J. Matrix Anal. Appl.*, 12(4), 726–9.

Tijs, S. H., Parthasarathy, T., Potters, J. A. M., and Rajendra Prasad, V. (1984): Permutation games: another class of totally balanced games, *OR Spektrum*, 6, 119–23.

Tilanus, C. B. (1976): Where short-term budget meets long-term plan. In *Quantitative Methods in Budgeting*, C. B. Tilanus ed., pp. 159–67, Martinus Nijhoff Social Sciences Division, Leiden.

van den Essen, A. (1990): *Polynomial maps and the Jacobian conjecture*, Technical Report 9034, Dept. of Mathematics, Catholic University at Nijmegen, The Netherlands.

van den Essen, A. (1992): *The exotic world of invertible polynomial maps*, Technical Report 9204, Dept. Mathematics, Catholic University at Nijmegen, The Netherlands.

van der Sluis, A. (1970): Condition, equilibration, and pivoting in linear algebraic systems, *Numer. Math.*, 15, 74–86.

van de Waerden, B. L. (1926): Aufgabe 45, *Jber. Deutsch. Math.-Verein*, 35, 117.

Varga, R. S. (1962): *Matrix Iterative Analysis*, Prentice-Hall, Englewood Cliffs, N. J.

Visick, G. (1995): A weak majorization involving the matrices $A \circ B$ and AB, *Linear Algebra Appl.*, 223/224, 731–44.

von Neumann, J. (1937): Uber ein ö konomisches Gleichungsystem und ein Verallgemeinung des Brouwerschen Fixpunktsatzes, *Ergebnisse eines Mathematischen Kooloquiums*, 8, Vienna (in German); translated in *Rev. Economic Studies*, 13, 10–18, (1945–46).

von Neumann, J. (1945): A model of general economic equilibrium, *Rev. of Economic Studies*, 13, 1–9.

von Neumann, J. (1953): A certain zero-sum two-person game equivalent to an optimal assignment problem. *Ann. Math. Studies*, 28, 5–12.

Welsh, D. J. A. (1976): *Matroid Theory*, Acadmic, London.

White, N. (ed.) (1986): *Theory of Matroids*, Cambridge University Press, Cambridge.

Widder, D. V. (1946): *The Laplace Transform*, Princeton University Press, Princeton, NJ.

Wielandt, H. (1950): Unzerlegbare nicht negative Matrizen, *Math. Z.*, 52, 642–8.

Wilkinson, J. H. (1963): *Rounding Errors in Algebraic Processes*, Her Majesty's Stationary Office, England.

Zagier, D. (1992): Realizability of a model in infinite statistics, *Comm. Math. Phys.*, 147, 199–210.

INDEX

absorbing, 48
accessible, 47
adjacency matrix, 209
Alexandroff inequality, 92, 188, 189, 195, 196, 203, 236, 238
Alexandroff inequality for mixed discriminant, 203
algebraic multiplicity, 5, 215
analytic function, 24
aperiodic, 48
arithmetic mean-geometric mean inequality, 80, 127, 137, 156, 268
assignment game, 96
associated with a matrix, 4, 236
average weight, 220

balancing algorithm, 262
base, 207
basic class of an isolated block, 38
Bernstein's Theorem, 176
binomial distribution, 185
Birkhoff von–Neumann Theorem, 59, 63, 64, 67, 71, 76, 82, 97, 106, 107, 111, 126
bisubmodularity, 112
Bondareva-Shapley Theorem, 95
boolean algebra, 225
boolean matrix, 225
Brouwer's fixed point theorem, 5, 53, 184, 272
budget allocation problem, 239

c.n.d. matrix, 161
c.p.d. matrix, 161
Carathéodory's Theorem, 56, 64, 111
Cartan matrix, 214
Cauchy–Binet formula, 235
Cauchy–Schwarz Inequality, 147, 180, 235
chainable entries, 67
characteristic function, 94

circuit, 220
circuit geometric mean, 130
circuit matrix, 124
circuit mean, 220
circuit product, 130
class, 38
closed system, 287
coalition, 94
coefficients of production, 278
Cohen's Theorem, 134
communicating states, 47
complementary slackness, 32, 250, 281, 282, 289, 310
completely mixed game, 10
completely mixed strategy, 10
completely monotonic function, 176
completely reducible matrix, 125
conditionally negative definite matrix, 161
conditionally positive definite function, 186
conditionally positive definite matrix, 161
cone, 4, 51, 58
constant returns to scale, 278
contingency table, 243
contracting model, 288
contraction mapping Theorem, 259
contraction ratio of Birkhoff, 255
convex, 4
convex cone, 251
convex function, 165
convex set, 4, 5, 53, 59, 60, 98, 165, 166, 193, 293
cooperative game, 94
core, 94
coxeter graph, 213
coxeter group, 213
cyclic matrix, 40

$D_1 A D_2$ Theorem, 118
dependent set, 197
Descarte's rule of signs, 194

AUTHOR INDEX